Research in Congenital Hypothyroidism

NATO ASI Series

Advanced Science Institutes Series

A series presenting the results of activities sponsored by the NATO Science Committee, which aims at the dissemination of advanced scientific and technological knowledge, with a view to strengthening links between scientific communities.

The series is published by an international board of publishers in conjunction with the NATO Scientific Affairs Division

A	**Life Sciences**	Plenum Publishing Corporation
B	**Physics**	New York and London
C	**Mathematical and Physical Sciences**	Kluwer Academic Publishers Dordrecht, Boston, and London
D	**Behavioral and Social Sciences**	
E	**Applied Sciences**	
F	**Computer and Systems Sciences**	Springer-Verlag
G	**Ecological Sciences**	Berlin, Heidelberg, New York, London,
H	**Cell Biology**	Paris, and Tokyo

Recent Volumes in this Series

Volume 157—Plasma Membrane Oxidoreductases in Control of Animal and Plant Growth
edited by Frederick L. Crane, D. James Morré, and Hans Löw

Volume 158—Biocompatibility of Co-Cr-Ni Alloys
edited by Hartmut F. Hildebrand and Maxime Champy

Volume 159—Biologically Based Methods for Cancer Risk Assessment
edited by Curtis C. Travis

Volume 160—Early Influences Shaping the Individual
edited by Spyros Doxiadis

Volume 161—Research in Congenital Hypothyroidism
edited by F. Delange, D. A. Fisher, and D. Glinoer

Volume 162—Nematode Identification and Expert System Technology
edited by Renaud Fortuner

Volume 163—Leishmaniasis: The Current Status and New Strategies for Control
edited by D. T. Hart

Series A: Life Sciences

Research in Congenital Hypothyroidism

Edited by

F. Delange

University of Brussels
Brussels, Belgium

D. A. Fisher

University of California, Los Angeles
Harbor-UCLA Medical Center
Torrance, California

and

D. Glinoer

University of Brussels
Brussels, Belgium

In association with

J. H. Dussault

Université Laval
Sainte-Foy, Québec, Canada

A. M. Ermans

University of Brussels
Brussels, Belgium

M. Irie

Toho University
Tokyo, Japan

Plenum Press
New York and London
Published in cooperation with NATO Scientific Affairs Division

Proceedings of a NATO Advanced Research Workshop on
Research in Congenital Hypothyroidism,
held May 16-18, 1988,
in Brussels, Belgium

Under the sponsorship of
North Atlantic Treaty Organization (NATO, Brussels)
Commission of the European Communities (CEC, Brussels)
Fonds National de la Recherche Scientifique (FNRS, Belgium)
Loterie Nationale (Belgium)
Ministère de l'Education Nationale (Belgium)

Library of Congress Cataloging in Publication Data

NATO Advanced Research Workshop on Research in Congenital Hypothyroidism (1988: Brussels, Belgium)
Research in congenital hypothyroidism / edited by F. Delange, D. A. Fisher, and D. Glinoer.
p. cm.—(NATO ASI series. Series A, Life sciences; v. 161)
"Proceedings of a NATO Advanced Research Workshop on Research in Congenital Hypothyroidism, held May 16-18, 1988, in Brussels, Belgium"—T.p. verso.
"Published in cooperation with NATO Scientific Affairs Division."
Includes bibliographies and indexes.
ISBN 978-1-4684-7582-1 ISBN 978-1-4684-7580-7 (eBook)
DOI 10.1007/978-1-4684-7580-7
1. Congenital hypothyroidism—Congresses. I. Delange, F. II. Fisher, Delbert A., 1928- . III. Glinoer, Daniel. IV. North Atlantic Treaty Organization. Scientific Affairs Division. V. Title. VI. Series.
[DNLM: 1. Hypothyroidism—congenital—congresses. WK 250 N279r 1988]
RJ274.7.N37 1988
616.4′44043—dc19
DNLM/DLC 89-3716
for Library of Congress CIP

Softcover reprint of the hardcover 1st edition 1989
A Division of Plenum Publishing Corporation
233 Spring Street, New York, N.Y. 10013

RESEARCH IN CONGENITAL HYPOTHYROIDISM

Proceedings of the NATO Advanced Research Workshop
Brussels, Belgium May 16-18, 1988

Program Committee

F. Delange (Brussels)
J.H. Dussault (Québec)
A.M. Ermans (Brussels)
D.A. Fisher (Torrance)
M. Irie (Tokyo)

Local Organizing Committee

F. Delange
A.M. Ermans
D. Glinoer
P. De Nayer
G. Van Vliet
J.P. Chanoine

Université Libre de Bruxelles et
Université Catholique de Louvain

Financial grants in support of this meeting have been received from the following contributors and were deeply appreciated :

Abbott Diagnostics Division Belgium
ACEC, Belgium
Amersham Belgium
Baxter Limited, Japan
Becton Dickinson Belgium
Boehringer Pharma Belgium
Chiba Corning Diagnostics KK, Japan
Christiaens, Belgium
Daiichi Radioisotope Laboratories, Ltd, Japan
Dainabot Co., Ltd, Japan
Eiken Chemical Co., Ltd, Japan
Fonteyn Medical Books, Belgium
Fujirebio Inc., Japan
IRE Medgenics Benelux
Kabivitrum Belgium
Labaz-Sanofi Belgium
Mallinckrodt Diagnostica Belgium
Medatom Belgium
Merck Sharp-Dohme Belgium
Nestlé-Guigoz Belgium
Nippon DPC Corporation, Japan
Pfizer Belgium
Pharmacia Belgium
Nutricia Belgium
Special reference Laboratories, Japan
Tomakomai Clinical Laboratories, Japan
Toshiba Medical Systems Belgium

PREFACE

Systematic screening for congenital hypothyroidism in the newborn was introduced some 15 years ago. The main objective was the prevention of mental retardation due to thyroid hormone deficiency during the early months of life. During the past decade screening programs have become routine throughout most of the industrialized world and many questions relating to implementation, organization and quality control of such programs have been largely resolved. Preliminary IQ and neurological data have indicated that screening and early treatment do, in fact, prevent mental retardation. However, a number of scientific questions related to congenital hypothyroidism remain unanswered and extensive research activities are ongoing in the field. The objective of the organizers of the Brussels workshop was to focus almost exclusively on these current research aspects of the screening programs.

This workshop is the third international conference specifically devoted to neonatal thyroid screening. The first was held in La Malbaie in Quebec in the fall of 1979. That meeting was well organized and highly productive. Its proceedings constitute a bible in the field. After the Quebec meeting, we witnessed major and rapid advances in our understanding of neonatal thyroid physiology as well as screening methodology, organization, data management, the significance of an approach to false negative and false positive results, patient follow-up, and assessment of follow-up and treatment, and the psychoneurological evaluation of affected infants. Some of these aspects were further developed during a second highly productive international conference organized in Tokyo in 1982.

About two years ago, the members of the Program Committee began planning a third workshop on congenital hypothyroidism to focus on current research in the field which would benefit from in depth discussion by a working group of about 100 members. Several topics were selected as indicated by the program, but there was the inevitable frustration of having to omit other topics of great potential interest such as calcium metabolism in congenital hypothyroidism, fetal growth in the absence of thyroid hormones, the critical analysis of the false negative and missed cases, etc...

The Program Committee is grateful to the 25 workshop speakers who accepted the invitation to participate and share with us their unique experience in the field. The Committee also appreciates the interest of others attending the meeting who, because of their experience in the field, and by poster presentations, contributed importantly to the success of the workshop.

The Local Organizing Committee acknowledges the support of the several National and International Agencies which co-sponsored the meeting. These include :

The North Atlantic Treaty Organization (NATO)
The Commission of the European Communities (CEC)
Fonds National de la Recherche Scientifique (FNRS) of Belgium,

Ministère de l'Education Nationale of Belgium.
Loterie Nationale, Belgium.

The proceedings of the workshop will be included among the Advanced Research Workshops of NATO.

Finally, important parts of the meeting were made possible by contributions of several European and Japanese companies.

These proceedings include articles resulting from the plenary lectures and their discussions prepared by the editors from the tapescripts of the workshop, abstracts of poster presentations, and the final conclusions and recommendations of the participants. The editors considered that this volume constitutes a significant step in the accumulation in knowledge on neonatal thyroid screening.

François Delange

Delbert Fisher

Daniel Glinoer

CONTENTS

PART 1. PHYSIOPATHOLOGY

PART 2. ETIOLOGY

PART 3. DIAGNOSIS

PART 4. THERAPY AND NEUROPSYCHOINTELLECTUAL DEVELOPMENT

PART 5. ABSTRACTS OF THE POSTERS SESSION

Moderators: C. Beckers and G. Giovannelli

PART 1. PHYSIOPATHOLOGY

PLACENTAL TRANSFER OF SUBSTANCES FROM MOTHER TO FETUS AFFECTING FETAL PITUITARY-THYROID FUNCTION

Lewis E. Braverman

University of Massachusetts School of Medicine

Worcester, MA

INTRODUCTION

Over the past years, especially since screening programs for the detection of congenital hypothyroidism have been established, a great deal of interest has centered on the placental transfer of hormones and drugs which might affect fetal thyroid function. Drs. Morreale de Escobar and Emèrson will discuss in detail the possible transfer of thyroid hormones from mother to fetus and the role of the placenta in fetal thyroid homeostasis. The present discussion will focus on studies carried out in collaboration with Dr. Elio Roti from the University of Parma, Parma, Italy and investigators in my laboratory, all related to the transplacental passage of a variety of substances known to affect adult hypothalamic-pituitary-thyroid function (Table 1).

TABLE 1. PLACENTA TRANSFER

1. Iodide
2. TRH
3. SRIF
4. Metoclopramide
5. Bromocriptine
6. PTU-MMI
7. TSI
8. Thyroid Blocking Ab
9. Beta Blocking Drugs

Iodide

The development of hypothyroidism, with or without goiter, has been described after iodine excess in patients with underlying thyroid disorders such as Hashimoto's thyroiditis, previous surgical or RAI treatment for Graves' disease, hemithyroidectomy for benign lesions, enzymatic defects in hormone synthesis and in patients treated with drugs such as lithium, phenazone and sulfisoxazole and, in some instances, in subjects with an apparently normal functioning thyroid gland (1,2). It has also been reported that exposure of the fetal thyroid gland to large quantities of iodine may induce goiter and hypothyroidism in the newborn (1). The fetal thyroid may be exposed to large amounts of iodine after amniofetography employing X-ray contrast agents or secondary to maternal exposure to oral or

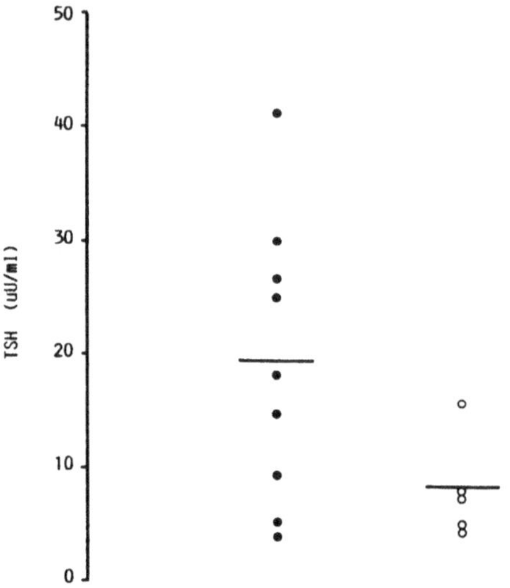

Fig. 1 - Individual cord blood thyrotropin (TSH) concentration in iodine exposed (●) and control (○) newborns. The horizontal lines represent the mean of the two groups. A significant difference, $p<0.005$, was observed (Student's test).

topical medications containing iodine. A significant increase in serum thyrotropin (TSH) concentrations has been observed in women who douched for 2 weeks with polyvinylpyrrolidone-iodine (Povidone-Iodine, PVP-I) (3). PVP-I, including vaginal douche and gel preparations, can be obtained without medical prescription and may be used during pregnancy. We evaluated in Parma, Italy, thyroid function in neonates at birth and in their mothers who used PVP-I as a vaginal douche or gel creme during the last trimester of pregnancy (4). These results were compared to those observed in newborns and mothers not exposed to PVP-I. This area of north Italy is moderately iodine deficient.

Maternal serum thyroxine (T_4), triiodothyronine (T_3), reverse triiodothyronine (rT_3) and TSH concentrations at delivery were not significantly different between the two groups of pregnant women. Cord blood thyroid hormone concentrations in the newborns of iodine exposed mothers were not significantly different from those in control newborns. In contrast, cord blood TSH concentrations in the neonates of mothers exposed to PVP-I during the last trimester of pregnancy were significantly higher than values in control neonates ($p<0.005$) (Figure 1). These data confirm that the fetal thyroid gland, even in the last trimester of pregnancy, does not adapt completely to the inhibitory action of iodine on thyroid hormone synthesis and/or release.

Administration of large doses of iodine to normal subjects results in a transient inhibition of thyroid hormone synthesis, the acute Wolff-Chaikoff effect (5). However the thyroid resumes normal hormone production, despite the continuous administration of pharmacological doses of iodine. This event is termed escape from or adaptation to the acute Wolff-Chaikoff effect (6,7). However, in subjects with underlying thyroid disorders or patients treated with compounds with weak antithyroid activity, the escape phenomenon might not occur and hypothyroidism could develop (2). The fetal thyroid is able to trap iodide by the tenth week of pregnancy (8) and is particularly prone to the inhibitory effect of iodine on hormonogenesis since the fetal thyroid develops the ability to escape from the inhibitory action of iodine only during the last 4 weeks of term pregnancy (9). The fetal thyroid may be exposed to large amounts of iodine from the maternal circulation since the placental is freely permeable to iodine (10). In this regard, Delange and his coworkers have reported that daily douching with PVP-I for 7 days prior to abortions carried out at 16-26 weeks of gestation resulted in a marked increase in amniotic fluid and fetal thyroid iodine content (11). Our present results are in agreement with previous studies reporting increased neonatal TSH concentrations after maternal iodine exposure. Neeb et al (12) observed a transient increase in blood TSH concentrations in 9 of 120 neonates whose mothers had used vaginal PVP-I as prophylaxis against vaginal infections. Screening 30,000 neonates in Lower Saxony, West Germany, Sander et al. detected 7 neonates with transient elevations of serum TSH concentration related to vaginal use of PVP-I (13).

Use of PVP-I for perineal preparation during labor and delivery resulted in a significant increase in neonatal serum TSH concentrations measured on days 3 and 5 of perinatal life (14). However, a more recent study did not reveal elevated cord blood TSH concentrations following the use of PVP-I perineal prep during delivery, despite increased maternal and serum total iodine concentrations (15). Transient hypothyroidism has also been reported in neonates treated with topical PVP-I preparations, especially when applied to the skin frequently and when the skin was denuded (4). It has been suggested that iodine induced hypothyroidism following topical application of PVP-I is more common in premature infants (16), perhaps due to the increased skin permeability present in premature infants. Furthermore, we have reported in the rat that escape from the acute inhibitory effect of iodine on thyroid hormone synthesis occurs in the late neonatal period which corresponds to the last few weeks of term human fetal life (17). In contrast with the majority of these studies, others have reported that iodine exposure during fetal and perinatal life is not accompanied by significant changes in thyroid function (18,19).

It is of interest that the vast majority of studies demonstrating transient perinatal hypothyroidism secondary to excess iodine exposure emanate from continental Western Europe where mild to moderate iodine deficiency is present. Whether this is due to the iodine deficiency or to the increased use of iodine containing medications during pregnancy and the perinatal period is not known. Another possible example of the difference in the neonatal response to excess iodine, depending upon ambient iodine intake, has recently been reported. At this meeting, Delange et al. reported that iodine overload in severely ill newborns results in transient hypothyroidism in Brussels (moderate iodine deficiency) while we have observed little or no change in blood TSH concentrations in very ill neonates exposed to excess iodine in Worcester, MA, USA (iodine sufficient) (Brown, R.S. and Braverman, L.E., unpublished observations).

Since the fetus and neonate appear to be at greater risk for the development of iodine induced thyroid hypofunction, it is advisable to avoid iodine containing medications during this more vulnerable period.

<u>Thyrotropin Releasing Hormone (TRH)</u>

TRH has been demonstrated to stimulate the fetal pituitary in the rat (20), rhesus monkey (21), and lamb (22). In many of these reports it was evident that TRH crosses the placenta, since increased TSH secretion occurred after maternal administration of TRH. No information was available as to whether TRH is capable of crossing the human placenta or stimulating the human thyrotroph antenataly.

Evidence has been presented in the ovine species that under appropriate experimental conditions, the neonatal T_3 surge can be partially disassociated from the neonatal TSH surge (23). Therefore, it was also of interest to determine the T_3 and T_4 responses of the perinatal thyroid to

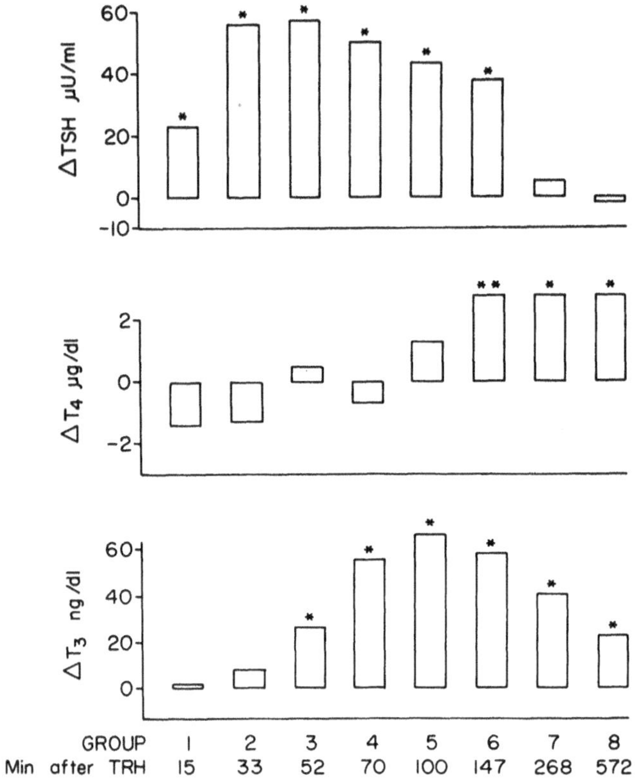

Fig.2 Differences between CB TSH, T_4, and T_3 concentrations in TRH-injected groups compared to those in group 0 which received no TRH. The data are plotted in order of the duration between maternal TRH injection and CB sampling. *, P<0.01 vs. group 0; **, P<0.05 vs. group 0.

endogenous TSH secreted as a result of TRH stimulation.

TRH or saline was administered to 214 term pregnant women at various time intervals (8-820 min) before delivery. Cord blood (CB) was obtained, and plasma TSH, T_4, T_3, rT_3, and thyroglobulin concentrations were measured by RIA (Figure 2). CB TSH was significantly elevated within 20 min after TRH administration and remained elevated for 180 min. CB T_3 rose significantly by 60 min and remained elevated for 820 min. CB T_4 was significantly increased from 120 to 820 min after TRH administration. There was no significant change in the CB thyroglobulin concentration. These findings demonstrated for the first time that TRH crosses the human placenta, that the fetal pituitary is responsive to TRH, and that endogenous TSH stimulates the fetal thyroid (24).

It had been suggested that measurement of amniotic fluid (AF) TSH could detect primary fetal hypothyroidism and permit the intrauterine treatment of hypothyroidism prior to birth (25). The usefulness of such a maneuver is questionable since thyroid hormone therapy for neonatal hypothyroidism after birth is so successful. Although we could not detect a rise in AF TSH following TRH administration to the mother during labor (26), AF TSH has been found to be elevated in fetal hypothyroidism using a sensitive TSH assay (27).

In normal subjects, TRH administration induces increases in serum TSH and PRL concentrations. However, a number of pathological states, including acromegaly and primary hypothyroidism, are associated with an enhanced serum GH response to TRH (28). Thyroid function in the term normal fetus is characterized by normal free T_4, low T_3, and relatively elevated TSH concentrations in the serum.

In view of the possible mild hypothyroidism which exists during fetal life, the effect of 400 ug TRH administered to mothers during labor on the fetal GH response was evaluated. TRH was administered iv to 59 pregnant women just before delivery. Cord blood (CB) samples were obtained at delivery and assigned to 6 groups, depending upon the duration of time between TRH injection and CB sampling. The control group comprised 79 pregnant women who received saline. A progressive rise and then a fall in CB GH concentrations were observed after TRH administration. Values were significantly elevated 61-90 min after TRH administration compared to values in saline-treated subjects (19.3 ± 3.1 vs 13.1 ± 0.9 ng/ml; $P<0.05$), demonstrating that TRH stimulates GH release in the fetus (28).

Somatostatin (SRIF)

In adults, SRIF administration reduces basal and stimulated secretion of GH and TSH. Studies were carried out in women during labor to determine whether SRIF crosses the human placenta and, if so, whether it would affect fetal pituitary function. SRIF (500 ug) was infused over 30 min to 68

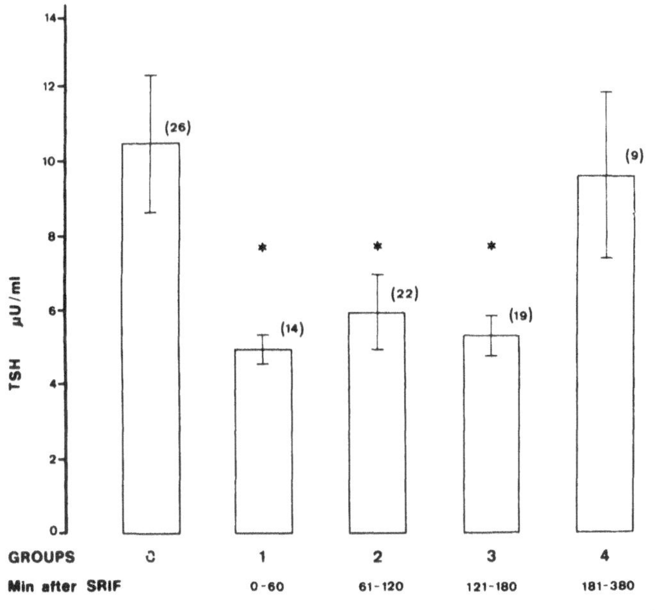

Fig. 3 Cord blood (CB) thyrotrophin (TSH) concentration in neonates whose mothers were treated with saline, group 0, and somatostatin (SRIF), groups 1 to 4. Number in parenthesis indicates the number of subjects in each group. The bars represent the mean values and the brackets the SE. Min after SRIF represent the elapsed time between the end of the SRIF infusion to the mother and CB samples at parturition. *P<0.05 vs group 0 by Duncan's test.

pregnant women during labor and, as a control, saline was infused into 26 pregnant women. There was a marked decrease in CB TSH from 0 to 180 min following SRIF administration (Figure 3), but no effect on serum T_4, T_3 and rT_3 concentrations was observed. CB GH was also markedly reduced from 0 to 120 min after SRIF infusion. This study strongly suggests that SRIF readily crosses the placenta and transiently suppresses fetal anterior pituitary TSH and GH secretion (29). Although no change in CB glucose concentration was observed, SRIF has been reported to be effective in treating neonatal hypoglycemia without adverse side effects (30).

Metoclopramide (MET) and Bromocriptine

The role of dopamine (DA) in the regulation of TSH secretion has been suggested by the observations that pharmacological quantities of infused DA and L-dopa lower basal TSH secretion in normal and hypothyroid subjects (31,32). Moreover, the administration of L-dopa and DA produced a marked decrease in the TSH response to exogenous TRH. These observations led to the postulate of an inhibitory dopaminergic effect on TSH secretion. This conclusion was further substantiated by the observation that the administration of DA agonists resulted in a decrease in basal and TRH-stimulated serum TSH concentrations in normal subjects and hypothyroid patients (33). The physiological role of the dopaminergic inhibitory effect on TSH secretion was also documented by the finding that the administration of MET, a DA receptor-blocking agent, almost always resulted in increased circulating TSH concentrations (34). An increased dopaminergic inhibitory tone on TSH secretion has been observed at night, accompanied by greater TSH release after MET administration (35). Sowers et al. (36) reported that changes in dopaminergic activity do play a role in the circadian variations in TSH secretion. In the few studies that have been carried out during pregnancy and the puerperium, the importance of dopaminergic control of TSH secretion is less evident.

MET, a potent dopamine receptor-blocking drug which readily crosses the placenta, or saline, were administered to 125 term pregnant women at various time intervals (5-412 min) before delivery. Maternal serum was obtained before and after MET injection. Cord blood was obtained at delivery in MET-treated and saline-treated (control group) women. No significant changes in serum TSH, T_4, T_3, or rT_3 concentrations were observed in maternal serum or cord blood after MET administration (37).

These results suggest that, in contrast to euthyroid nonpregnant women and men, MET administration does not induce a rise in serum TSH concentration in term pregnant women or in the term fetus. Thus, the dopaminergic inhibitory effect on anterior pituitary TSH secretion may not be an important factor in TSH regulation during pregnancy or in the fetus, or the dose of MET employed may be unable to overcome the dopamine inhibitory effect.

These findings were further confirmed by the failure of the dopamine agonist, bromocriptine, 5 mg administered orally to women during labor, to decrease CB TSH concentrations although maternal serum TSH concentrations were decreased (38).

Tubero-infundibular DA is generally accepted as the major inhibitor of Prl secretion in man and animals. During pregnancy, markedly elevated serum Prl concentrations are usually present by the first trimester and continue elevated throughout pregnancy. Dopaminergic control of Prl secretion during pregnancy is suggested by the observations that the administration of antidopaminergic drugs to pregnant women significantly increases serum Prl concentrations (39) whereas the administration of dopaminergic drugs markedly decreases Prl secretion (40).

In our 2 studies cited above, bromocriptine administration markedly inhibited both maternal and fetal Prl concentrations. In contrast, MET administration during labor failed to increase fetal serum Prl concentration. However, these latter observations do not negate the role of DA in the regulation of fetal Prl secretion but rather suggest that Prl secretion is maximal in the term fetus and cannot be further stimulated.

Propylthiouracil (PTU), Methimazole (MMI) and Carbimazole (CBZ)

The thionamide antithyroid drugs, PTU, MMI, and CBZ, are widely used in the therapy of Graves' disease and toxic nodular goiter. Many studies conducted in animals have reported the development of fetal goiter and hypothyroidism after maternal treatment with these antithyroid agents. These observations have provided good evidence for the transplacental passage of the thionamide drugs.

Treatment of hyperthyroidism with these drugs during pregnancy is effective in controlling maternal hyperthyroidism. Since they readily cross the placenta, the smallest effective dose should be employed in order to avoid fetal thyroid insufficiency. Cooper has recently summarized some of the pharmacologic features of the antithyroid drugs and it is evident that MMI and CBZ traverse the placenta and enter breast milk far more readily than does PTU (41). This strongly suggests that PTU is the antithyroid drug of choice in the treatment of hyperthyroidism during pregnancy and during the post partum period. Fetal goiter resulting from maternal antithyroid drug treatment has been observed (10). Doses ranging from 300 to 800 mg PTU daily may induce fetal hypothyroidism. In a study in which pregnant women with Graves' disease were treated with the apparent lowest dose of PTU which maintained euthyroidism in the mother, a mild decrease in the serum T_4 concentration with a concomitant increment of serum TSH was observed in the newborn for the first 4 postpartum days (42). Unfortunately, 7 of the 11 patients also received thyroid hormone therapy and even lower doses of PTU might have been used in these 7 patients if exogenous thyroid hormone had not been employed. These findings also suggest that the placenta is not freely permeable to the administered thyroid hormones since they failed to prevent the rise in fetal TSH induced by PTU administration.

Thyroid Stimulating and Blocking Antibodies

It is beyond the scope of this presentation to discuss the pathophysiology of neonatal Graves' disease. The etiology of this almost always self limited disorder is the transplacental passage of thyroid stimulating immunoglobulins (TSI) from mothers with Graves' disease to their fetuses. Indeed, the occurrence of neonatal Graves' disease can often be predicted by the detection of high titers of TSI in the sera of pregnant women during the last trimester. Most recently, the placental transfer of maternal TSH receptor, TSI, and thyroid growth blocking antibodies has been suggested to be casually related to the occurrence of sporadic, congenital hypothyroidism. Although such antibodies may be associated with congenital hypothyroidism in some babies whose mothers have primary, autoimmune myxedema, Brown and her colleagues and others at this meeting strongly suggest that maternal blocking antibodies rarely cause sporadic congenital hypothyroidism. Further discussion of this problem will be carried out later today in the autoimmunity session.

Beta Blocking Drugs (10)

The beta blocking drugs, especially propranolol, have been successfully employed in the treatment of hyperthyroidism and are effective in relieving those symptoms related to catecholamine activity such as tremor, sweating, nervousness, tachycardia and increased cardiac output.

Propranolol crosses the placenta and its concentration in cord blood is at times even higher than that observed in maternal blood and may increase even further during the neonatal period. Reduced metabolism of the drug or a reduced volume of distribution during perinatal life may account for these findings. Propranolol has been reported to delay cervical effacement and dilation, thereby prolonging labor, and to decrease Apgar scores in the newborn.

Fetal endocrine and metabolic effects induced by the maternal administration of propranolol have not been comprehensively studied. Neonatal hypoglycemia has been reported. The effects of maternal or neonatal propranolol treatment on fetal iodothyronine metabolism or hypothalamic-pituitary function during the perinatal period has not beem carefully studied.

Acknowledgements

I am indebted to Ms. Gloria Ober for her excellent secretarial assistance. This work was supported in part by Grant AM 18919 NIDDK, NIH, Bethesda, MD.

References

1. J. Wolff: Iodide goiter and the pharmacologic effects of excess iodide. Am. J. Med. 47:101, (1969).
2. A.G. Vagenakis, L.E. Braverman: Adverse effects of iodide on thyroid function. Med. Clin. North Am. 59:1075, (1975).
3. M. Safran, L.E. Braverman: Effect of chronic douching with polyvinylpyrrolidone-iodine on iodine absorption and thyroid function. Obstet. Gynecol, 60:35, (1982).
4. G. Robuschi, M. Montermini, A. Alboni, E. Borciani, G. Cersosimo, L. Negrotti, A. Gnudi, M. Safran, L.E. Braverman, and E. Roti: Cord blood iodothyronine and thyrotropin concentrations in newborns of mothers exposed to povidone iodine in the last trimester. J. Endocrinol. Invest., 10:183, (1987).
5. J. Wolff, I.L. Chaikoff: Plasma inorganic iodide as a homeostatic regulator of thyroid function. J. Biol. Chem., 174:555, (1984).
6. J. Wolff, I.L. Chaikoff, R.C. Goldberg, J.R. Meier: The temporary nature of the inhibitory action of excess iodide on organic iodide synthesis in the normal thyroid. Endocrinology, 45:504, (1949).
7. L.E. Braverman and S.H. Ingbar: Change in thyroidal function during adaptation to large doses of iodide. J. Clin. Invest., 42:1216, (1963).
8. T.H. Shepard: Onset of function in the human fetal thyroid: biochemical and radioautographic studies from organ culture. J. Clin. Endocrinol. Metab., 27:945, (1967).
9. D.A. Fisher, A.H. Klein: Thyroid development and disorders of thyroid function in the newborns. N. Engl. J. Med., 304:702, (1981).
10. E. Roti, A. Gnudi, and L.E. Braverman: The placental transport, synthesis and metabolism of hormones and drugs which affect thyroid function. Endocrine Reviews, 4:132, (1983).
11. I. Mahillon, W. Peers, A.M. Ermans and F. Delange: Effect of vaginal douching with povidone-iodine during early pregnancy on the iodine supply to mother and fetus. Ann. d'Endocrinologie, 48:167, (1987).

12. U. Neeb, I. Caesar, E. Krause: Passegere TSH-Erhohung des neugeborenen nach vaginaler Polyvinylpyrrolidon-Jod-Application sub partu. Geburtshilfe u Frauenheilk, 39:973, (1979).
13. J. Sander, Ch. Niehaus: Konnatale hypothyreose. Munch. Med. Wochenschr, 122:309, (1980).
14. D. L'Allemand, A. Gruters, P. Heidemann, P. Schurnbrand: Iodine-induced alterations of thyroid function in newborn infants after prenatal and perinatal exposure to povidone iodone. J. Pediatr., 102:935, (1935).
15. L.K. Bachrach, G.N. Burrow: Maternal-fetal absorption of povidone-iodine. J. Pediatr., 104:158, (1984).
16. R.S. Brown, L.E. Braverman: Transient hypothyroidism in a premature infant after skin preparation with iodine. Pediatr. Res., 14:475, (1980).
17. T. Theodoropoulos, L.E. Braverman, A.G. Vagenakis: Iodide-induced hypothyroidism:a potential hazard during perinatal life. Science, 205:502, (1979).
18. N. Etling, F. Gehin-Fouque, J.P. Viehl, J.P. Gautray: The iodine content of amniotic fluid and placenta transfer of iodinated drugs. Obstet. Gynecol., 53:376, (1979).
19. S.P. Pyati, R.S. Ramamurthy, M.T. Krauss, R.S. Pildes: Absorption of iodine in the neonate following topical use of povidone iodine. J. Pediatr., 91:825, (1977).
20. A. Kajihara, A. Kojima, T. Onaya, Y. Takemura, T. Yamada: Placenta transport of thyrotropin releasing factor in the rat. Endocrinology, 90:592, (1972).
21. M. Azukizawa, Y. Murata, T. Ikenoue, C.B. Martin, Jr., J.M. Hershmann: Effect of thyrotropin-releasing hormone on secretion of thyrotropin, prolactin, thyroxine, and triiodothyronine in pregnant and fetal rhesus monkeys. J. Clin. Endocrinol. Metab., 43:1020, (1976).
22. A.H. Klein, D.A. Fisher: Thyrotropin-releasing hormone-stimulated pituitary and thyroid gland responsiveness and 3,5,3'-triiodothyronine suppression in fetal and neonatal lambs. Endocrinology, 106:697, (1980).
23. J. Sack, M. Beaudry, P.V. DeLamater, W. Oh, D.A. Fisher: Umbilical cord cutting triggers hypertriiodothyroninemia and non-shivering thermogenesis in the newborn lamb. Pediatr. Res., 10:169, (1976).
24. E. Roti, A. Gnudi, L.E. Braverman, G. Robuschi, R. Emanuele, P. Bandini, L. Benassi, A. Pagliani, and C.H. Emerson: Human cord blood concentrations of thyrotropin, thyroglobulin, and iodothyronines after maternal administration of thyrotropin-releasing hormone. J. Clin. Endocrinol. and Metab., 53:813, (1981).
25. I.A. Kourides, C.V. Heath, F. Ginsberg-Fellner: Measurement of thyroid-stimulating hormone in human amniotic fluid. J. Clin. Endocrinol. Metab., 54:635, (1982).
26. G. Robuschi, L.E. Braverman, R. Emanuele, L. d'Amato, E. Gardini, M.S. Foscolo, C. Gualerzi, L. Benassi, A. Gnudi, E. Roti: Amniotic fluid thyrotropin (TSH) following maternal administration of thyrotropin releasing hormone. J. Perinat. Med., 13:219, (1985).
27. K. Yoshida, T. Sakurada, T. Takahashi, N. Furuhashi, K. Kaise and Y. Yoshinaga: Measurement of TSH in human amniotic fluid: diagnosis of fetal thyroid abnormality in utero. Clin Endocrinol. 25:313, (1986).

28. E. Roti, A. Gnudi, G. Robuschi, R. Emanuele, L. Benassi, and L.E. Braverman: Response of growth hormone to thyrotropin-releasing hormone during fetal life. J. Clin. Endocrinol. and Metab., Vol 54, 6:1255, (1982).
29. E. Roti, G. Robuschi, Alessandro Alboni, R. Emanuele, L. d'Amato, E. Gardini, M. Salvi, E. Dall'AGlio, A. Gnudi, and L.E. Braverman: Inhibition of foetal growth hormone (GH) and thyrotrophin (TSH) secretion after maternal administration of somatostatin. Acta Endocrinologica, 106:393, (1984).
30. H.F. Kitson, R.B. McCrossin, M. Jimenez, A. Middleton, and M. Silink: Somatostatin treatment of insulin excess due to B cell adenoma in a neonate. J. Pediatr., 96:145, (1980).
31. F. Massara, F. Camanni, V. Vergano, L. Belforte, G.M. Molinatti: Inhibition of thyrotropin and prolactin secretion by dopamine in man. J. Endocrinol. Invest. 1:25, (1978).
32. C.M. Feek, J.S.A. Sawers, N.S. Brown, J. Seth, W.J. Irvine, A.D. Toft: Influence of thyroid status on dopaminergic inhibition of thyrotropin and prolactin secretion: evidence for an additional feedback mechanism in the control of thyroid hormone secretion. J. Clin. Endocrinol. Metab., 51:585, (1980).
33. K. Miyai, T. Onishi, M. Hosokawa, K. Ishibashi, Y. Kumahara: Inhibition of thyrotropin and prolactin secretion in primary hypothyroidism by 2-Br-ergocryptine. J. Clin. Endocrinol. Metab. 39:391, (1974).
34. J.R. Sowers, R.W. McCallum, J.M. Hershman, H.E. Carlson, R.A.L. Sturdevant, N. Meyer: Comparison of metoclopramide with other dynamic tests of prolactin secretion. J. Clin. Endocrinol. Metab. 43:679, (1976).
35. M.F. Scanlon, A.P. Weetman, M. Lewis, M. Pourmand, M.D. Rodriquez-Arnao, D.R. Weightman, R. Hall: Dopaminergic modulation of circadian thyrotropin rhythms and thyroid hormone levels in euthyroid subjects. J. Clin. Endocrinol. Metab., 51:1251, (1980).
36. J.R. Sowers, R.A. Catania, J.M. Hershman: Evidence for dopaminergic control of circadian variations in thyrotropin secretion. J. Clin. Endocrinol. Metab., 54:673, (1982).
37. E. Roti, G. Robuschi, R. Emanuele, L. d'Amato, A. Gnudi, M. Fatone, L. Benassi, M.S. Foscolo, C. Gualerzi, and L.E. Braverman: Failure of metoclopramide to affect thyrotropin concentration in the term human fetus. J. Clin. Endocrinol. and Metab., 56:1071, (1983).
38. E. Roti, G. Robuschi, A. Alboni, L. d'Amato, M. Montermini, E. Gardini, M. Salvi, E. Borciani, E. Dall'Aglio, S. Bisi, G. Zammarchi, R. Lasagni, A. Gnudi, and L.E. Braverman: Human foetal prolactin but not thyrotropin secretion is decreased by bromocriptine. Acta Endocrinologica, 112:35, (1986).
39. A. Kauppila, S. Kivinen, and O. Ylikorkala: Metoclopramide increases prolactin release and milk secretion in puerperium without stimulating the secretion of thyrotropin and thyroid hormones. J. Clin. Endocrinol. Metab., 52:436, (1981).
40. E.E. Muller, A.R. Genazzani, S. Murru, and P. Fioretti: Effect of piribedil on plasma prolactin levels in women with puerperal or pathological hyperprolactinaemia. Acta Endocrinologica. (Copenh). 86:33, (1977).
41. D.S. Cooper: Antithyroid drugs. N. Engl. J. Med., 311:1353, (1984).
42. R.G. Cheron, M.M. Kaplan, P.R. Larsen, H.A. Selenkow, J.F. Crigler, Jr.: Neonatal thyroid function after propylthiouracil therapy. N. Engl. J. Med., 364:525, (1981).

DISCUSSION

PINCHERA (Pisa) : My question is related to the regulation of TSH secretion and the importance of the dopaminergic pathway. How do you reconcile the effect observed in pregnant women with bromocriptin and the lack of effect of metoclopramide on TSH secretion (compared to prolactin secretion) with the effects observed in non pregnant women?

BRAVERMAN (Worcester) : The discrepancy is perhaps dose-related. In the pregnant woman, the dose of metoclopramide employed was 10 mg (in bolus i.v. injection) which induced a decrease in prolactin but had no effect on TSH. In the newborn, however, there was no effect on either prolactin or TSH. This is in contrast to what one sees if you give metoclopramide to a non pregnant woman, suggesting perhaps that during pregnancy, at least as far as metoclopramide (the dopamine antagonist) is concerned, the effect is not as distinct as it is in the non pregnant woman. Bromocriptin on the other hand (it may again be a dose-related phenomenon) seems to affect both prolactin and TSH during pregnancy.

PINCHERA (Pisa) : What is your view on the possible effect of the administration of iodized oil (given to prevent cretinism) during the last trimester of pregnancy to promote - instead of prevent - hypothyroidism in the fetus?

BRAVERMAN (Worcester) : The administration of iodized oil does not commonly induce hypothyroidism in the newborn in areas of the world where there is marked iodine deficiency. That is a paradox because there is certainly no question that amniofetography during pregnancy (using Bethadine preparations at the time of birth) can indeed induce elevations in fetal TSH. It may be that in the parts of the world where iodized oil has been administered, the low ambient intake is such that the mother traps much more the iodine, so that it may be less available to the fetus.

CZERNICHOW (Paris) :Do you think that there might be a different regulation of TSH and prolactin during the early minutes of life to explain the acute rise of TSH and also of prolactin (described by Sack and Fisher 12 years ago) and that bromocriptin could play a role on this acute stimulation ?

BRAVERMAN (Worcester) : I do not think that the dopominergic pathway plays a role on the TSH rise because the changes observed with bromocriptin are very small and probably represent a pharmacological - and not a physiological - effect.

CZERNICHOW (Paris) : On your slides concerning vaginal douching, I noticed that in most babies the amniotic fluid iodide concentration was low and not significantly different from the non-douching pregnant women. Is this an indication that the transfer of iodide from vagina to fetus comes through the blood and not by direct transfer to the uterine cavity ?

BRAVERMAN (Worcester) : Yes, I think that is correct.

TRANSFER OF THYROID HORMONES FROM THE MOTHER TO THE FETUS

Gabriella Morreale de Escobar, María Jesús Obregón
and Francisco Escobar del Rey

Unidad de Endocrinología Experimental, Instituto
de Investigaciones Biomédicas, C.S.I.C., Facultad de
Medicina, Arzobispo Morcillo 4, 28029 Madrid (Spain)

INTRODUCTION

Impairment of thyroid function early in life results in developmental defects. The prevention of the CNS damage caused by congenital hypothyroidism is the aim which is being successfully achieved by neonatal thyroid screening programs. But CNS damage may also result from maternal hypothyroxinemia alone. Especially severe CNS damage is found in the endemic cretin of the neurological type born to iodine deficient mothers: these women are hypothyroxinemic throughout pregnancy, and their fetus has an impaired thyroid function caused by the low iodine supply. The CNS damage increases in severity from that caused by maternal hypothyroxinemia alone, congenital hypothyroidism alone, or the combined maternal and fetal impairment of thyroid function found in areas of marked iodine deficiency. But whereas the mental retardation caused by congenital hypothyroidism may still be prevented by early treatment of the newborn, the CNS damage caused by maternal hypothyroxinemia is prevented by adequate treatment during pregnancy, and the birth of neurological cretins requires adequate iodization of the mothers before conception, or very early in pregnancy. These associations between thyroid failure and CNS damage, and the experimental models used to study them, have been previously reviewed by us [1-4]. These reviews contain pertinent bibliography of the abundant work on the subject carried out by many investigators.

The above considerations have centered our attention on the problem of maternal-fetal thyroid hormone relationships. In view of the main interest of the present meeting, namely research in congenital hypothyroidism, we will give special attention to the possibility that maternal thyroid hormones might protect the developing brain of the hypothyroid fetus.

MATERNAL TO FETAL TRANSFER OF THYROID HORMONES BEFORE ONSET OF FETAL THYROID FUNCTION

Taking the day of appearance of the vaginal plug as gestational day (dg) = 0, uterine implantation occurs at 6 dg, and birth occurs 22.7 ± 0.2 (SEM) dg. By injecting radioiodide into pregnant rats, which were thyroidectomized (T) a few hours earlier, Nataf and Sfez [5], and Geloso [6] showed that the fetal thyroid starts secreting labeled T4 after 17.5-18 dg. Weiss and Noback [7] described changes in bone development of fetuses obtained at 16 dg from goitrogen-treated rats, that is, before onset of fetal thyroid secretion.

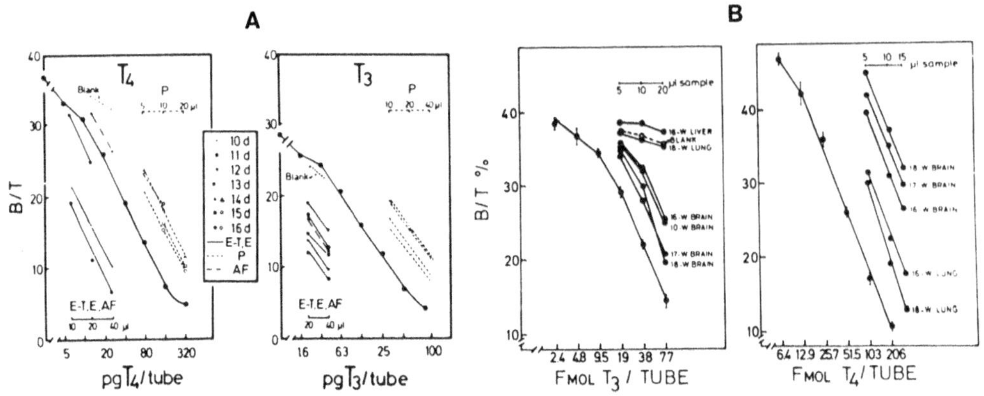

Figure 1 A and B.- 1A compares the T3 and T4 standard curves with serial dilutions of eluates obtained from embryo-trophoblasts (E-T), embryonic (E) placental (P) and amniotic fluid (AF) samples obtained in the rat at different gestational ages, shown in the inset. 1B shows similar curves with extracts from human fetal brains, between 10 and 16 weeks of gestation. From Obregón et al. [15] and Bernal and Pekonen [19] .

Sweney and Shapiro [8] showed labeled T4 and T3 in amniotic fluid, liver and palatal areas of 13 and 14-dg-old embryos from rats injected with labeled T4. Results from both studies clearly suggested the possibility that maternal thyroid hormones are available to the rat embryo before onset of fetal thyroid function, despite opinions to the contrary [9-13].

Having developed specific and sensitive RIAs [14], and extensive extraction and purification procedures which permitted the determination of very small concentrations of T4 and T3 in tissues [14-16], we decided to re-investigate this possibility by measuring steady-state levels of both iodothyronines in embryonic and fetal tissues obtained before onset of fetal thyroid function. Different types of samples were obtained from the embryonic compartment, at different stages of development, and after perfusion of the dams with phosphate buffered saline. As summarized in Fig. 1 A, all samples obtained from the fetal compartment before onset of fetal thyroid function contained both T4 and T3 [15]. They were found in molar ratios which were quite different from those of maternal plasma, thus excluding contamination with maternal blood. An independent study by Woods et al. [17] prompted by theoretical considerations [18], showed transfer of labeled iodothyronines from rat dams to their embryos very early in gestation. The results obtained in rats appear to be pertinent to early phases of human development. Bernal & Pekonen [19] have shown the presence of T3 in the brain of 10-weeks old human embryos Fig. 1 B). T4 was later also found (personal communication by J. Bernal) in a whole 7 week-old human embryo; the amount of tissue was too small to attempt the detection of T3. Secretion of T4 by the human fetal thyroid is very low until 16-18 weeks of gestation, and that of T3 until the last trimester [11].

Studies by Pérez-Castillo et al. [20] have shown the presence of nuclear receptors for T3 in rat embryos as early as 13 dg, in brain by 14 dg, and in liver, heart and lung from 16 dg onwards. Data for the rat brain are shown in Fig. 2 A. The nuclear T3 receptor was also found in the brain of 10 week-old human fetuses [19] (Fig. 2 B), and in a 7 week old whole embryo. Therefore, both the hormone and its receptor are present before onset of fetal thyroid function, and it does not appear far-fetched to suggest that a biological effect might ensue, though this possibility has not yet been confirmed nor the possible effect defined.

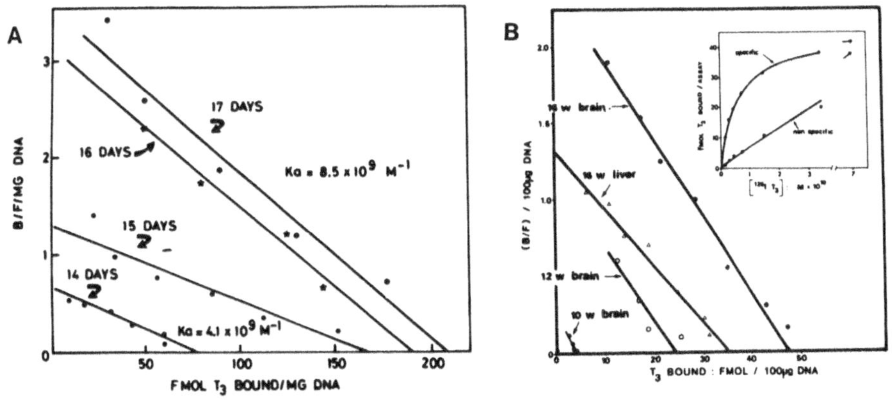

Figure 2. Scatchard plots for nuclear T3 binding obtained with rat fetal brain (Panel A) at different gestational ages, and with human fetal brain and liver (panel B) at 10, 12, and 16 weeks for brain, and 16 weeks for liver. From Perez-Castillo et al.[20], and Bernal and Peconen[19].

To confirm the maternal origin of the iodothyronines found in early embryonic samples in the rat, we determined their concentrations in fetal tissues obtained from normal (C) mothers and from dams which had been thyroidectomized (T) a few months before mating [16]. Data obtained from C dams extended and confirmed the previous ones regarding the presence of both T4 and T3 in all embryonic samples obtained well before the onset of fetal thyroid function. In contrast, when the embryonic samples had been obtained from T mothers, concentrations of both T4 and T3 were below the limits of detection in all tissues, placentas included, at least up to 17 dg (Fig. 3).

Starting at 18 dg, the thyroid was dissected out of the fetus before homogenization for the determination of T4 and T3. The extrathyroidal concentrations thus obtained are shown in Fig. 3 on a logarithmic scale. There is a marked increase in fetal T4 and T3 concentrations between 17 and 18 dg, in agreement with the timing previously reported [5,6] for onset of fetal thyroid secretion in the rat. T4 and T3 found in fetuses before and up to 17 dg are obviously of maternal origin.

Therefore, fetuses from T mothers develop in a thyroid hormone-deficient condition, at least up to 17.5-18 dg. Their number is markedly reduced (from 11.1 ± 0.4 to 6.8 ± 0.3), and the individual weight of the fetuses is smaller. Before onset of fetal thyroid function, fetal weight is reduced by maternal T to 59 % of C values [16]. After fetal thyroid function has started,

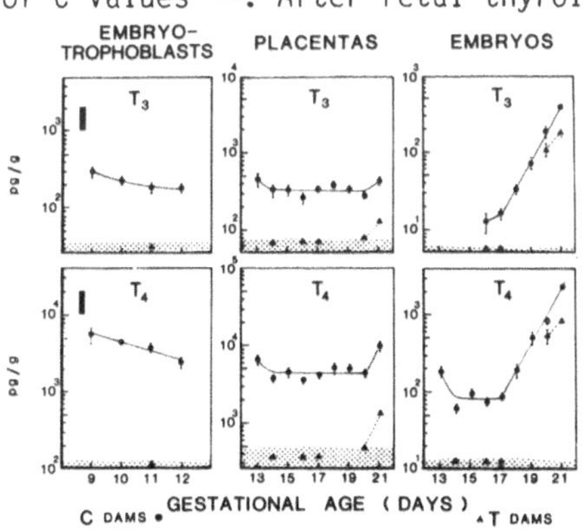

Figure 3.- T4 and T3 concentrations in embryo-trophoblasts (9-12 dg), placentas and fetuses (13-21 dg) obtained from C and T dams. Shaded areas correspond to the limit of detection. All differences between samples from C and T dams were statistically significant. From Morreale de Escobar et al.[16]

there appears to be a catch-up in growth, but even at 21 dg (near term) the body weight of fetuses from T dams is only 80 % of normal [16,21]. Bonet and Herrera [22] have recently described that maternal T impairs the physiological increase of maternal body weight (BW) which takes place during gestation. The BW of the 21-dg-old fetuses was also reduced. Treatment of the T dams with 1.8 μg T4 / 100 g BW / day throughout gestation prevented these alterations. Interestingly, treatment with the same T4 dose only during the first 12 days of pregnancy was enough to prevent these changes, whereas treatment from day 12 to 21 did not. The fetal pituitary growth hormone (GH) content was reduced to 31 % of that of fetuses from C dams. As indicated for the increment in maternal BW, and for the fetal BW, the decrease in fetal pituitary GH was prevented by treating the T dams with T4 during the first half of gestation; treatment only during the second half did not restore fetal pituitary GH. They also showed [23] that T dams left untreated during the first half of pregnancy have an impaired capacity to sustain fetal metabolic demands when food is withheld. These results clearly show the adverse effects of maternal hypothyroidism during a period of development when she is the only source of thyroid hormones, both for herself and for the developing conceptus. Later correction of both maternal hypothyroidism by T4 treatment, and of fetal thyroid hormone deficiency by onset of fetal thyroid function and possible maternal to fetal transfer of T4, cannot compensate for this initial damage. At present it is not possible to assess the degree to which different alterations caused by maternal T contribute to poor fetal development: maternal impairment in building up endogenous metabolic stores later needed for fetal development, deficiency of thyroid hormones in the embryonic tissues themselves, or both. Whichever the mechanism (s) involved, the adult progeny of T rats suffer from permanent behavioural defects [24]. These are not avoided by the treatment of the dams near term with GH, despite the fact that such a treatment reverses many of the metabolic alterations found in fetuses from T dams near term [25,26].

There are similarities between the results obtained in rats and those described in man. We have already indicated that both the hormone and the nuclear T3 receptor are found in the human fetal brain several weeks before onset of active thyroid secretion, and during a phase of forebrain development when active neuroblast division is starting in the forebrain (for reviews, see [1-4]). After 10 weeks of gestation, and coinciding with the active phase of neuroblast division, there was a rapid increase in cerebral DNA; receptor concentration increased ten-fold by 16 weeks of gestation [19]. Present information thus shows that there is maternal transfer of thyroid hormone in man before the beginning of the second trimester of pregnancy. Therefore, maternal hypothyroidism could result in decreased availability of thyroid hormone during the initial phases of development and during the spurt in forebrain neuroblast division. It is known that maternal hypothyroxinemia results in lowered I.Q. of the progeny, even in areas without iodine deficiency [27,28]. Although an incapacity of the hypothyroxinemic woman to maintain an adequate metabolic milieau for the fetus might play an important role, defects in their progeny can no longer be attributed only to this [29], and the possibility should also be considered that a deficiency of thyroid hormones in the embryo itself might play and adverse role. This has received very little attention, because it was believed that maternal thyroid hormones were not available to the developing embryo before midgestation.

MATERNAL TO FETAL TRANSFER OF THYROID HORMONES AFTER ONSET OF FETAL THYROID FUNCTION

We have summarized evidence that maternal thyroid hormones cross the placenta into the fetal compartment early in gestation. Once the fetal thyroid starts functioning, this transfer might decrease and eventually stop, or it might continue until term. It is experimentally difficult to investigate these possibilities. Labeled-iodine containing hormones found in the fetal compartment after their injection into the mother do not necessarily indicate transfer, as the fetal thyroid might have synthesized them from labeled iodine deiodinated from the hormone administered to the dam. Determinations of T4 and T3 in fetal tissues by RIA do not permit to distinguish between maternal and fetal contributions. Either the maternal or the fetal thyroid function has to be blocked, or both. But conclusions may then only be drawn for maternal-fetal relationships where either the mother, the fetus, or both, are hypothyroid, and cannot be extrapolated to normal conditions.

1) Normal maternal and fetal thyroid status

As reviewed up to 1967 by Geloso [6], if labeled T4 is injected into pregnant females of different species (rat, mouse, guinea pig, rabbit, sheep, pig, monkey) labeled iodothyronines are always found in the fetal plasma, in varying concentrations depending on the species and on the gestational phase, except in sheep and pig. There was less information regarding passage of labeled T3 from mother to fetus, which appeared to be important in the rat, and minimal in the guinea pig. As pointed out [6], some of the experiments were carried out with perfused guinea pigs placenta left in situ, and in monkeys where the specific activity of the thyroxine in fetal plasma and fetal thyroid were measured. Results from such experiments strongly suggested that the labeled iodothyronines found in the fetal compartment were mostly derived from transplacental transfer, not from fetal thyroid synthesis.

More recent experiments carried out in the rat [30,31] have approached the problem by determining T4 and T3 kinetics in pregnant rats at the end of gestation, using labeled iodothyronines and giving potassium iodide to block radioiodide recycling. Placental transfer rates were calculated from labeled iodothyronine concentrations in maternal and fetal plasma. Placental transfer of the hormones were detected, but considered to represent a minimal part of total fetal production rates. As the authors themselves pointed out, the calculations were based on several assumptions regarding fetal T4 and T3 kinetics which could not be tested experimentally. The extensive investigations carried out in the sheep model have been reviewed by others [10-12,32,33], and indicate virtual placental impermeability to T4 and T3 in the maternal to fetal direction.

It might be pertinent to note here that Artiodactyla (among which sheep and swine) have epitheliochorial placentas, quite different from those of Anthropoidea (among which man and monkey) and most Rodentia (among which rat, mouse and guinea pig)which are of the same type, namely hemochorial. The first type of placenta contains maternal endothelium, reticulum and epithelium, and fetal trophoblast and endothelium, whereas the latter type does not contain the maternal tissues.

As regards transfer in normal women at the end of uncomplicated pregnancies, both Grumbach and Werner [34] and Kearns and Hutson [35] found that significant amounts of labeled T4 and T3 were transferred from the mother to the newborn, although at a slow rate. In both studies T3 was more easily transferred than T4. Recycling by the fetal thyroid of radioiodine deiodinated from the labeled iodothyronine given to the mother was avoided by the administration of potassium iodide, and the lack of fetal thyroidal uptake of the label was confirmed at birth.

The use of labeled iodothyronines was later avoided, and the problem was investigated by giving normal women near term relatively large doses of either T4 [36], or T3 [37,38]. The end-points measured to assess maternal to fetal transfer were usually related to cord-blood T4, T3, or suppression of fetal TSH. The use of high doses of T4 or T3 might have altered the maternal-fetal transport system, either by enhancing transfer, or by activating placental deiodinations [13], thus actually decreasing the proportion of hormone transferred. Despite these possibilities, results obtained in uncomplicated pregnancies suggest that transfer of both iodothyronines occurs near term, though neither T4 nor T3 cross the placental barrier freely. Fisher et al. [36] found much smaller increases in cord-blood BEI (butanol extractable iodine) than in maternal plasma BEI after administration of a single very large dose of T4 (0.5-8 mg) 3 to 52 hr before delivery, and concluded that there was virtually no transfer of T4 in normal women near term. From the data obtained, Fisher later [39] calculated that as much as 36 μg T4 would be transferred to the fetus under normal conditions, and considered that this might be important for the protection of an athyreotic fetus. As measured by increased cord-blood T3 concentrations, transfer of T3 was observed in women given relatively low doses of T3 (50 μg / day for the last three weeks of pregnancy) [38], inadequate even to suppress maternal T4 levels (through suppression of maternal TSH) [37]. However, up to 300 ug T3 per day were needed to observe a decrease in cord-blood T4 [37,38]. This finding lead to the conclusion that appreciable transfer of T3 only occurs when high doses are administered to women. However, as we will later show for the rat, it is also possible that sensitivity of the fetal pituitary near term to T3 (but not to T4) is still very low as compared to adults. As reviewed by Fisher [32] there seems to be a progressive maturation of T3 negative feed-back control of pituitary TSH release with advancing gestational age in fetal sheep and monkeys, and in neonatal rats.

In summary, existing evidence strongly suggests that both T4 and T3 cross the human placenta near term in normal pregnancies, though not freely. It is impossible at present to assess how much maternal transfer contributes to total fetal thyroid hormone economy, although it probably represents a smaller proportion than the fetal contribution.

2) Maternal and / or fetal thyroid failure

i) Impairment of maternal thyroid function. If maternal transfer of thyroid hormones continues after onset of fetal thyroid function, some change would be expected to occurr in fetal thyroid hormone concentrations or TSH secretion if the usual maternal contribution were decreased. This was indeed the case when total extrathyroid T4 and T3 concentrations were determined in fetuses from T dams (see data at 20 and 21 dg, Fig. 3). It was also found that not all tissues were equally affected: by 21 dg the brain was spared from T4 and T3 deficiency [20]. We suggested that this might involve a compensatory increase in the secretion of hormone by the fetal thyroid.

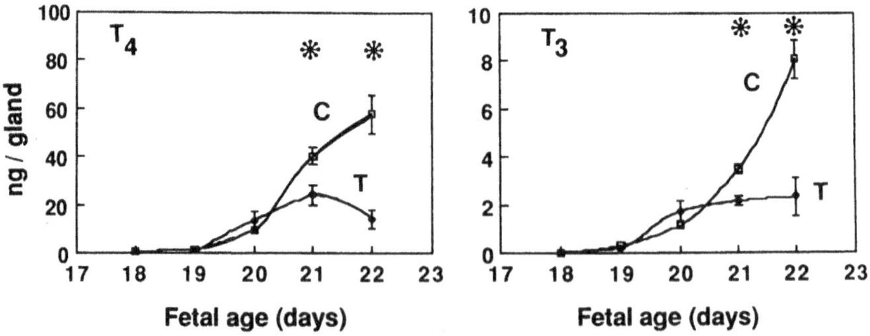

Figure 4. Total T4 and T3 contents in proteolytic digests (Pronase) of fetal thyroid glands obtained from C and T mothers between 18 and 22 dg. Asterisks identify significant differences between fetuses from C and T dams. From Ruiz de Oña et al. [40].

Gray and Galton [41] found that thyroidal T4 content was decreased in rat fetuses from T dams near term, and interpreted their data as indicating increased fetal thyroidal secretion. We confirmed this also also for the thyroidal T3 content of 21-dg-old fetuses from T dams [21]. Figure 4 shows the changes in total T4 and T3 contents of the fetal thyroid from 18 until 22 dg, as measure by RIA in proteolytic digests. Total thyroidal content of T4 increased from 0.66 $\pm$ 0.07 (SEM) ng / gland at 18 dg to 57.7 $\pm$ 8.2 at 22 dg; data for T3 were 0.021 $\pm$ 0.005 ng / gland and 8.07 $\pm$ 0.78. The increases were about a hundred-fold for T4, and fourhundred-fold for T3. The T4 to T3 ratio was about 30 : 1 at 18 dg and 7 : 1 at 22 dg. When the mother had been T, however, the fetal T4 and T3 contents were the same as those of fetuses from C dams up to 20 dg, but then stopped increasing. By 22 dg the thyroidal T4 and T3 contents in fetuses from T dams were 24 % and 28 %, respectively, of those of fetuses from C dams. This finding confirms the previous observations at 21 dg [21] and support the suggestion that the thyroid of the fetuses from T mothers is stimulated to secrete increasing amounts of hormones to compensate for the absence of the maternal contribution. Both the data and their interpretation agree with those of Gray and Galton [41], and are further supported by the finding of increased plasma TSH levels in fetuses from T dams [22].

It is likely that the fetal thyroid would not be able to compensate completely for the lack of maternal contribution if it were faced with conditions impairing its function. This is the case when the iodine intake is low [42], as summarized in Figure 5. In such a situation normal T4 and T3 levels might not be attained in any of the fetal tissues, the brain included.

In summary, data from experiments in rats near term show that the fetus senses maternal thyroid failure, and the fetal thyroid reacts to this situation with compensatory mechanisms which attempt to maintain normal thyroid hormone concentrations. A plausible explanation is that the stimulus for activation of these mechanisms is directly related to the decreased maternal contribution of thyroid hormones to fetal economy, though it cannot be excluded that the signal is related to other metabolic alterations of the T dam.

ii) Fetal hypothyroidism. Models of fetal hypothyroidism alone are difficult to achieve in small laboratory animals. Geloso and Bernard [43] decapitated fetuses at 17.5 dg and compared their BEI (isotope equilibration method) at 21 dg with that of age-paired normal fetuses, and that of decapitated fetuses from T mothers. Data are summarized in Table 1.

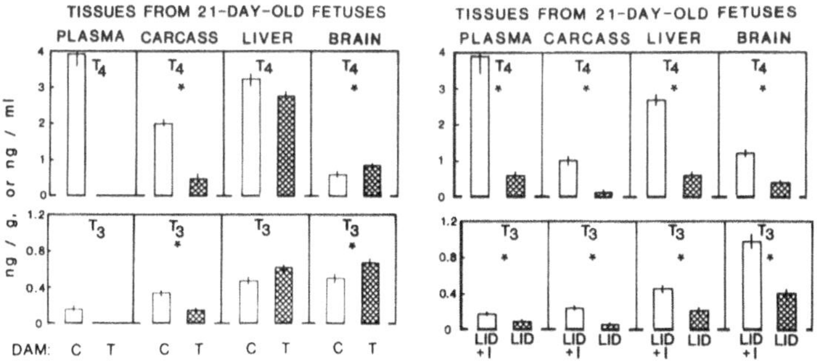

Figure 5.- Concentrations of T4 and T3 in plasma and tissues of fetuses from C and T dams, and from rats on a low iodine diet (LID) or the same diet supplemented with KI (LID+I). Asteriscs identify statistically significant differences. Data are from Morreale de Escobar et al. [16] and from Escobar del Rey et al. [42]. Reproduced from Morreale de Escobar et al., Frontiers in Thyroidology, Plenum Press (1986).

Table 1- Fetal plasma butanol extractable iodine (BEI), by isotopic equilibrium (means ± 95% confidence limits)[43].

Mothers	Fetuses	ng BEI / ml	ng T4§ / ml
Normal	Normal	2.6 ± 0.6	3.9 ± 0.9
Normal	Decapitated	0.9 ± 0.2	1.4 ± 0.3
Thyroidectomized	Normal	2.0 ± 0.7	3.0 ± 1.1
Thyroidectomized	Decapitated	0.3 ± 0.1	0.5 ± 0.2

§Assuming the BEI was composed mainly of T4: BEI x 1.5
The corresponding BEI values for the normal and thyroidectomized mothers were 6.2 ± 1.5 and 0.7 ± 0.5 ng /ml; T4, 9.3 ± 2.3 and 1.1 ± 0.8 ng / ml.

These results indicate that at the end of gestation the rat fetal thyroid is the main source of the fetal plasma T4, as decapitation of the fetus results in a 60 % decrease of the fetal plasma BEI. But it is not the only source, as maternal T further decreases the fetal plasma T4 levels. The authors concluded that a certain amount of maternal thyroid hormones cross the placenta, but that the transfer is too limited to compensate for the absence of the fetal thyroid [43].

Apart from the technical difficulties involved, this approach does not permit to study the effects of possible maternal transfer of thyroid hormones on the fetal hypothyroid brain, and fetal TSH secretion. We therefore recurred, as many others [44-46], to the use of goitrogens which cross the placenta and block the production of fetal thyroid hormones. This treatment obviously also blocks maternal thyroid function so that the experimental design includes groups of MMI-treated dams on a replacement dose of thyroid hormone, given by constant infusion to avoid supraphysiological peaks of thyroid hormone levels.

Table 2 outlines the experimental design which we used. Methimazole (MMI) was used to block maternal and fetal thyroid function. Some of the MMI-treated dams received a replacement dose of T4 (1.8 µg / 100 g BW / day) by constant infusion, as described in detail [21] elsewhere. The main findings are summarized in Figure 6, which shows the T4 and T3 concentrations in fetal thyroid digests, carcass and brain. The concentrations of T4 and T3 in the fetal thyroid were markedly decreased by MMI treatment, irrespective of

Table 2. Experimental design for experiments A, B and C

Exper :	Group of dams :	MMI # (0.02%)	T4 infused § (µg / 100 g)	T3 infused § (µg / 100 g)
A	C (or T)	-	-	-
	C (or T) + MMI	+	-	-
	C (or T) + MMI + T4	+	+(1.8)	-
B	C (or T)	-	-	-
	C (or T) + MMI	+	-	-
	C (or T) + MMI + T4	+	+(1.8)	-
	C (or T) + MMI + T3	+	-	+(0.45)
C	C	-	-	-
	C + MMI	+	-	-
	C + MMI + T4 (3 doses)	+	+(1.8;2.4;3.6)	-
	C + MMI + T4 (3 doses)	+	-	+(0.5;1.5;4.5)

Given as drinking water, from 16-21 dg (Exp. A) or from 14-21 dg (Exp. B and C)

§ In µg T4 or T3 / 100 g BW of dam / day, from 17-21 dg (Exp. A) or from 15-21 dg (Exp. B and C).

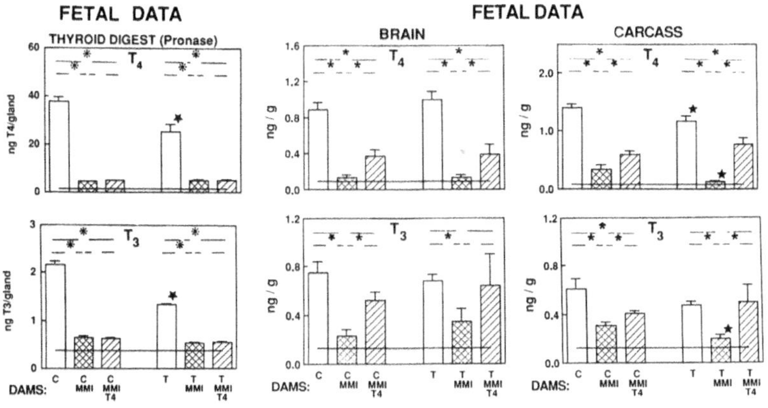

Figure 6.- Concentrations of T4 and T3 in fetal thyroid digests, and in the brain and carcass of fetuses from the six groups of Experiment A (Table 2). Asteriscs between data bars identify statistically significant differences. Horizontal lines indicate limits of detection. From Morreale de Escobar et al. [21] as illustrated in [4].

the infusion of T4 into the MMI-treated dams. Hormone levels in tissues were also decreased by the MMI treatment. Comparison of the hormone levels in fetuses from C + MMI + T4 (or T + MMI + T4) dams with those from C + MMI (or T + MMI) mothers shows that infusion of T4 into the mothers not only ameliorates fetal deficiency of T4, but also of T3. This occurs without an increase of fetal plasma T3 (not shown) [21], indicating that the fetal brain derived its T3 from local deiodination of T4. Our conclusion that some maternal T4 crosses the placenta near term, at least if the fetal thyroid is impaired, is in agreement with the data and conclusions by others [22,41], who injected physiological T4 doses.

When the rat fetal thyroid is impaired (as would be the case in human congenital hypothyroidism) maternal T4 has a protective effect on the fetal brain and on other fetal tissues. This might mitigate adverse effects of fetal thyroid failure on the developing brain. It appeared important to determine whether, or not, T3 is also transferred from the mother to the fetus near term, and also mitigates the effects of fetal hypothyroidism. Experiment B was carried out (Table 2) similar to the one summarized above, but with several modifications. MMI was started two days earlier, at 14 dg, and the infusion of hormone into the dams was started at 15 dg. Several MMI-treated dams were again infused with the same dose of T4 used for Experiment A, and other MMI-treated dams were infused with T3 at a dose (0.45 µg / 100 g BW. day) equivalent for adult rats to that of T4. Data for fetuses from dams infused with T4 confirmed those of the previous

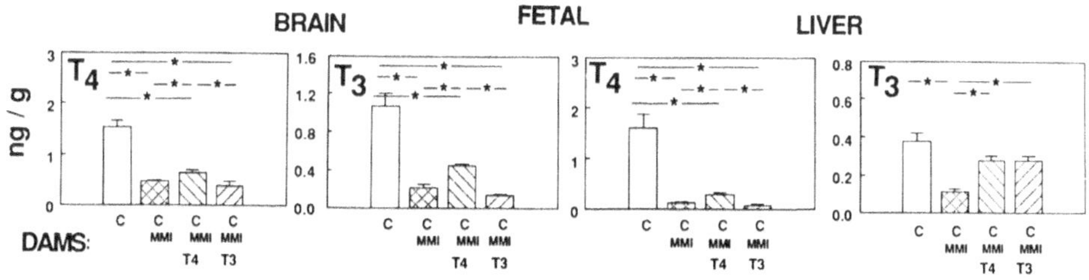

Figure 7.- Concentrations of T4 and T3 in the brain and lung of fetuses from four of the groups of Experiment B (Table 2), corresponding to the intact (C) dams. Asteriscs placed mid-way between groups identify statistically significant differences. From Morreale de Escobar et al [47].

experiment, illustrated in Figure 6 as regards the transfer of T4 and its effects on the T4 and T3 deficiency of fetal tissues, the brain included. The new finding [47] was that infusion of T3 into the MMI-treated dams increased the concentration of T3 in fetal plasma, and in all fetal tissues studied, except the brain (Fig. 7). MMI treatment had increased both the maternal and fetal plasma TSH levels (not shown). Infusion of both T4 and T3 decreased plasma TSH to normal levels in the mothers. Infusion of T4 decreased fetal TSH, although it remained elevated as compared to normal fetuses. In contrast, the infusion of T3 had no effect on fetal plasma TSH. Data shown in Figure 7 correspond to the groups from C dams only; results from the four groups of fetuses from T dams were very similar.

Previous data summarized here suggested an important role of local generation of T3 from T4 in providing the developing brain with increasing amounts of T3. The results of this last experiment show that the brain of rat fetuses near term is entirely dependent on T4 availability, as T3 does not enter the brain, at least at physiological plasma T3 levels. 5' D-II activities were measured in the cerebral cortex of the fetuses corresponding to Figure 7. Activity increased with MMI treatment, in response to thyroid hormone deficiency. Infusion of T4 into the mothers decreased enzyme activity markedly, although not to control levels, whereas infusion of T3 did not, showing that the changes in thyroid hormone level were accompanied by a biological effect. Data are shown and discussed in more detail in chapter 6 by Obregón et al.

Preliminary results have been obtained in fetuses from Experiment C (Table 2). The intact (C) dams were infused with saline; the MMI-treated dams were infused with 3 different T4 doses (1.8, 2.4 and 3.6 μg/ 100 g BW / day) and 3 different T3 doses (0.5, 1.5 and 4.5 μg /100 g BW / day), chosen on the basis of the results described by Knobil and Yasimovich [44]. The lowest of the T4 and T3 doses already reversed maternal plasma TSH to C values. The 1.8 μg T4 dose ↕ was insufficient to raise the maternal hepatic T3 level and alfa-glycerophosphate dehydrogenase (a-GPD) activity of the MMI-treated dams to normal. A dose of 2.4 μg T4 reversed both variables to C values. The 3.6 μg T4 dose resulted in liver T3 levels and a-GPD activites which were higher than C values, although not higher than values for non-pregnant females. The lowest of the T3 doses already reversed liver T3 and a-GPD activity to C values, the higher doses resulting in dose-related increases of hepatic T3 and a-GPD activity.

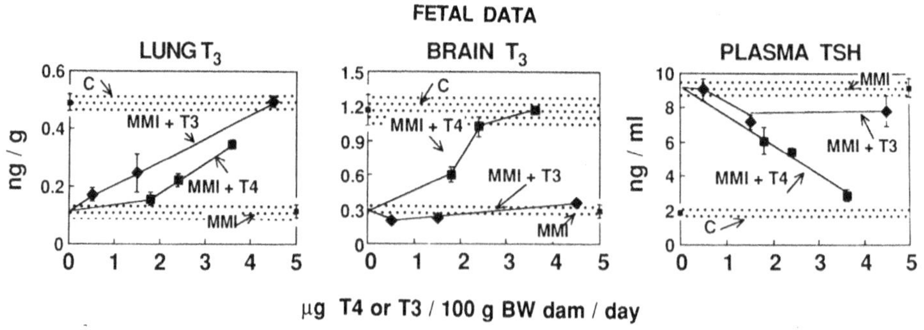

Figure 8.- "Dose-response" curves between fetal concentrations of T3 in lung and brain, or fetal plasma TSH, and the dose of T4 or T3 infused into the mothers. Groups are those of Experiment C (Table 2). Dotted areas correspond to means ± SEM for C or MMI-treated groups. From Calvo et al. [48].

↕ The doses indicated from this point onwards are the amount infused /100 g BW / day, the BW being that of the dams at 15 dg.

Findings in fetal tissues depended on the iodothyronine used for infusion, on the dose, on the end-point measured in the fetus (T4, T3, TSH), and on the tissue being considered. An example is shown in Figure 8, for the fetal lung and brain T3, and plasma TSH. The results obtained in fetal lung (and, although not shown, in fetal plasma and fetal liver) demostrate that when T3 is infused, there is a dose-related transfer of T3, accompanied by a dose-related increase of T3 in several fetal tissues. But, as already found at a single T3 dose in Experiment B, fetal plasma TSH was scarcely affected by the infusion of this iodothyronine. This occurred even at the highest T3 dose. Again, there was no increase in cerebral T3 above the level found in MMI-treated fetuses with any of the T3 doses used, the highest being 9 times the replacement dose for the adult rat. In contrast, the maternal infusion with 2.4 ug T4 raised cerebral T3 levels to C values. It also depressed plasma TSH. We wish to point out that the dose of T4 which normalized cerebral T3 levels in the fetus was lower than the dose needed to normalize fetal pasma TSH (> 3.6 ug T4, Fig.8) and fetal plasma T4 (not shown), and was also the dose normalizing maternal liver T3 and a-GPD.

Recent studies on the role of the choroid plexus in the transport of T4 and T3 into the brain show that after injection of labeled thyroid hormones, the choroid plexus and the brain avidly accumulate T4, but not T3 [49]. Such a mechanism might afford an explanation for the lack of effect of T3 on fetal brain T3 concentrations.

In summary, results obtained with this experimental design show that maternal thyroid hormones also cross the placenta late in gestation, at least when the fetal thyroid is impaired, and despite active deiodinations by the placenta [13]. Maternal T4 has a more important role, as it mitigates T3 deficiency of the brain, whereas maternal T3 does not. Neither does maternal T3 decrease fetal TSH secretion. We believe it is quite interesting that the fetal brain already posesses mechanism(s) (see chapter 6 by Obregón et al.) which achieve maintainance of normal cerebral T3 concentrations at levels of maternal T4 which are within the physiological range for the mother, and which are not sufficient to increase fetal plasma T4 (or to decrease fetal TSH) to normal values, nor to normalize T3 concentrations in other fetal tissues. If the mother is hypothyroxinemic, the fetal brain is not protected, even if maternal T3 levels are normal.

We wish to point out that the conclusion that the human placenta at term is virtually impermeable to the iodothyronines partly based on studies [39,40] where normal women were given T3 near term, and some parameter related to fetal TSH secretion was measured. The data obtained in the rat suggest that this may be a misleading end-point to assess transfer of T3. Moreover, the results should not be extrapolated to T4.

If experimental data obtained in the rat are pertinent to human babies with congenital hypothyroidism, they strongly support the use of T4 for early treatment, or in those cases where the condition is diagnosed in utero. They also indicate that T3 might actually be harmful, if given in doses which depress maternal T4 levels. On the other hand, normal maternal T3 levels might protect tissues other than the brain from T3 deficiency and thus avoid the appearance of overt signs of hypothyroidism at birth, even if their plasma TSH is high. The possible relevance of these findings to human endemic cretinism has been discussed elsewhere [4].

But are any of these findings pertinent to human congenital hypothyroidism? We have already discussed more extensively the possibility that they are [21]. Autonomy of the pituitary-thyroid feedback system in human fetuses during the third trimester of pregnancy is well established, but the finding that fetal thyroid failure is accompanied by high TSH levels at birth does not exclude that TSH would have been higher (and cord-blood T4 lower) if there had been no transfer of hormones from the mother to the hypothyroid fetus. Transfer of maternal thyroid hormones to athyreotic fetuses would explain that in some of them T4 levels are clearly detectable in cord-blood, and then decrease rapidly [50]. It would also explain the variability with which athyreotic babies present signs of hypothyroidism at birth, as this might be related to the maternal thyroid hormone levels and the adequacy of placental transfer, which might vary for different pregnancies [39]. These same causes for variability might account for differences in permanent brain damage presented by athyreotic babies. It has been established that the babies with the lowest plasma T4 levels at birth present more frequently signs of having suffered from hypothyroidism in utero, as assessed from delayed maturation of prenatal ossification centers; these babies tend to have the lowest IQ (although within the normal range) even if treated promptly after birth [51].

FINAL COMMENTS

A comprehensive program with the aim of preventing CNS damage caused by thyroid failure should include: i) neonatal thyroid screening, treatment as soon as possible after birth (or even before birth) being the principal aim; ii) iodine prophylaxis to avoid maternal hypothyroxinemia combined with fetal thyroid failure and iii) careful monitoring of plasma T4 and free T4 levels during pregnancy to detect and treat cases of maternal hypothyroxinemia, with or without clinical hypothyroidism.

REFERENCES

1. G. Morreale de Escobar and F. Escobar del Rey, Brain damage and thyroid hormones, in:"Neonatal Thyroid Screening", G. N. Burrow, ed., Raven Press, New York (1980).
2. G. Morreale de Escobar, A. Ruiz-Marcos, and F. Escobar del Rey, Thyroid hormones and the developing brain, in: "Congenital Hypothyroidism", J. H. Dussault and P. Walker, eds., Marcel Dekker, New York (1983).
3. G. Morreale de Escobar, M. J. Obregón, and F. Escobar del Rey, Fetal and maternal thyroid hormones, Horm Res 26:12 (1987).
4. G. Morreale de Escobar, C. Ruiz de Oña, M. J. Obregón, and F. Escobar del Rey, Models of fetal iodine deficiency, in: "Iodine in Neurodevelopment", G. R. DeLong and J. Robbins, eds., Plenum Press, New York, in press.
5. B. Nataf and M. Sfez, Debut du fonctionnement de la thyroide foetale du rat, Compt ren Soc Biol 156:1235 (1961).
6. J. P. Geloso, Fonctionnement de la thyroide et correlations thyreohypophysaires chez le foetus de rat, Annales d'Endocrinologie 28:1 (1967).
7. R. M. Weiss and C. R. Noback, The effects of thyroxine and thiouracil on the time of appearance of ossification centers of rat fetuses, Endocrinology 45:389 (1949).
8. L. R. Sweney and B. L. Shapiro, Thyroxine and palatal development in the rat, Dev Biol 42:19 (1975).
9. M. Hamburgh, The role of thyroid and growth hormone in neurogenesis, in: "Current Topics in Developmental Biology", A. A. Moscona and A. Monroy, eds., Academic Press, New York, vol 4:109 (1969).
10. D. A. Fisher, J. H. Dussault, J. Sack, and I. J. Chopra, Ontogenesis of hypothalamic-pituitary-thyroid function in man, sheep and rat, Rec Progr Horm Res 33:59 (1977).
11. D. A. Fisher, and A. K. Klein, Thyroid development and disorders of thyroid function in newborn, New Engl J Med 304:702 (1981).
12. D. A. Fisher, The unique endocrine milieu of the fetus, J Clin Invest 78:603 (1986).
13. E. Roti, A. Gnudi, and L. E. Braverman, The placental transport, synthe-

sis and metabolism of hormones and drugs which affect thyroid function, Endocr Rev 4:131 (1983).
14. M. J. Obregón, G. Morreale de Escobar, and F. Escobar del Rey, Concentrations of triiodothyronine in the plasma and tissues of normal rats as determined by radioimmunoassay: Comparison with results obtained by an isotopic equilibration technique, Endocrinology 103:2145 (1978).
15. M. J. Obregón, J. Mallol, R. Pastor, G. Morreale de Escobar, and F. Escobar del Rey, Thyroxine and triiodothyronine in rat embryos before onset of fetal thyroid function, Endocrinology 114:305 (1984).
16. G. Morreale de Escobar, R. Pastor, M. J. Obregón, and F. Escobar del Rey, Effects of maternal hypothyroidism on the weight and thyroid hormone content of rat embryonic tissues, before and after onset of fetal thyroid function, Endocrinology 117:1890 (1985).
17. R. J. Woods, A. K. Sinha, and R. Ekins, Uptake and metabolism of thyroid hormones by the rat fetus in early pregnancy, Clin Sci 67:359 (1984).
18. R. Ekins, Roles of thyroxine-binding proteins and maternal thyroid hormones in development, Lancet i:1129 (1985).
19. J. Bernal and F. Pekonen, Ontogenesis of the nuclear 3,5,3'-triiodothyronine receptor in the human fetal brain, Endocrinology 114:677 (1984).
20. A. Pérez-Castillo, J. Bernal, B. Ferreiro, and T Pans, The early ontogenesis of thyroid hormone receptor in the rat fetus, Endocrinology 117:2457 (1985).
21. G. Morreale de Escobar, M. J. Obregón MJ, C. Ruiz de Oña, and F. Escobar del Rey, Transfer of thyroxine from the mother to the rat fetus near term: Effects on brain 3,5,3'-triiodothyronine deficiency, Endocrinology 122:1521 (1988).
22. B. Bonet and E. Herrera, Different responses to maternal hypothyroidism during the first and second half of gestation in the rat. Endocrinology 122:450 (1988).
23. E. Herrera and B. Bonet, Effect of maternal hypothyroidism during two phases of gestation on fetal growth and metabolic response to starvation in the rat, Annales d'Endocrinologie 44:17A, abstract (1983).
24. C. E. Hendrich, W. J. Jackson, and S. P. Porterfield, Behavioral testing of progenies of Tx (hypothyroid) and growth hormone-treated Tx rats: an animal model for mental retardation, Neuroendocrinology 38:429 (1984).
25. S. P. Porterfield and C. E. Hendrich, The effects of growth hormone treatment of thyroid-deficient pregnant rats on maternal and fetal carbohydrate metabolism, Endocrinology 99:786 (1976).
26. C. E. Hendrich, S. P. Porterfield SP, and V. A. Galton, Pituitary-thyroid function of fetuses of hypothyroid and growth hormone treated hypothyroid rats, Horm Metab Res 11:362 (1979).
27. G. W. Greenman, M. O. Gabrielson, J. Howard-Flanders, and M. A. Wessel, Thyroid function in pregnancy. Fetal loss and follow-up evaluation of surviving infants, New Engl J Med 267:426 (1962).
28. E. B. Man and S. A. Serunian, Thyroid function in human pregnancy. IX. Development or retardation of 7-year- old progeny of hypothyroxinemic women, Am J Obst Gynecol 125:949 (1976).
29. A. J. McMichael, J. D. Potter, and B. S. Hetzel, Iodine deficiency, thyroid function and reproductive failure, in: " Endemic Goiter and Endemic Cretinism", J. B. Stanbury and B. S. Hetzel, eds., John Wiley & Sons, New York, (1980).
30. J. D. Dubois, A. Cloutier, P. Walker, and J. H. Dussault, Absence of placental transfer of L-triiodothyronine in the rat, Pediat Res 11:116 (1977).
31. J. H. Dussault and P. Coulombe, Minimal placental transfer of L-thyroxine in the rat, Pediat Res 14:228 (1980).
32. D. A. Fisher, Ontogenesis of hypothalamic-pituitary-thyroid function in the human fetus, in: "Pediatric Endocrinology", F. Delange, D. A. Fisher, and P. Malvaux , eds., Karger, Basel, (1985).
33. D. A. Fisher, Development of fetal thyroid function and its control, in:

"Iodine in Neurodevelopment", G. R. DeLong and J. Robbins, eds., Plenum Press, New York, in press.
34. M. M. Grumbach and S. H. Werner, Transfer of thyroid hormone across the human placenta at term, J Clin Endocrinol 16:1392 (1956).
35. J. E. Kearns and W. Hutson, Tagged isomers and analogues of thyroxine (Their transmission across the human placenta and other studies), J Nuclear Med 4:453 (1963).
36. D. A. Fisher, H. Lehman and C. Lackey, Placental transfer of thyroxine, J Clin Endocrinol Metab 24:393 (1964).
37. S. Raiti, G. B. Holzman, R. L. Scott, and R. M. Blizzard, Evidence for the placental transfer of tri-iodothyronine in human beings. New Engl J Med 277:456 (1967).
38. J. H. Dussault, V. V. Row, G. Lickrish, and R. Volpé, Studies of serum triiodothyronine concentration in maternal and cord blood: Transfer of triiodothyronine across human placenta. J Clin Endocrinol 29:595 (1969).
39. D. A. Fisher, Panel discussion on hyperthyroidism in the pregnant woman and neonate, J Clin Endocrinol Metab 27:1639 (1967).
40. C. Ruiz de Oña, M. J. Obregón, G. Morreale de Escobar, and F. Escobar del Rey, Thyroid hormone economy in fetal rats after onset of fetal thyroid function, 17th Meeting of the European Thyroid Association, Montpellier, (1988).
41. B. Gray and V. A. Galton, The transplacental passage of thyroxine and foetal thyroid function in the rat, Acta Endocrinol (Kbhvn) 75:725 (1974).
42. F. Escobar del Rey, R. Pastor, J. Mallol, and G. Morreale de Escobar, Effects of maternal iodine deficiency on the L-Thyroxine and 3,5,3'-Triiodo-L-thyronine contents of rat embryonic tissues before and after onset of fetal thyroid function, Endocrinology 118:1259 (1986).
43. J. P. Geloso and G. Bernard, Effets de l'ablation de la thyroide maternelle ou foetale sur le taux des hormones circulantes chez le foetus de rat, Acta Endocrinol (Kbhvn) 56:561 (1967).
44. E. Knobil and J. B. Josimovich, Placental transfer of thyrotropic hormones, thyroxine, triiodothyronine and insulin in the rat, Ann NY Acad Sci 75:895 (1959).
45. S. K. Roy and Y. Kobayashi, Placental transfer of L-thyroxine and tri-iodo-L-thyronine in rats during late pregnancy, Proc Soc Exper Biol Med 110:699 (1962).
46. F. Comite, G. N. Burrow, and E. C. Jorgensen, Thyroid hormone analogs and fetal goiter, Endocrinology 102:1670 (1978).
47. G. Morreale de Escobar, M. J. Obregón, C. Ruiz de Oña, Comparison of maternal to fetal transfer of T3 versus T4 in rats, as assessed by T3 levels in fetal tissues, Annales d'Endocrinologie 48:178A, abstract (1987).
48. R. Calvo, C. Ruiz de Ona, M. J. Obregón, F. Escobar del Rey, and G. Morreale de Escobar, Differential effects of maternal T4 and T3 on hypothyroid rat fetal tissues, 17th Meeting of the European Thyroid Association, Montpellier,(1988).
49. P. W. Dickson, A. R. Aldred, J. G. T. Menting, P. D. Marley, W. H. Sawyer, and G. Schreiber, Thyroxine transport in choroid plexus, J Biol Chem 262:13907 (1987).
50. D. A. Price, R. M. Erlich, and P. G. Walfish PG, Congenital hypothyroidism. Clinical and laboratory characteristics in infants detected by neonatal screening, Arch Dis Child 56:845 (1981).
51. J. Letarte and S. La Franchi, Clinical features of congenital hypothyroidism, in: "Congenital Hypothyroidism", J. H. Dussault and P. Walker, eds., Marcel Dekker, New York, (1983).

DISCUSSION

EMERSON (Worcester) : It certainly seems reasonable that very early in gestation a lack of thyroid hormone from the mother might have adverse effects. I want to make a comment on work from our laboratory where we found that thyroidectomizing the mother at approximately 7 days had no effect on fetal serum TSH concentrations. This is somewhat in contrast to what you presented. I wonder if it has to do with the timing of maternal thyroidectomy in relation to mating ?

MORREALE (Madrid) : Our experiments were performed before mating. In other experiments, thyroidectomy has been performed on the day of mating, that is much earlier than at day 7 after mating. Those first few days are extremely important and, therefore, the experimental models are different and our results cannot be compared.

EMERSON (Worcester) : In our experiments there was no difficulty in seeing a rise in TSH in the fetus with methimazole treatment.

MORREALE (Madrid) : Well, naturally, with methimazole you block the fetal thyroid and you do get increased TSH. This is totally different from maternal thyroidectomy alone.

KLEIN (Hanover) : We were able to retrieve the amniotic fluid from 2 infants (at 15 and 16.5 weeks of gestation) who were born with congenital hypothyroidism, one with dyshormonogenesis and the other with an atrophic gland. The fluid was analyzed by Dr. I. Kourides (New York) who found high levels of TSH at that early stage. What is your comment about this finding ?

MORREALE (Madrid) : I have no personal comment but I think Dr. Fisher will tell us more later on about the early rise in TSH during gestation.

ROLE OF THE PLACENTA IN FETAL THYROID HOMEOSTASIS

Charles H. Emerson

University of Massachusetts School of Medicine
Worcester, MA

INTRODUCTION

The placenta is usually regarded as a conduit through which maternal nutrients pass to the fetus, or as a barrier that allows for fetal development in an insulated environment. Thus, with respect to thyroid function, it is well known that the placenta transfers iodide to, and excludes thyrotropin from, the fetal circulation. It is less recognized that the placenta is a metabolically active organ, and that this activity is important for the conversion of fetal substrates. In order to relate this aspect of placenta physiology to fetal thyroid homeostasis it is necessary to first review thyroid hormone profiles and pharmacokinetics in the fetus.

THYROID HORMONE PROFILES AND IODOTHYRONINE PHARMACOKINETICS IN THE FETUS.

There are important differences in the plasma thyroid hormone profiles of the fetus and the adult. Figure 1, which depicts the ontogenesis of plasma thyroid hormones in the guinea pig (Castro et al., 1986), illustrates this. Similar data have been reported in other species including man and the sheep (Fisher et al., 1973; Fisher et al., 1972; Dussault et al., 1972; Nwosu et al., 1978; El-Zaheri et al., 1981; Isaac et al., 1979; Chopra, Sack and Fisher, 1975; Fisher and Klein, 1981). Early in gestation fetal plasma thyroxine (T4) and 3,5,3'-triiodothyronine (T3) concentrations are very low. As the fetus develops, plasma T4 concentrations increase. In many species, including the guinea pig, the increase is such that, at term, total and free T4 concentrations are higher in fetal than in maternal plasma. In contrast to T4, total and free T3 concentrations are lower in fetal plasma than in maternal plasma throughout gestation. Finally, whereas 3,3'5'-triiodothyronine (reverse T3,rT3) is difficult to detect in maternal plasma, it is easily measurable in fetal plasma. For example, as shown in Figure 1, rT3 is detectable in fetal plasma as early as 45 days of gestation. By term, fetal plasma rT3 concentrations are strikingly elevated. These differences are initially related to an inability of the fetal thyroid to secrete iodothyronines. As fetal thyroid function matures, they become more a result of dissimilarities, between the fetus and the adult, in the peripheral conversion of T4, and in the clearance of T3 and rT3.

Chopra et al. (Chopra, Sack and Fisher, 1975) have studied the production rate and clearance of fetal iodothyronines. Their data are summarized in Tables 1 and 2. Table 1 demonstrates that by the third trimester thyroidal secretion of T3 is, relative to T4 secretion, only

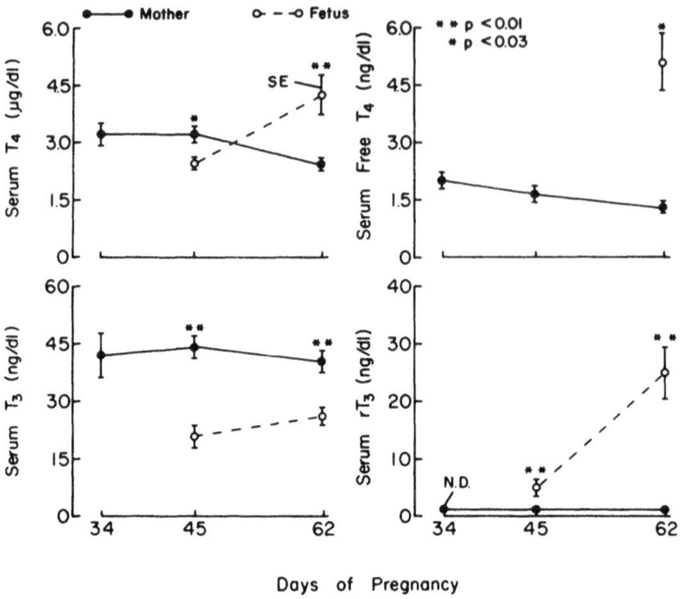

Figure 1. Plasma thyroid hormone concentrations in the guinea pig fetus and mother (Reproduced from Castro et al., 1985a)

slightly lower in fetal sheep than in adult sheep. In contrast, whereas the adult sheep produces similar amounts of T3 from peripheral conversion of T4 as it does from thyroidal secretion, the fetus produces almost no T3 from peripheral conversion. Consequently the net production rate of T3 in the fetus is lower than in the adult. Moreover, the metabolic clearance rate (MCR) of T3 in the fetus is about twice that of the adult. Therefore, the relatively low concentrations of T3 in the fetus are due to both to diminished production and accelerated clearance. Table 2 shows data for rT3. It is apparent that the peripheral production rate of this iodothyronine is almost three times greater, and MCR is nearly fourfold lower, in the fetus than in the the adult. Therefore, fetal plasma rT3 concentrations are high because of increased production and low clearance.

Studies of broken cell preparations of fetal tissues (Harris et al.1978; Wu et al., 1978; Borges, LaBourene and Ingbar, 1980) indicate that they contain relatively little type I iodothyronine 5′-deiodinase activity. This probably accounts for the low peripheral production and clearance in the fetus, of T3 and rT3 respectively. Immaturity of type I 5′-deiodinase systems does not, however, account for the increased clearance of T3 and increased production of rT3. In fact, this is probably not related to any organs of the fetus proper. Rather it may be due to the placenta, an organ which is part of the fetoplacental unit and shares in the fetal circulation. This concept is based on in vitro and in situ studies showing that the placenta is able to deiodinate, convert, and clear certain iodothyronines. These properties of the placenta conform, to a remarkable degree, with what is known about the pharmacokinetics of iodothyronines in the fetus.

IODOTHYRONINE DEIODINASE ACTIVITIES IN PLACENTA HOMOGENATES

Several years ago we demonstrated that T4 and T3 underwent deiodination when incubated with rat or human placenta homogenates (Roti et al., 1981; Roti et al., 1982a). In these studies, the analytic methods were such that at least 2 percent deiodination of the substrate was required in order to detect iodothyronine degradation. In addition, deiodination could not be

Table 1. The Production Rate of T3 as a percent of the production rate of T4 in the adult and fetal sheep.

Source of T3	T3 Production, % of T4 production	
	Adult	Fetus
Thyroid Secretion	9.6	6.7
Peripheral Conversion	9.6	UD

Adapted from Chopra et al., 1975

detected if the incubation tubes were devoid of reducing agents such as dithiotheritol (DTT). When T3 was the substrate, 3,3'-diiodothyronine (3,3'-T2) and 3'-monoiodothyronine (3'-T1) were generated. This suggested that T3 was deiodinated in the inner ring, first to 3,3'-T2, which then underwent inner ring deiodination to 3'-T1. When T4 was the substrate, rT3 but not rT3 degradation products, was generated. When rT3 was the substrate, no deiodination was noted.

These studies and work from other laboratories (Yoshida et al., 1985; Suzuki et al., 1986) indicate that placenta homogenates contain abundant inner ring iodothyronine deiodinase activity. Interestingly, the deiodinase is active against both inner ring iodine atoms only when a single iodine is present in the outer ring, as in the case of T3. When two iodine atoms are present in the outer ring, as in T4, the deiodinase is active against only one of the two inner ring iodine atoms. The reason that rT3 is a poor substrate for the placenta inner ring deiodinase is not known. However, it is likely that if this was not the case, fetal plasma rT3 concentrations would be much lower. Another important characteristic of this enzyme is that, at similar substrate concentrations, T3 is more rapidly deiodinated than T4. Thus, in placenta homogenates, the order of susceptibility of iodothyronines to the inner ring deiodinase is T3 > T4 > rT3. This differs from most outer ring deiodinase enzymes.

Recently it was reported that human placenta and cultures of human chorion and/or decidual cells also contain outer ring deiodinase activity directed against T4 and rT3 (Hidal and Kaplan, 1985; Kaplan and Shaw, 1984). In contrast to placenta inner ring deiodinase, T3 is a poor substrate for placenta outer ring deiodinase. At the present time, the function of the placenta outer ring deiodinase is unclear. It is difficult to demonstrate outer ring deiodination when T4 and rT3 is perfused through the fetal side

Table 2. The Production Rate and Metabolic Clearance Rate of rT3 in the Adult and Fetal Sheep.

	Adult	Fetus
rT3 Production Rate, $mg/M^2/day$	0.038 ± 0.005	0.102 ± 0.008
rT3 MCR $L/M^2/day$	74 ± 18	19 ± 2

Adapted from Chopra et al., 1975

of the intact placenta (vide infra). In addition, we have recently shown in rat placenta homogenates (Emerson et al., 1988) that outer ring deiodination proceeds at a much slower rate than inner ring deiodination. Moreover, the studies showing that the peripheral production rate of T3 is very low in the fetus (Chopra, Sack and Fisher, 1975) are not compatible with the concept that the placenta generates substantial amounts of T3 from fetal T4. Possibly some T3 is generated from T4 in the placenta. However, the T3 produced may then be rapidly degraded by placenta inner ring deiodinase.

SIMILARITY OF FETAL IODOTHYRONINE PHARMACOKINETICS AND IODOTHYRONINE DEIODINATION AND CLEARANCE IN THE PLACENTA.

To obtain further information on the role of placenta deiodination, we (Castro et al., 1985; Emerson and Braverman, 1984) and others (Cooper et al.1983) have performed studies with an in situ perfused guinea pig placenta model. In this model the maternal relationship to the placenta is preserved and the fetus is removed in order to cannulate the umbilical vessels. To prepare this model a near term pregnant guinea pig is anesthetized and one of the fetuses is surgically exposed. One umbilical artery and the umbilical vein are cannulated and the remaining umbilical artery is ligated. The fetal side of the placenta is then perfused through the umbilical artery with Krebs-Henseleit buffer containing labeled or stable iodothyronines. The placenta effluent is collected in timed fractions from the umbilical vein cannula (one pass system), or is returned to the perfusion reservoir (recirculating system). In the one-pass system aliquots of the placenta effluent are collected at intervals and analyzed. In the recirculating system the aliquots are removed from the perfusion reservoir at various times during the perfusion. In addition to studying deiodination on the fetal side of the placenta, information regarding fetal-maternal transfer across the placenta can be obtained by sampling the maternal circulation at a time in which iodothyronines are perfused into the fetal side of the placenta. To evaluate maternal-fetal transfer, iodothyronines can be perfused into the maternal circulation, and the placenta effluent collected and analyzed.

Figure 2 shows studies in which 125{I}, 125{I}-rT3, and 125{I}-T3 were recirculated through the fetal side of the guinea pig placenta. When 125{I} was recirculated there was a rapid decline in the radioactivity of the

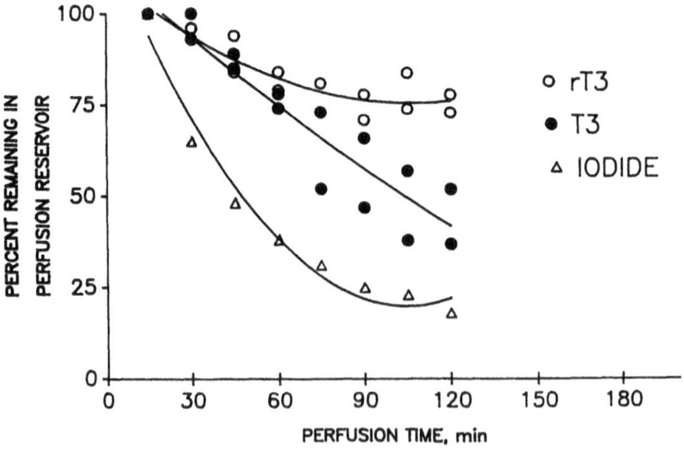

Figure 2. Loss of rT3, T3, and iodide from the fetal side of the guinea pig placenta in a recirculating perfusion system.

Table 3. Estimates of deiodination, loss of radioactivity, and clearance of 125(I) labeled iodothyronines during a single pass through the perfused pig guinea pig placenta.

	Iodothyronine		
	Concentration Perfused, nM	Percent Deiodinated	Percent Cleared
T4	44.00	0.4	< 5
T3	0.14	27.3	43
T3	140.00	15.4	31
rT3	0.14	UD	< 5

Estimates are based on published (Cooper et al., 1983, Emerson and Braverman, 1984, and Castro et al., 1985) and unpublished studies. Cleared refers to the loss of iodothyronine from the fetal side of the placenta due to deiodination and to transfer of the iodothyronine into the placenta or to the maternal circulation. Calculations are based on data obtained after one hour of perfusion to reach equilibrium.

perfusion reservoir. This was expected since iodide is known to cross the placenta relatively rapidly (Roti, Gnudi and Braverman, 1983). When 125(I)-rT3 was recirculated, the radioactivity in the reservoir decreased during the first 30 to 60 minutes, probably representing an equilibration phase. Following this there was little or no change in the concentration of radioactivity in the reservoir. In addition, HPLC analysis revealed that all of the radioactivity in the reservoir represented 125(I)-rT3. Similar results have been obtained by Cooper et al. (Cooper et al., 1983). These studies strongly suggest that transfer of rT3 between the fetal and maternal circulations is very limited, and that the placenta does not degrade rT3. This concept is also supported by studies in the rat from our laboratory. In these studies (El-Zaheri et al., 1981) rT3 was infused for 5 days into the maternal circulation of pregnant rats. Despite the fact that maternal plasma rT3 concentrations increased from 9 ± 6 (mean $\pm$ SD) to 315 ± 69 ng/dl, fetal plasma rT3 concentrations only increased from 19 ± 3 to 35 ± 11 ng/dl.

When 125(I)-T4 is recirculated through the fetal side of the placenta there is also very little decrease in radioactivity after the initial equilibration phase. However, whereas 125(I)-rT3 degradation cannot be detected during recirculation through the fetal side of the placenta, a fraction of the 125(I)-T4 that is recirculated is deiodinated to 125(I)-rT3. As reported by Cooper et al. (Cooper et al.1983), the fractional rate of 125(I)-T4 deiodination is approximately 0.4 percent during a single pass through the placenta.

When 125(I)-T3 is recirculated through the fetal side of the placenta the results are considerably different from those obtained for 125(I)-rT3 and 125(I)-T4. As shown in Figure 2, there is a steady decline in radioactivity throughout the perfusion period. In this and other studies the decrease in radioactivity in the reservoir was associated with the appearance of radioactivity in the maternal circulation. Moreover, as noted when T3 is incubated with placenta homogenates, the 125(I)-T3 is degraded sequentially. For example, in a typical study, the reservoir contained only 125(I)-T3 at the beginning of the perfusion. After one hour the amount of radioactivity in the reservoir that consisted of 125(I)-T3 had decreased to approximately 27 percent of the total radioactivity as judged by high pressure liquid chromatography (HPLC) analysis. At this time, 44 percent of the radioactivity consisted of 125{I}-T2, and 29 percent consisted of 125{I}-T1. After two hours no 125{I}-T3 was detected, the 125{I}-T2 had decreased to 25 percent of the total radioactivity and the 125{I}-T1 had increased to 75 percent of the total radioactivity. These data strongly suggest that the fetal side of the placenta disposes of T3 by clearing and degrading this iodothyronine. In some species, the placenta probably also transfers T3 or T3 metabolites to the maternal circulation. In view of the rapid deiodination of T3 on the fetal side of the placenta, it seems likely that the bulk of the T3 is deiodinated before it is transferred to the maternal circulation.

Table 3 summarizes data regarding the deiodination and clearance of iodothyronines in the perfused guinea pig placenta. It is apparent that there are many similarities between the in situ perfused placenta and the fetus with respect to iodothyronine clearance and metabolism. These similarities take on added significance by the fact that between 40 to 50 percent of the fetal cardiac output circulates through the placenta (Rudolph, 1969). This is far greater than the blood flow through most fetal organs (Rudolph, 1969). This high blood flow would amplify the ability of the placenta to deiodinate and clear T3 from the fetal circulation. Similarly, by augmenting the amount of T4 available to the placenta, it would increase rT3 production in the fetoplacental unit.

In summary, there are a number of studies with placenta homogenates and the perfused placenta which support the hypothesis that this organ has a major role in maintaining low T3 and high rT3 concentrations in fetal plasma. In addition to this work, other reports are relevant to this discussion. In a study by Sack et al. (Sack et al., 1976), serum T3 concentrations were monitored in newborn sheep during delivery, usually by caesarean section. One group of newborns was delivered into room air, a cold stress that produces a surge in TSH secretion. In this group the umbilical cord was immediately cut. In this group, as in other groups (vide infra), care was taken to maintain the maternal placenta relationships until the umbilical cord was cut. Another group of newborns were delivered into room air. In this group umbilical cord cutting was delayed for one hour after delivery. A third group was delivered into a 39 C water bath in order to prevent the cold induced surge in TSH secretion. In this group cord cutting was also delayed for one hour. In all groups an abrupt surge in plasma T3 concentrations was observed. This surge was consistently observed after umbilical cord cutting, and not always after cold stress. Among the possible causes of the surge in plasma T3 was that it was due to an increase in thyroid hormone secretion. This seems unlikely because plasma T4 concentrations did not increase after umbilical cord cutting. Another possibility is that the surge was due to an abrupt increase in peripheral 5'-deiodinase activity, resulting in increased production of T3 from T4. This also seems unlikely because, if there was an increase in peripheral 5'-deiodinase activity, plasma rT3 concentrations should also decline rapidly in the newborn. However, the decline in plasma rT3 concentrations after delivery is quite gradual (Chopra, Sack and Fisher, 1975). The third possibility seems the most likely. That is that umbilical cord cutting was associated with a sudden decrease in the MCR of T3 due to loss of placenta T3 clearing and deiodinating activity. Because of the inverse reciprocal relationship between the MCR and the plasma concentrations of a substance this would be associated with an increase in plasma T3.

An observation that is difficult to reconcile with this hypothesis is found in the studies of Penny et al. (Penny et al., 1986) and Abuid et al. (Abuid, Stinson and Larsen, 1973). In these studies thyroid hormone concentrations were measured in sera obtained, at the time of vaginal delivery, from the umbilical arteries and umbilical vein of human newborns. In the study of Penny et al. (Penny et al., 1986) it was found that rT3 concentrations were slightly higher in umbilical vein plasma than in umbilical artery plasma. This finding seems to support an active role for placenta deiodinase in generating rT3 across the placenta bed. However as previously noted, T3 is more rapidly deiodinated in the placenta than T4. Therefore it is not clear why, if the deiodinase was active, there was not also a decrease in T3 concentrations across the placenta bed. Placenta inner ring deiodinase activity does decline during the latter part of gestation (Roti, et al., 1985b, Yoshida et al., 1985). However this activity is still easily detectable when placenta homogenates obtained at delivery are incubated in the presence of DTT (Roti et al. 1981). It may be that conditions during delivery are such as to severely limit placenta deiodinase activity. In fresh homogenates of rat or human placenta, deiodination is not apparent in the absence of reducing agents. These presumably act to regenerate the active enzyme from the sulfenyl iodide intermediate (Visser, 1980). Possibly the endogenous cofactors that serve this function are not available during the later stages of vaginal delivery, a time at which the placenta is separating from the uterine wall. If this were the situation, T3 deiodination and clearance by the placenta would be minimal.

PLACENTA DEIODINASE ACTIVITY AND THE TRANSFER OF IODOTHYRONINES BETWEEN THE MATERNAL AND FETAL CIRCULATIONS (DEIODINATION-LIMITED TRANSFER, DLT)

The relatively large concentration gradient for iodothyronines between the maternal and fetal circulations is evidence that the transfer of these substances across the placenta is limited. There are probably two reasons for this. The major reason is undoubtedly that placenta membranes are relatively impermeable to iodothyronines. Species with epitheliochorial placentation have even less permeability than those with hemochorial placentation because there are added cell layers between the fetal and maternal circulations (Ramsey, 1982). This is supported for thyronines by the reports suggesting that 3,5,-dimethyl-3'-isopropyl-l-thyronine crosses the placenta in species with hemochorial placentation (Comite, Burrow and Jorgensen, 1978; Tamagna, Hershman and Jorgensen, 1979; Ballard et al., 1980), but not the sheep (Bachrach et al., 1982), a species with epitheliochorial placentation. It also seems likely that the permeability of the placenta for some iodothyronines is greater than for others. For example, as previously noted, the loss of radioactivity when 125(I)-T3 is recirculated through the fetal side of the placenta is much greater than when 125(I)-rT3 is recirculated.

There may be a second reason, referred to here as deiodination-limited transfer (DLT), for the relatively poor transfer of iodothyronines from the maternal to the fetal circulations. The deiodination-limited transfer hypothesis states that some maternal iodothyronine molecules do cross the placenta membranes separating the maternal and fetal circulations. However, most of them then are deiodinated in the inner ring before their transfer across the placenta membranes is complete. The magnitude of deiodination-limited transfer (DLT) for the maternal-fetal transfer of a particular iodothyronine can be expressed in quantitative terms by the following equation.

$$DLT = rDI - rDN$$

The terms rDI and rDN refer respectively, to the net rate of maternal-fetal transfer of the iodothyronine under conditions in which placenta deiodination is completely inhibited, and the net rate of maternal-fetal transfer of the iodothyronine under conditions in which deiodination is not inhibited. From the previous discussion it is apparent that DLT would be greater for those iodothyronines which are permeable in placenta membranes and are good substrates for the placenta deiodinase (i.e. T3). DLT would also be more of a factor limiting transfer across the placenta in species with a thin layer of cells between the maternal and fetal circulations. This

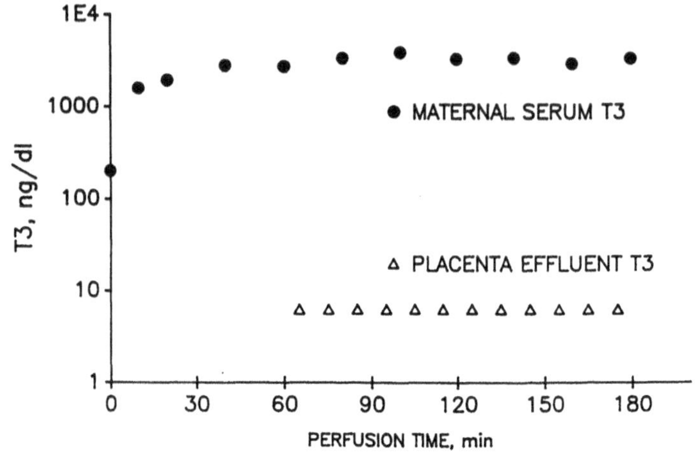

Figure 3. Maternal serum and placenta effluent T3 concentrations during T3 infusion into the maternal circulation of the guinea pig. The least detectable concentration of T3 in the T3 radioimmunoassay was 8 ng/dl. When T3 was undetectable, the least detectable value was plotted.

consists primarily of species with hemochorial placentation (human, rat, guinea pig) and would probably exclude species with epitheliochorial placentation (sheep). However, even amongst species with similar placentation there may be differences in the magnitude of DLT. In the rat, a species with hemochorial placentation, approximately half of the rT3 in the fetal circulation may be generated in placenta membranes from maternal T4 (Emerson and Braverman, 1984). The guinea pig also has a hemochorial placenta but in this species, at least during the latter stages of gestation, little rT3 appears to be generated from maternal T4 (Cooper et al.1983). However, in preliminary studies we have been able to demonstrate some DLT in the guinea pig with respect to T3. In these studies supraphysiological amounts of T3, 478 ng/min, were infused into the maternal circulation. T3 concentrations were measured in maternal plasma and placenta effluent fractions collected in a one-pass system. As demonstrated in Figure 3, T3 was not detected in placenta effluent fractions despite the fact that extremely high T3 concentrations were achieved in the maternal circulation. Figure 4 shows a study employing a similar protocol except that, after one hour, sodium iopanoate was added to the buffer perfusing the placenta in order to inhibit placenta deiodinase activity. The fact that sodium iopanoate inhibits inner ring deiodinase activity was established in studies with placenta homogenates (Fay et al., 1984) and the in situ perfused placenta (Castro et al. 1985). As can be seen, the addition of sodium iopanoate to the perfusion buffer was associated with the prompt appearance of T3 in the placenta effluent. These studies document that T3 transfer across the guinea pig placenta is severely limited. Some of this limitation appears to be related to DLT since inhibition of placenta deiodinase was associated with increased transfer of T3 across the placenta. However, even when deiodination was completely inhibited, the T3 gradient between the maternal circulation and the placenta effluent was still large, indicating that DLT can only partially account for limited T3 transfer.

DEIODINATION-LIMITED TRANSFER AS A FACTOR IN MODULATING FETAL IODOTHYRONINE CONCENTRATIONS

The most important aspect of the DLT hypothesis is its prediction that fetal plasma T3 and rT3 concentrations would be perturbed, in situations in which the magnitude of DLT is high, if placenta deiodinase activity becomes

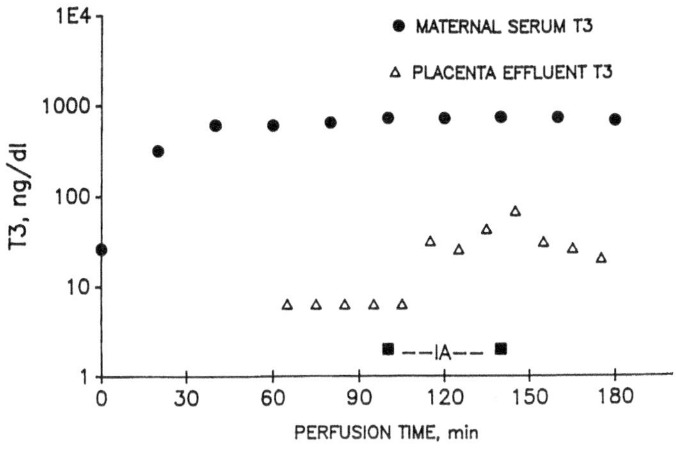

Figure 4. Maternal serum and placenta effluent T3 concentrations during T3 infusion into the maternal circulation of the guinea pig, and sodium iopanoate (IA) infusion into the fetal side of the placenta. The figure indicates time that IA was infused.

inhibited. Since the concentration gradient of T3 across the placenta favors maternal-fetal transfer, inhibition of placenta deiodinase activity would increase maternal-fetal T3 transfer. This would lead to increased T3 concentrations in fetal plasma. Similarly, because the DLT hypothesis predicts that maternal T4 is deiodinated to rT3 which is then transferred to the fetus, fetal plasma rT3 concentrations would decrease if placenta deiodinase activity was inhibited. As previously noted placenta deiodinase activity is probably, by virtue of its ability to degrade fetal T3 of fetal origin, and to convert T4 of fetal origin to rT3, one of the major reasons why fetal plasma T3 concentrations are low and rT3 concentrations are high. Therefore, regardless of whether the substrates for the placenta deiodinase are derived from the maternal or the fetal circulation, inhibition of deiodinase activity would result in similar changes in fetal iodothyronine concentrations.

In early gestation, at a time when fetal thyroid function is poorly developed, the major source of substrates for the placenta deiodinase may be maternal iodothyronines. This is much more likely to be the case for species with hemochorial placentation and may not be a factor for those with epitheliochorial placentation. Accordingly, in species with hemochorial placentation, inhibition of placenta deiodinase activity in early gestation could perturb fetal plasma iodothyronine concentrations. In the later stages of gestation, when the ability of the fetal thyroid to secrete hormones is more fully developed, fetal iodothyronines probably constitute the bulk of the substrate for the placenta deiodinase molecules that generate products which enter the fetal circulation. Therefore in all species, regardless of their type of placentation, inhibition of placenta deiodinase activity in the later stages of gestation could alter fetal serum T3 and rT3 concentrations.

It is not known under what circumstances or how often placenta deiodinase activity is impaired. It is also not clear if increases in fetal plasma T3 concentrations, or decreases in rT3 concentrations, would affect fetal function or development. Because T3 and not rT3 has thyromimetic properties, a rise in fetal plasma T3 is more likely to be deleterious than a reduction in plasma rT3 concentrations. rT3 is, however, relatively potent in inhibiting brain type II 5′-deiodinase (Visser et al. 1982; Kaiser,

Goumaz and Burger, 1986) and it is conceivable that the high serum rT3 concentrations in the fetus may modulate the activity of this enzyme. Elsewhere in this monograph Dr. Fisher discusses the ontogenesis of fetal responses to T3. This information and future studies in this area should be helpful in efforts to determine if the unique profile of thyroid hormones in the fetal circulation, which appears to be partially dependent on placenta deiodinase activity, has functional significance.

ACKNOWLEDGEMENT

Supported by Research Grant DK87250 from the NIH, Bethesda, MD, U.S.A.

REFERENCES

Abuid, J.A., Stinson, A. and Larsen, P.R., 1973, Serum triiodothyronine and thyroxine in the neonate and the acute increase in these hormones following delivery, J. Clin. Invest., 52:1195.

Bachrach, L.K., Kudlow, J.E., Silverberg, J.D.H., Kent, G. and Burrow, G.N., 1982, Treatment of ovine cretinism in utero with 3,5-dimethyl-3'-isopropyl-l-thyronine, Endocrinology, 111:132.

Ballard, P.L., Benson, B.J.,, Brehier, A., Carter, J.P., Kriz, B.M. and Jorgensen, E.C., 1980, Transplacental stimulation of lung development in the rabbit by 3,5-dimethyl-3'-isopropyl-L-thyronine, J. Clin. Invest., 65:1407.

Borges, M., LaBourene, J. and Ingbar, S.H., 1980, Changes in hepatic iodothyronine metabolism during ontogeny of the chick embryo, Endocrinology, 107:1751.

Castro, M.I., Alex, S., Young, R.A., Braverman, L.E. and Emerson, C.H., 1986, Total and free serum thyroid hormone concentrations in fetal and adult pregnant and nonpregnant guinea pigs, Endocrinology, 118:533.

Castro, M.I., Braverman, L.E., Alex, S., Wu, C.F. and Emerson, C.H., 1985, Inner-ring deiodination of 3,5,3'-triiodothyronine in the in situ perfused guinea pig placenta, J. Clin. Invest., 76:1921.

Chopra, I.J., Sack, J. and Fisher, D.A., 1975, 3,3',5'-triiodothyronine (reverse T3) and 3,3',5-triiodothyronine (T3) in fetal and adult sheep: studies of metabolic clearance rates, production rates, serum binding, and thyroidal content relative to thyroxine, Endocrinology, 97:1080.

Comite, F., Burrow, G.N. and Jorgensen, E.C., 1978, Thyroid hormone analogs and fetal goiter, Endocrinology, 102:1670.

Cooper, E., Gibbons, M., Thomas, C.R.T., Lowy, C. and Burke, C.W., 1983, Conversion of thyroxine to 3,3',5'-triiodothyronine in the guinea pig placenta, Endocrinology, 112:1808.

Dussault, J.H., Hobel, C.J., DiStefano, J.J., Erenberg, A. and Fisher, D.A., 1972, Triiodothyronine turnover in maternal and fetal sheep, Endocrinology, 90:1301.

El-Zaheri, M.M., Vagenakis, A.G., Hinerfeld, L., Emerson, C.H. and Braverman, L.E., 1981, Maternal thyroid function is the major determinant of amniotic fluid 3,3',5'-triiodothyronine in the rat, J. Clin. Invest., 67:1126.

Emerson, C.H. and Braverman, L.E., 1984, Peripheral deiodination of thyroid hormones in placenta and fetal membranes, Horm. Met. Res. (Suppl.), 14:56.

Emerson, C.H., Bambini, G., Alex, S., Castro, M.I., Roti, E. and Braverman, L.E., 1988, The effect of thyroid dysfunction and fasting on placenta inner ring deiodinase activity in the rat, Endocrinology, 122:809.

Fay, M., Roti, E., Fang, S.L., Wright, G.., Braverman, L.E. and Emerson, C.H., 1984, The effects of propylthiouracil, iodothyronines, and other agents on thyroid hormone metabolism in human placenta, J. Clin. Endocrinol. Metab., 58:280.

Fisher, D.A., Dussault, J.H., Erenberg, A. and Lam, R.W., 1972, Thyroxine and triiodothyronine metabolism in maternal and fetal sheep, Pediatr. Res., 6:894.

Fisher, D.A., Dussault, J.H., Hobel, C.J. and Lam, R., 1973, Serum and thyroid gland triiodothyronine in the human fetus, J. Clin. Endocrinol. Metab., 36:397.

Fisher, D.A. and Klein, A.H., 1981, Thyroid development and disorders of thyroid function in the newborn, N. Engl. J. Med., 304:702.

Harris, A.R.C., Fang, S.L., Prosky, J., Braverman, L.E. and Vagenakis, A.G., 1978, Decreased outer ring monodeiodination of thyroxine and reverse triiodothyronine in the fetal and neonatal rat, Endocrinology, 103:2216.

Hidal, J.T. and Kaplan, M.M., 1985, Characteristics of thyroxine 5'-deiodination in cultured human placental cells. Regulation by iodothyronines, J. Clin. Invest., 76:947.

Hultberg, B. and Sjoblad, S., 1977, Lysosomal enzymes in medium from cultured skin fibroblasts from normal individuals and patients with lysosomal diseases, Clin. Chim. Acta, 80:79.

Isaac, R.M., Hayek, A., Standefer, J.C. and Eaton, R.P., 1979, Reverse triiodothyronine to triiodothyronine ratio and gestational age, J. Pediatr., 94:477.

Kaiser, C.A., Goumaz, M.O. and Burger, A.G., 1986, In vivo inhibition of the 5'-deiodinase type II in brain cortex and pituitary by reverse triiodothyronine, Endocrinology, 119:762.

Kaplan, M.M. and Shaw, E., 1984, Type II iodothyronine 5'-deiodination by human and rat placenta in vitro, J. Clin. Endocrinol. Metab., 59:1808.

Nwosu, U.C., Kaplan, M.M., Utiger, R.D. and Delivoria-Papadopulos, M., 1978, Surge in fetal plasma triiodothyronine before birth in sheep, Am. J. Obstet. Gynecol., 132:489.

Penny, R., Sims, M.E., Campbell, W.G., Spencer, C.A. and Nicoloff, J.T., 1986, Thyroid indices arterial and venous cord blood: significantly greater levels of reverse triiodothyronine in venous blood than in arterial blood, Metabolism, 35:645.

Ramsey, E.M. . In: The Placenta:Human and Animal, edited by New York,NY: Praeger, 1982, p. 54.

Roti, E., Fang, S.L., Green, K., Emerson, C.H. and Braverman, L.E., 1981, Human placenta is an active site of thyroxine and 3,3',5-triiodothyronine tyrosyl ring deiodination, J. Clin. Endocrinol. Metab., 53:498.

Roti, E., Fang, S.L., Braverman, L.E. and Emerson, C.H., 1982, Rat placenta is an active site of inner ring deiodination of thyroxine and 3,3',5-triiodothyronine, Endocrinology, 110:34.

Roti, E., Braverman, L.E., Fang, S.L., Alex, S., and Emerson, C.H., 1982 Ontogenesis of placental inner ring thyroxine deiodinase and amniotic fluid 3,3',5'-triiodothyronine concentration in the rat, Endocrinology, 111:959.

Roti, E., Gnudi, A. and Braverman, L.E., 1983, The placental transport, synthesis and metabolism of hormones and drugs which affect thyroid function, Endocr. Rev., 4:131.

Rudolph, A.M., 1969, The course and distribution of the foetal circulation. in: Foetal Autonomy, edited by Wolstenholme, G.E.W. and O'Connor, M. London: J. & A. Churchill,Ltd., 1969, p. 147-161.

Sack, J., Beaudry, M., DeLamater, P.V., Oh, W. and Fisher, D.A., 1976, Umbilical cord cutting triggers hypertriiodothyroninemia and nonshivering thermogenesis in the newborn lamb, Pediat. Res. 10:169.

Suzuki, M., Yoshida, K., Sakurada, T., Kaise, N., Kaise, K,, Fukazawa, H., Nomura, T., Itagaki, Y., Yonemitsu, K., Yamamoto, M, Saito, S. and Yoshinaga, K., 1986, Effect of changes in thyroid state on metabolism of thyroxine by rat placenta, Endocrinol. Japon., 33:37.

Tamagna, E.I., Hershman, J.M. and Jorgensen, E.C., 1979, Thyrotropin suppression by 3,5-dimethyl-3'-isopropyl-L-thyronine in man, J. Clin. Endocrinol. Metab., 48:196.

Visser, T.J., 1980, Mechanism of inhibition of iodothyronine-5'-deiodinase by thioureylenes and sulfite, Biochim. Biophys. Acta, 611:371.

Visser, T.J., Leonard, J.L., Kaplan, M.M. and Larsen, P.R., 1982, Kinetic evidence suggesting two mechanisms for iodothyronine 5'-deiodination in rat cerebral cortex, Proc. Natl. Acad. Sci. USA, 79:5080.

Wu, S.Y., Klein, A.H., Chopra, I.J. and Fisher, D.A., 1978, Alterations in tissue thyroxine-5'-monodeiodinating activity in perinatal period, Endocrinology, 103:235.

Yoshida, K., Suzuki, M., Sakurada, T., Shinkawa, O., Takahashi, T., Furuhashi, N., Kaise, N., Kaise, K., Kitaoka, H., Fukazawa, H., Nomura, T., Itagaki, Y., Yamamoto, M., Saito, S. and Yoshinaga, K., 1985, Human placental thyroxine inner ring monodeiodinase in complicated pregnancy, Metabolism, 34:535.

DISCUSSION

<u>FISHER (Torrance)</u> : One interesting observation that has been difficult to explain over time is the fact that reverse T3 concentrations after cord-cutting remain fairly similar in the human and the sheep : reverse T3 decreases progressively in humans over several weeks after birth and in the sheep over a period of a couple of weeks. So, reverse T3 is clearly coming from non placental sources, at least in part.

<u>EMERSON(Worcester)</u> : Yes, I agree with you. My opinion is that the clearance of reverse T3 in the fetus is very low whether the placenta is there or not (that is, the placenta is not clearing reverse T3). We must therefore assume that the fetus is producing some reverse T3 and that its clearance remains low, even after delivery.

ROLE OF THE MATERNAL CARRIER PROTEINS IN THE SUPPLY OF THYROID HORMONES TO THE FETO-PLACENTAL UNIT: EVIDENCE OF A FETO-PLACENTAL REQUIREMENT FOR THYROXINE

Roger Ekins, Arun Sinha, Maria Ballabio, Mark Pickard, Michael Hubank, Zaidan al Mazidi, and Mohamed Khaled

Department of Molecular Endocrinology
University College and Middlesex School of Medicine
University of London, London W1N 8AA, United Kingdom

INTRODUCTION

For a number of years, the consensus view amongst endocrinologists has been that thyroid hormones of maternal origin neither cross the placenta in significant amount[1], nor are implicated in the development of the fetus in general, or of the fetal CNS in particular[2]. Nevertheless, for a number of reasons (some of which have been discussed in[3]), the validity of this view appeared to us to be open to doubt; we therefore initiated studies (ca 1980) specifically intended to verify or disprove some of the experimental evidence and theoretical postulates on which this view was based.

Results of our preliminary studies[4] immediately confirmed the suspicion that previous work on thyroid hormone (TH) transport from mother to fetus was either irrelevant to the events occurring in early pregnancy (having been confined to animal studies late in pregnancy - ie following development of the fetal thyroid gland) - or, in the case of the relatively few studies conducted during early pregnancy, had been misinterpreted or ignored.

Aside from the questionable nature of past experimental evidence relating to this issue, the principal explanation offered for the supposed absence of placental TH transport appeared equally unsound, albeit superficially persuasive. This relied on the proposition that thyroxine binding globulin (TBG) and the other TH binding proteins (thyroxine-binding pre-albumin (TBPA) and albumin) present in maternal serum prevent passage of TH from maternal to fetal circulations[5]. This concept reflects a view widely held amongst endocrinologists, ie that TBG and other specific hormone binding proteins in serum serve to minimise hormone loss from the vascular compartment and, indeed, that this constitutes their principal physiological role[6].

The serum concentration of TBG rises approximately 2 - 3 fold in human pregnancy; the concentrations of some of the specific steroid binding proteins likewise rise, albeit - in certain cases - to a greater factorial extent. Speculation on these proteins' physiological role has therefore inevitably centered on the possibility that they are in some way implicated in governing hormone transport during gestation, the notion that the rise in TBG "protects" the fetus from maternal TH[5] being a prime example of such concepts. Thus although, in the context of this Symposium, the principal questions of interest are: do thyroid hormones traverse the placenta and, if so, are they of importance to the full maturation of the fetal brain?, the role of TBG and other thyroid binding proteins in TH transport - particularly during pregnancy - clearly constitutes an important complementary issue. Coincidentally it is one which, because of its wider relevance to endocrinology and reproductive physiology, has attracted much recent interest and controversy.

In this presentation we hope to clarify the confusion that centers on the role of binding proteins in hormone transport, showing inter alia that, contrary to common opinion, binding proteins facilitate, and do not inhibit, hormone efflux from the vascular compartment. Indeed we suggest that TBG may have evolved specifically to ensure adequacy of a maternal thyroxine supply to the feto-placental unit, possibly forming part of a complex feed-back system governing the thyroid hormone environment to which the (early) fetus is exposed. However, we also present data relating to the more general hypothesis[3] that maternal thyroxine not only traverses the placenta, but is of crucial importance to the development of the fetal CNS.

Hormone transport to target tissues: the free hormone hypothesis

The "free hormone hypothesis" of hormone delivery is widely accepted amongst endocrinologists. It enshrines the concept that, in the case of hormones which exist in blood (largely) in protein-bound form, the serum free hormone concentration constitutes the essential determinant of hormone action. A corollary is that only hormone in the free state is able to traverse target tissue capillary walls and penetrate target cells. These concepts reflect the observation that, in circumstances in which serum protein-bound hormone levels differ significantly from normal (eg during pregnancy, or in consequence of genetic abnormality), endocrine status appears to correlate with the serum free concentration, not the bound. The thyroid/pituitary/hypothalamic feedback system provides the classic example of these propositions, the system appearing to operate in such a way as to maintain free thyroid hormone concentrations at a (near) normal level in the face of wide differences in the concentrations (or compositions) of the thyroid binding proteins present in blood. Similar concepts apply in the case of the steroid hormones.

Such observations underlie the view that measurements of serum free hormone are diagnostically more valuable than measurements of bound (or total) concentrations. Clearly they also imply that both the bound hormone moiety and the binding proteins themselves are physiologically irrelevant, the latter view being sustained by the fact that no physiological consequence attributable to an absence or elevation of these proteins has ever been observed. Nevertheless, the notion that the hormone binding proteins in serum possess no physiological role is difficult to accept. Consequently endocrinologists have frequently sought explanations of these proteins' existence and biological function.

A conventional view - the "vascular retention hypothesis" - has been that they provide an intravascular "buffer store" of hormone, attenuating transient surges of secretion or peripheral demand[6,7], and restricting hormone entry into "sensitive" cells[6]. Illustrative of this concept is Osorio and Myant's suggestion that TBG prevents TH transport from mother to fetus[5], thus serving to isolate or "protect" the fetus from the maternal endocrine system.

Hormone transport to target tissues: "bound hormone" hypotheses

Other investigators have sought explanations which challenge the free hormone hypothesis. For example, following observation of the differential effects of corticosteroids on the induction of hepatic and pancreatic aminotransferases in rats with altered serum CBG levels, Keller, Richardson and Yates suggested that certain tissues are permeable to specific binding proteins[8], and thus accessible to protein-bound hormone per se. They therefore hypothesised that CBG "increases the specificity of the adrenocortical system by targeting corticosteroids according to features of micro-circulation". Though Keller et al's hypothesis has subsequently been largely disregarded, similar ideas have more recently been proposed by Siiteri et al[9] who claim to have observed intra-cellular localization of CBG and SHBG, and have suggested that, following their structural modification, these proteins convey hormone directly into the cell nucleus. However, a major difficulty arising with all such concepts is that physiological manifestations of abnormality in binding-protein levels would be expected to be evident; moreover they offer no specific explanation of the characteristic changes of binding protein levels seen in pregnancy.

The ideas of Pardridge and his colleagues in this area[10,11,12] have commanded particular attention during the past decade, and are currently the subject of considerable controversy. They center on the proposition that the equilibrium constant governing the protein binding of hormone is altered in the microvasculature of target organs, causing the intracapillary free hormone concentration to be elevated in such tissues, and large amounts of (dissociated) bound hormone to be selectively transported to them. Thus Pardridge claims that "the function of plasma protein binding is the selective delivery of ligands to tissues in a way that varies from organ to organ"[11], implying - like Keller et al - that variation in binding protein levels alters the distribution of hormone throughout the body. However, though the notion that binding proteins affect the delivery of hormone in this way is common to all recent challenges to the free hormone hypothesis (including our own), we totally reject Pardridge's ideas, believing them to derive from an oversimplified theoretical analysis of the kinetics of hormone efflux from target organ capillaries, coupled with crucial misinterpretations of experimental data[13]. However, in part because of the interest Pardridge's views continue to provoke, but more particularly because of the relevance of some of Pardridge's experimental data - when correctly interpreted - to the subject of this Symposium, this issue merits more detailed discussion here.

Pardridge's challenge of the validity of the free hormone hypothesis originally derived from a fundamental misunderstanding of the hypothesis itself[14]. It must be emphasized that this does not propose that the amount of hormone delivered to an individual tissue is restricted to the amount initially present in the free state in the afferent blood supply. Illustrative of misunderstanding on this point, the observation that the "splanchnic extraction of testosterone or estradiol, or the brain extraction of progesterone, is on the order of 30-50%, which is 10-fold the percentage of free hormone in serum"[11] was originally viewed by Pardridge as contradicting the hypothesis. This misconception constituted the principal foundation of Pardridge's proposal of an "apparent", in vivo, dissociation constant (K_D^{app}) deviating from the "absolute" constant (K_D) estimated in vitro[10], and whose value was supposedly given by $K_D e^{k_p t}$, where (t) is the capillary transit time and (k_p) the capillary wall permeation constant[10,11]. The claimed manifestation of this effect was elevation of the "apparent" in vivo free hormone level in tissues (eg the liver) characterised by "long" capillary transit times (i.e. times comparable with the dissociation half-time of the protein-hormone complex), this phenomenon supposedly accounting for the high rates of unidirectional (radiolabelled) hormone efflux observed in Oldendorf-type experiments. However, following criticism of these ideas and their mathematical basis[14], Pardridge and Landaw formulated, in 1984, a radically altered hypothesis[15], albeit the notion of a tissue-specific elevation of an "apparent" dissociation constant continues to feature in it. The revised hypothesis proposes that the "major factor leading to the rapid transport in vivo of protein-bound ligands into tissues such as brain is an endothelial-induced decrease in the affinity of the plasma protein for the ligand"[15] arising, for example, from "a conformational change in the plasma protein" as it transits the target tissue.

The basis for this new proposal differs from that underlying Pardridge's original hypothesis; it centers on the discrepancy between Pardridge and Landaw's experimental observations on radiolabeled hormone uptake in brain and other tissues, and theoretical predictions based on a revised hormone efflux equation, described by these authors as the "modified Kety-Renkin-Crone" equation[15], which takes the form:

$$FE = 1 - e^{-\frac{k_p t}{1 + K[P]}} \qquad (1)$$

where FE = fractional efflux of radiolabeled hormone during a single pass through the target organ
[P] = protein concentration
K = affinity constant
t = capillary transit time
k_p = capillary wall permeation rate constant

Pardridge's more recent views depend crucially on the demonstration that increase in the protein content of the injected labeled-hormone bolus used in his organ-perfusion experiments does not cause a reduction in tissue uptake of hormone of the magnitude predicted by Equation 1[15]. However, it is readily demonstrable that this simplified equation (which is esentially identical to one proposed in our critique of Pardridge's earlier theoretical analysis[14]) depends on the assumption that all kinetic events occurring within, and adjacent to, the capillary in the course of unidirectional hormone efflux proceed at infinite speed, i.e. that the only constraint on the rate of hormone efflux is the rate constant governing capillary wall hormone permeation.

The invalidity of this proposition can readily be demonstrated by reference to the more correct efflux equation applicable to Pardridge's experiments, assuming (for the sake of clarity) that the extracapillary free (labeled) hormone concentration remains essentially at zero as the tracer bolus transits organ capillaries, and that intracapillary free hormone mixing and diffusion are infinitely rapid (implying that the free hormone concentration across the entire capillary radius is uniform[13,14]):

$$FE = 1 - e^{-\frac{k_p t}{1 + K[P] + \frac{k_p}{k_d}}} \qquad (2)$$

where k_d = dissociation rate constant of bound hormone complex.

Clearly the simplified form of Equation 2 relied on by Pardridge reflects the assumption that $k_d = \infty$ (ie that the protein hormone complex dissociates infinitely rapidly), implying that $k_p/k_d = 0$. This assumption is clearly questionable. Moreover, inclusion of a term in the efflux equation reflecting bound hormone dissociation is sufficient to explain all Pardridge's experimental observations (see, for example, Figure 1). It is quite evident that the insignificant discrepancies between experimental results and those predicted from Equation 2 provide no justification for the postulation of hitherto unsuspected intracapillary hormone release mechanisms of the kind proposed by Pardridge.

A kinetic model of hormone transport

Pardridge's experimental results nevertheless confirm a proposition that we have frequently advanced, i.e. that circumstances *may* arise in a particular target organ in which bound hormone dissociation exerts rate limiting effects on hormone delivery[14,16]. In such circumstances, variation in the bound hormone level will affect hormone delivery to the organ concerned, as is implicit in Equation 2, and exemplified by the data shown in Figure 1. This proposition underlies our own challenge to the free hormone hypothesis, and our suggested explanation for the evolution and existence of serum binding proteins.

Though Equation 2 is capable of explaining Pardridge and Landaw's data, more complete theoretical analysis of the kinetics of hormone efflux from target tissue capillaries suggests that additional parameters affect the intracapillary free hormone concentration (or, more particularly, the concentration at the capillary wall, which governs the rate of hormone efflux into adjacent tissue).

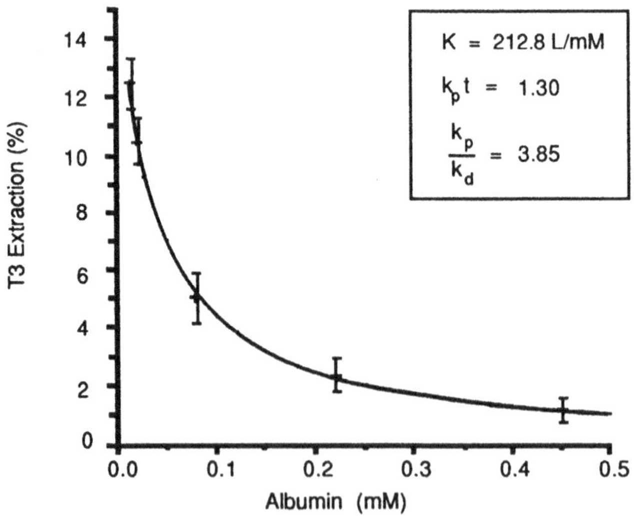

Figure 1. Unidirectional extraction of [125I] T3 by rat brain plotted versus arterial albumin concentration (data reproduced from Figure 5, Reference 18). Note the typically good fit of Equation 2 to the experimental data.

Target tissue uptake of hormone can be shown to be a complex function of a number of factors (including the extravascular free hormone concentration, the capillary wall permeation rate constant, the rate of intracapillary free hormone difusion and mixing, the intracapillary bound hormone level, and the dissociation rate of the bound hormone complex) some of which are "tissue specific". Our analysis reveals that serum binding proteins *assist* in sustaining a high free hormone concentration at the capillary wall in the face of hormone efflux, thus enhancing (not, as commonly thought, restricting) hormone efflux from the capillary. Moreover it is demonstrable that a rise in the serum concentration of a binding protein that carries two hormones (as does TBG) is potentially capable of specifically enhancing transport to certain tissues of one relative to the other in consequence of differences in their kinetic characteristics. Thus, in the case of T4 and T3, a rise in TBG is likely to increase T4 delivery to target tissues in which the rate of hormone efflux per unit area of capillary wall is high. This suggests that, although the free hormone concept may broadly apply throughout the body, binding proteins may nevertheless serve a subtle role in directing particular hormones to particular target organs in particular physiological circumstances (albeit for reasons quite different from those previously advanced by other investigators).

The marked changes in the concentrations of many binding proteins seen in pregnancy suggest that the physicochemical phenomena discussed above may specifically operate to regulate hormone supplies to tissues implicated in the reproductive process. Moreover, evidence that the feto-placental unit may be of special significance in this context is provided by the hitherto unexplained observation that a pregnancy-induced CBG rise is seen only in species characterized by hemochorial and hemoendothelial placentae[17]. The nature of barriers to hormone transport between maternal and fetal circulations may underlie this unexpected correlation, differences in the permeabilty of these barriers providing a possible explanation both for species differences in the spectrum of hormone binding proteins present in the blood, and in the alterations in the levels of these proteins arising in pregnancy. In the case of TBG, consideration of T4 and T3 binding kinetics suggests that a TBG rise results in preferential T4 delivery to any organ in which there is a high local rate of capillary efflux of TH. Meanwhile our results on placental transport of T4 indicate that fetal T4 uptake in the rat is relatively high in the first 9 - 10 days of fetal life, but thereafter falls to very low levels[4]. Placental T4 accumulation nevertheless remains high throughout pregnancy, the accumulated hormone being degraded to iodine (and reverse T3). Such degradation clearly provides a rich placental source of iodine for the developing fetal thyroid - a phenomenon likely to be of special importance when dietary iodine (and hence the iodide level in maternal blood) are low. These observations are clearly commensurate with the proposition that TBG is physicochemically designed to supply T4 to particular target tissues.

Nevertheless, although our own data and those subsequently obtained by others (eg[18]) have shown the belief that the feto-placental unit does not require T4 to be questionable, they do not directly confirm the proposition that the presence of large protein bound T4 pool in maternal blood serves to ensure adequacy of a feto-placental T4 supply. Nevertheless, the suggestion that this constitutes the primary role of TBG (and analogous serum binding-proteins in other mammalian species) accords with observations that high levels of protein bound T4 are seen only in mammals, that total TBG deficiency in man is essentially confined to males (and is without apparent physiological consequence), and that the bound T4 level in serum rises in pregnancy. Moreover, the presence in blood of a specific protein capable of enhancing T4 availability to the feto-placental unit might be expected to confer considerable evolutionary advantage, particularly in conditions of iodine scarcity, and in species such as man characterized by a long gestational period. Likewise, the finding that lack of maternal T4 in early fetal life is associated with irreversible neurological damage[19], and the recent observation that T4 alone is available to enter the fetal brain[20], are entirely commensurate with the idea that T4 is of special importance in regard to the differentiation of the fetal CNS.

In the following sections, we present further evidence supportive of this view.

PLACENTAL TRANSFER AND FETOPLACENTAL METABOLISM OF THYROID HORMONES AND IODIDE

Our initial experimental approach was to investigate the placental transfer of TH from mother to fetus, paying particular attention to early pregnancy[4,21]. In brief, rats were injected with [^{131}I]iodide and either L-[3'-^{125}I]T3 or L-[3',5'-^{125}I]T4 at various stages of pregnancy, and then killed after 1 hour. Radioactivity was determined in total, TCA soluble (iodide and other cleavage products) and TCA insoluble (iodothyronines) fractions of maternal blood and tissues, placentae and fetal tissues. Due to changes in fetal water content throughout pregnancy, concentrations of radioactivity were expressed as a function of the tissue protein content (radioactivity/g protein) and, in order to standardise results, all values were finally expressed as a percentage of the radioactivity associated with 1 g of maternal plasma protein.

At 10 days gestation, following injection of [^{125}I]T4, fetal concentrations of radioactivity in TCA insoluble and soluble fractions were 9 and 35% of the maternal plasma concentration respectively[21]. Placental concentrations were similar at the earliest gestational age at which they were examined (14 days), but whereas they increased throughout the remainder of pregnancy, fetal concentrations declined. Although fetal extraction of maternal [^{125}I]T3 was also observed in early pregnancy - the respective concentrations of radiolabel in acid insoluble and soluble fractions being 17 and 73% of the maternal plasma concentration - the absolute amounts of T3 transferred represent only a minor supply relative to T4, due to the 100-fold higher concentration of the latter in maternal blood. These findings have been confirmed by independent observations in a variety of animal species[18,22].

In order to illustrate the possible relevance of transferred maternal T4 to fetal development, concentrations of radioactivity in various maternal organs and fetus were compared[4,21]. At 10 days gestational age, the concentration in the fetal iodothyronine fraction was equivalent to, or greater than, that in maternal brain, ovary and heart, but less than that in maternal liver and kidney. The concentration in the fetal TCA soluble fraction was only less than that of maternal kidney, indicating efficient metabolism of accumulated T4. After mid gestation, the fetal iodothyronine concentration was comparable only with that of maternal brain, whereas TCA soluble metabolites were higher than those in maternal brain and heart. Thus, in early gestation, fetal concentrations were of the same order as those found in many maternal organs, implying that the supply of maternal T4 is sufficient to influence fetal development.

The presence of acid soluble radioactivity (^{125}I) in fetuses and placentae indicates metabolism of maternal T4. The ratio ^{125}I/^{131}I in the TCA soluble fraction of 10 day old fetuses was greater than that of maternal plasma, demonstrating local deiodination of maternal T4. Chromatography of fetal extracts[21] revealed that, in early pregnancy, 35% of transferred radiolabel was present as T3, 22% was in the form of rT3 and 43% remained as T4. In late pregnancy only 3% was recovered in the T3 fraction, whereas 66% co-migrated with rT3. In the placenta, relatively little T4 was converted to T3 following the onset of independent fetal thyroid function, most of the radiolabel (ca 70%) appearing as rT3. These findings are consistent with our observations of feto-placental 5'-deiodinase; activity of this enzyme is high in early pregnancy but declines after mid gestation[21]. The associated change in metabolic pattern may serve a dual role: to limit the supply of maternal T4, while at the same time ensuring adequacy of iodide supply for fetal T4 synthesis.

Table 1. Uptake of ^{131}I by maternal tissues and the feto-placental burden

Tissue	n	Ratio	% Dose of ^{131}I/organ
Liver	12	46 ± 4	3.43
Kidney	12	66 ± 5	0.63
Ovary	12	42 ± 12	0.06
Brain	12	6 ± 2	0.04
Heart	13	28 ± 7	0.13
Spleen	13	38 ± 4	0.12
Thyroid	12	6400 ± 4300	1.17
Maternal total			5.58
Fetoplacental burden	4		7.19

$$\text{Ratio} = \frac{\%\text{ dose /g wet weight}}{\%\text{dose /ml plasma}} \times 100$$

We have also investigated placental transfer of maternal iodide throughout gestation, demonstrating substantial ^{131}I accumulation in both fetus and placenta 1 hour after injection[21]. Iodide accumulation by the feto-placental unit increased throughout gestation, reaching a peak after establishment of autonomous fetal thyroid function. At 20 days gestation, the combined ^{131}I content of the total feto-placental burden was higher than that in all maternal organs studied (thyroid included) (Table 1). Under normal conditions, although an iodine deficiency state may prevail during pregnancy (due to increased renal clearance), the feto-placental unit is able to compete efficiently with maternal tissues for iodide.

Placental Control of Feto-Maternal Thyroid Hormone Economy

The particular relevance of TH with respect to the maturation of a variety of tissues and the organism as a whole, and the especially severe, potentially irreversible damaging effect of their deficiency on developing fetal brain and nervous system, constitute the basis for the concept of a specific placental control of maternal thyroid function. This proposition is consistent with certain indirect evidence suggestive of a nonpituitary regulation of TH concentrations in pregnancy and is centred on the hypothesis of a placenta functioning as an integrated and autonomous endocrine system, interconnected both with the fetus and the mother. As in the case of steroid hormone production (eg estriol) by the fetal adrenals, gonads and placenta, a complex function endocrine role may be postulated for the placenta, the organ controlling TH homeostasis through the synthesis and secretion of a placental thyro- tropin, independently from the maternal hypothalamus-pituitary axis. This putative placental control system would be expected to respond to different and specific feedback signals both from the fetal (T4, TSH ?) and the maternal (T4, TBG ?) compartments, fulfilling the critical role of monitoring the changing feto-placental requirements for T4 and ensuring an appropriate T4 environment for optimal fetal development (Figure 2).

Clinical evidence suggests that the increase in maternal thyroid activity consequent upon higher circulating TBG levels may not be mediated through pituitary TSH, whose circulating levels in early pregnancy have been described, in most reports, as lower or unchanged. The possibility of a non- pituitary control of TH secretion through a putative placental hormone has recently received renewed attention, speculation centering either on the existence of a specific chorionic thyrotropin (hCT) or, alternatively, a secondary biological activity of chorionic gonadotropin (hCG) (which is known to induce clinical hyperthyroidism in patients with hydatidiform mole or choriocarcinoma). While there is no agreement concerning the existence, or chemical and immunological properties, of hCT, the thyrotropic activity of hCG has been demonstrated under several in vitro conditions[23]. The molecular basis of this phenomenon may be found in the significant degree of structural homology with TSH. In our experience, the intrinsic thyrotropic activity of 1 IU of partially purified hCG is equivalent to 0.11 μU of TSH using, as a bioassay system, the functional parameter of iodide uptake in rat thyroid FRTL-5 cells[24]. This implies that hCG concentrations of 50-80 IU/ml normally found in the first trimester of pregnancy would correspond to physiologically significant amounts of TSH, able to suppress pituitary TSH secretion. Nevertheless, speculation regarding the physiological relevance of the thyrotropic

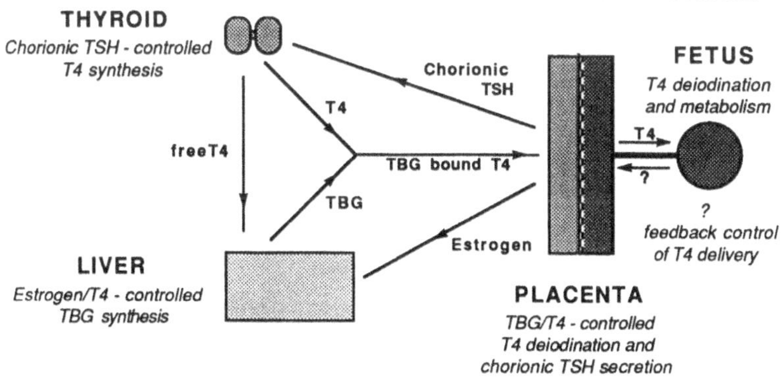

Figure 2. Putative feedback control system governing fetal T4 environment

activity measured in vitro must take into account, among other factors, the observation that hCG, like other glycoprotein hormones (eg TSH, FSH) encompasses a "family" of structurally-related isohormones with differing biological potencies and half-lives; the molecular composition of hCG can thus influence its interactions with the human thyroid. Although thyroid stimulating activity and hCG immunoactivity coelute in gel filtration[24], the occurrence of particular hCG variants (with higher degree of glycosylation or modifications in the carboxy-terminal region) might, in our view, specifically account for a regulation of thyroid function in pregnancy.

Distribution and Metabolism of Thyroid Hormones in Fetal Rat Brain

The transfer of T4 from the maternal circulation to the feto-placental unit indicates that fetal rat tissues are exposed to TH of maternal origin before the onset of an autonomous thyroid function. Comparison of the distribution pattern of radioactivity in fetal and maternal organs, following injection of dams with [^{125}I]T4, revealed localisation of activity in fetal brain in early pregnancy (10 days gestation). In the fetus, the concentration of radioactivity associated with the iodothyronine fraction was marginally higher in the liver relative to brain, whereas in the mother, the hepatic concentration was 12-fold higher than that in brain[21].

Current views regarding TH action centers on the proposition that these hormones mediate their effects by interaction with nuclear receptors. We therefore examined the subcellular localisation of transferred maternal T4 in fetal brain and liver. Large percentages of radiolabelled iodothyronine were recovered in the nuclear fractions of both liver and brain, with substantial amounts also present in the cytosol (Table 2).

Expression of concentrations of radioactivity in the brain relative to the corresponding liver values (brain/liver ratio) for mother and fetus, indicated a more favourable distribution of iodothyronine in subcellular fractions of fetal brain as compared to maternal brain (Table 3). The fetal brain appears both to accumulate T4 and to actively metabolise it to free iodide, T3 and rT3, as confirmed by chromato- graphic analysis of the subcellular fractions. These observations accord with the demonstration of 5'- and 5-deiodinase activities in fetal rat brain [20].

The presence of iodothyronines in organelles other than the nucleus, raises the possibility of extranuclear-mediated effects of TH, at least during development. Nevertheless, it should be emphasised that these observations must be viewed with caution at this stage, since a degree of translocation of TH, due to the subcellular fractionation procedure, cannot be ruled out.

Table 2. Percentage distribution of TCA soluble and insoluble radioactivity in subcellular compartments of fetal liver and brain.

	Liver		Brain	
Fraction	soluble	insoluble	soluble	insoluble
Nuclei	20 ± 19	39 ± 22	17 ± 15	28 ± 23
Lysosomes	9 ± 5	2 ± 1	6 ± 3	3 ± 2
Mitochondria	17 ± 28	24 ± 19	21 ± 13	16 ± 10
Microsomes	5 ± 5	3 ± 2	5 ± 8	5 ± 7
Cytosol	49 ± 25	28 ± 8	51 ± 20	48 ± 32

Table 3. Brain/liver ratios of TCA soluble and insoluble radioactivity in adult and fetal rats

	Adult		Fetus	
Fraction	soluble	insoluble	soluble	insoluble
Nuclei	0.02 ± 0.03	0.26 ± 0.06	1.75 ± 0.40	0.94 ± 0.07
Lysosomes	0.05 ± 0.04	0.05 ± 0.01	0.84 ± 0.07	0.70 ± 0.05
Mitochondria	0.16 ± 0.02	0.05 ± 0.02	2.83 ± 0.22	0.95 ± 0.04
Microsomes	0.31 ± 0.06	0.03 ± 0.02	0.85 ± 0.05	0.14±0.10
Cytosol	0.14 ± 0.09	0.04 ± 0.01	1.45 ± 0.12	0.94±0.07

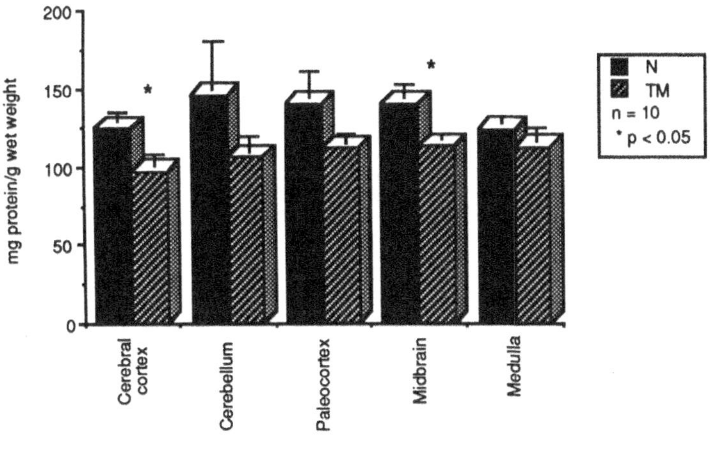

Figure 3. Brain protein concentration in adult progeny of N and TM dams

EFFECTS OF EARLY MATERNAL HYPOTHYROXINEMIA ON CELLULAR FUNCTION IN THE CNS OF THE PROGENY

Following our initial hypothesis, we investigated the consequences of maternal hypothyroxinemia in early pregnancy on brain function in progeny at various stages of postnatal life. Dams were partially thyroidectomised (parathyroid-spared) and maintained under normal animal house conditions with food and water ad libitum. The total daily iodine intake was similar to control dams, at 34 µg per day. These animals were severely hypothyroxinemic, but T3 levels were within the normal range. Control (N) and thyroidectomised (TM) dams were mated and pups were standardised to a constant litter size. Both control and experimental progeny were studied at different postnatal ages, from 14 days to 7 months. Serum levels of T4 and T3 in progeny of thyroidectomised animals were found to be within the normal range.

In 14 day old animals, body and brain weight were reduced by 15 and 35%, respectively[25]. Although preliminary data indicated no change in brain DNA content[25], more recent studies have demonstrated a 18% reduction. The total protein content of brain was reduced by 20%, and subfractionation revealed a severe depression (60%) in cytosolic protein, whereas insoluble protein fractions were unchanged[25]. The glycoprotein content of the cytosolic fraction was found to be increased by 46%, whereas that of particulate fractions was diminished. Acid phosphatase and ß-D-galactosidase activities were both reduced by 35%, and a less severe (17%) decrease was observed in aryl sulphatase activity.

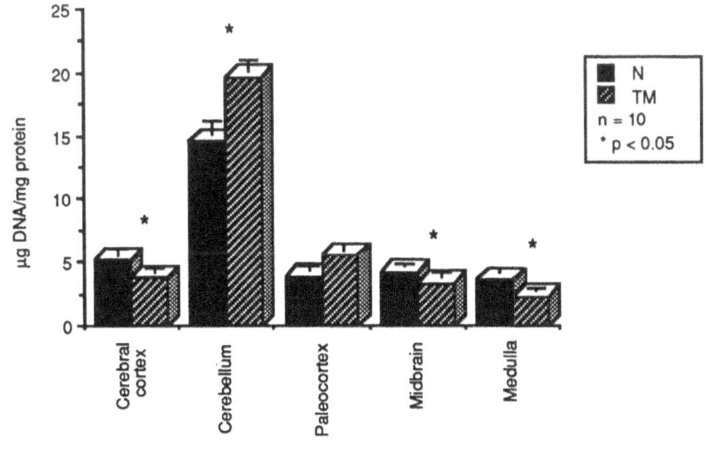

Figure 4. Brain DNA concentration in adult progeny of N and TM dams

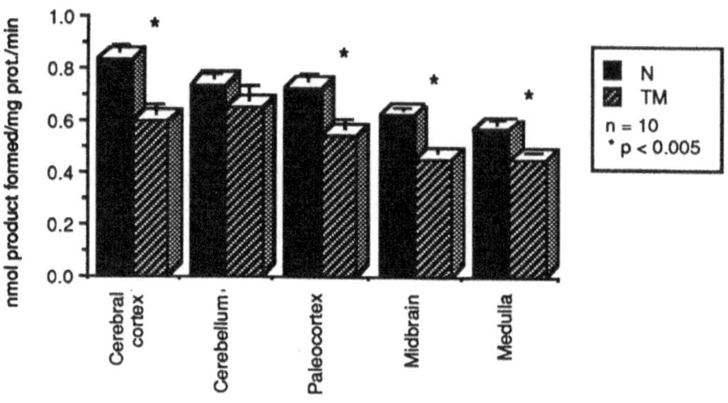

Figure 5. Brain LDH activity in adult progeny of N and TM dams

In adult animals (7 months), the significant reduction in brain/body weight ratio persisted, but the decrease in brain weight was less than that in 14 d old animals. Protein concentration was reduced in all brain regions studied, but only significantly so in cerebral cortex and midbrain (Figure 3). DNA concentration was diminished in cerebral cortex, midbrain and medulla (by 38, 25 and 55%, respectively) (Figure 4), while no change was observed in RNA concentrations. Protein/DNA ratio (an indicator of cell size) was increased in cerebral cortex, midbrain and medulla, but decreased in cerebellum and paleocortex. RNA/protein ratio (an indicator of protein synthetic capacity) was only significantly changed in paleocortex.

These observations in both young and adult animals are indicative of gross biochemical changes in brain. In particular, reductions in protein concentration may reflect compromised enzyme synthesis. This possibility was investigated in more detail in adult progeny by determination of specific activities of selected enzymes associated with a range of cellular functions.

Energy metabolism

Lactate dehydrogenase (LDH) activity was significantly reduced in all regions except cerebellum (Figure 5), and Na^+,K^+-ATPase activity was significantly decreased in paleocortex (Figure 6). In contrast, no changes were observed in either succinate dehydrogenase (a mitochondrial marker enzyme) or Mg^{2+}-ATPase activities.

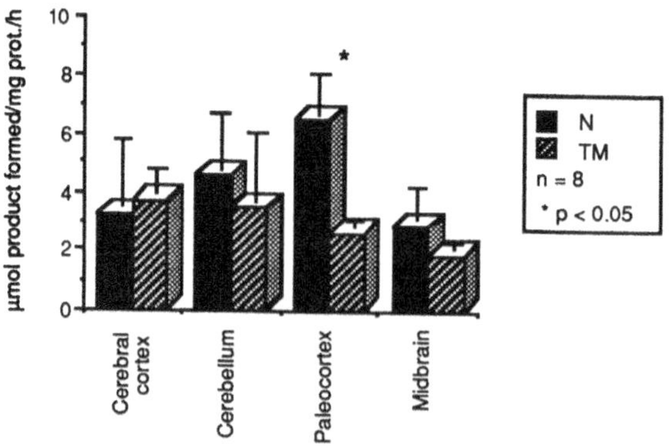

Figure 6. Brain Na^+,K^+-ATPase activity in adult progeny of N and TM dams

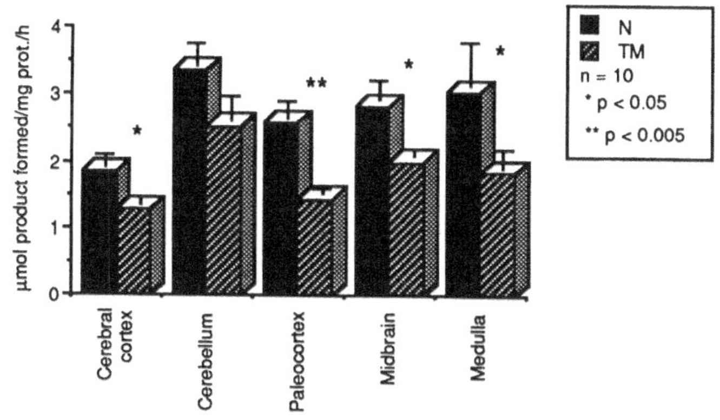

Figure 7. Brain ß-D-galactosidase activity in adult progeny of N and TM dams

Lysosomal enzymes

No significant change in both ß-D-glucosidase and aryl sulphatase B activities was observed. Specific activities of ß-D-galactosidase (Figure 7) and aryl sulphatase A (Figure 8) were significantly diminished in all regions, except cerebellum. Acid phosphatase activity was also significantly reduced in paleocortex and midbrain.

Neurotransmitter enzymes

Maternal hypothyroxinemia resulted in a differential effect on enzymes associated with neurotransmitter function; acetylcholine esterase (AChE) activity was reduced in cerebral cortex, paleocortex, midbrain and medulla (Figure 9), whereas no changes were observed in choline acetyl transferase activity. Aminergic transmitter function was unaffected, since monoamine oxidase activity remained within the normal range in all brain regions.

Calmodulin Calcineurin Systems

During early postnatal development, calcium-dependent, calmodulin-activated phosphatase activity (calcineurin) appeared to be diminished in TM dams at pH 4.0, whereas a significant increase was observed at pH 7.0. At seven months, a significant increase of the acidic component was observed in cerebral cortex. Therefore, a range of calcineurin dysgenesis at different age points are evident.

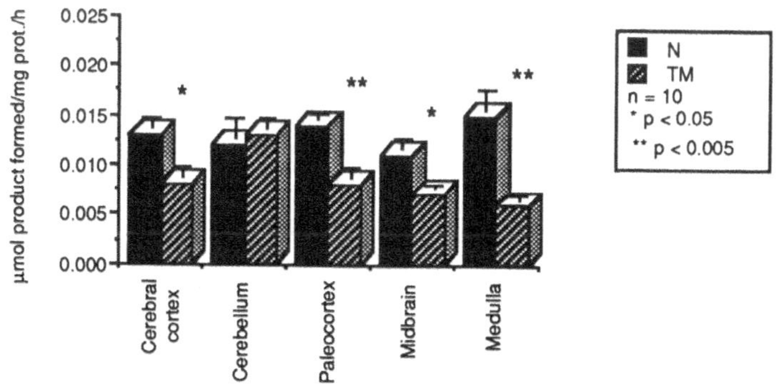

Figure 8. Brain aryl sulphatase A activity in adult progeny of N and TM dams

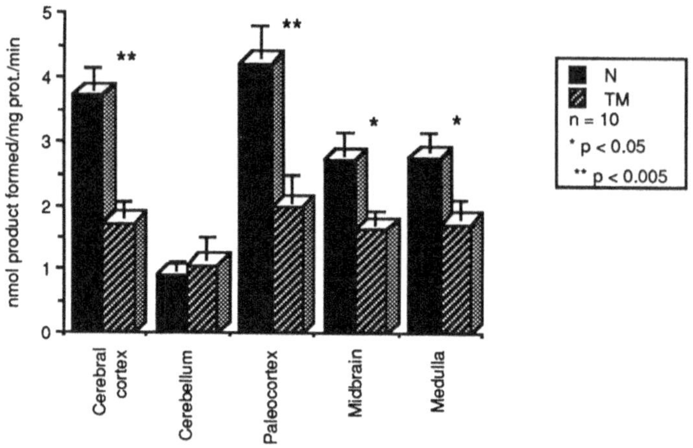

Figure 9. Brain AChE activity in adult progeny of N and TM dams

Tissue Culture Systems as Models

Studies on the progeny of thyroidectomised rat dams revealed both gross and specific biochemical deficits in the postnatal brain. However, the in vivo model presents restrictions which make difficult identification of the specific mechanisms responsible. Chief of these is the inability to specifically attribute any biochemical alteration to a lack of availability to the fetus of TH in utero.

In order to obviate these problems, in vitro techniques have also been employed. It is recognised that neural cells in culture behave similarly to cells in vivo in their responses to hormones, nuclear binding of T3, and developmental profiles. The use of tissue culture in a chemically defined medium eliminates interference from maternal and extraneural sources, permitting identification of the primary effects of TH. This approach has often been proposed as a useful model of neurological development and allows the intrinsic properties of neural cell types and the specificty of TH action to be investigated in a well controlled manner.

Initially, the effects of TH on protein synthesis in neurons were examined[26], revealing that T3 (1 - 30 nM) stimulated leucine uptake and subsequent incorporation into TCA-precipitable material. Stimulation of both uptake and incorporation was maximal (149 and 35% over control values, respectively) at 10 nM T3. Stimulation of leucine uptake was rapid (<1 hour), but the effect was abolished in the presence of inhibitors of protein synthesis, indicating that it was secondary to a pleiotrophic response of the cells. Although leucine incorporation was stimulated to similar degrees in soluble and insoluble (membrane-associated) protein fractions (respective values were +127 and 116% over controls), it is as yet unknown if this reflects a blanket stimulation of protein synthesis by T3.

Glucose is considered to be the primary energy substrate in the CNS; therefore, any changes in glucose utilisation may profoundly effect brain development. We have previously shown that T3 enhances 2-deoxyglucose uptake in neurons in primary culture, and have also observed similar effects in astrocytic cultures. Neuronal cultures were found to be unresponsive to insulin, indicating that TH may be important in regulation of energy metabolism during the period of neuronal proliferation and early differentiation in utero.

Much in vitro work has concentrated on neuronal, or mixed neural cell cultures, yet the importance of astrocytes, in terms of neuronal migration and survival, neurite outgrowth and differentiation, and in the control of oligodendroglial proliferation and trophic growth (specifically pertinent to early neurogenesis) has been acknowledged. TH are known to regulate the morphogenesis and cytoskeletal organisation of astrocytes, and to control glutamine synthetase activity, with consequent repercussions for inhibitory neurotransmitter availablity. Astrocytes are also known to migrate, though the region specific control of their production and mechanism of their migration remain unknown.

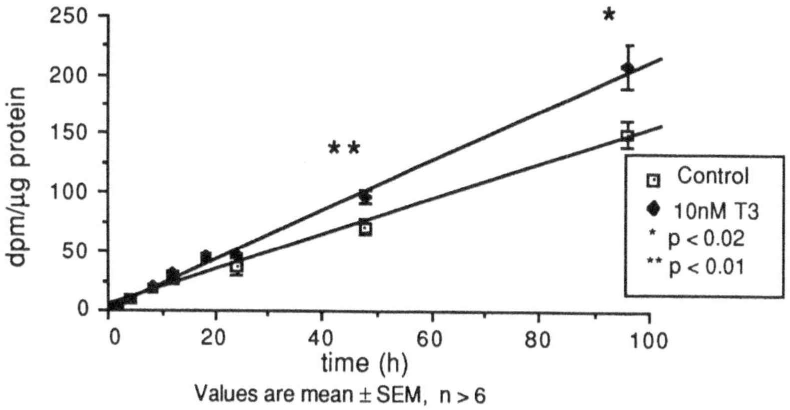

Figure 10. Effect of T3 on [3H]fucose incorporation into glycoproteins in astrocytes

Preliminary data obtained in vivo indicated differences in quantity and localisation of Con A binding glycoproteins in the brains of progeny from hypothyroxinemic rat dams. Alterations in the expression of important glycoprotein cell surface markers early in development may have long term consequences resulting from interference with normal cell development, connectivity and establishment of neural circuitry. In addition, many enzymes important in brain development are glycoproteins. We have therefore investigated the effects of TH on glycoprotein synthesis in cultured astrocytes, using the incorporation of ^{3}H-fucose as a marker.

Addition of T3 (10 nM) resulted in significant stimulation of fucose incorporation at 48 and 96 hours (32 and 39% over control values, respectively) (Figure 10). No immediate difference in free cytosolic ^{3}H-fucose was noticed, implying a lack of a direct effect of T3 on uptake, nor was the effect due to a blanket increase in protein synthesis.

The stimulation of fucose incorporation into glycoproteins may be accounted for by an increase in the production of a peptide precursor or precursors, an increase in production of the enzyme responsible for fucosylation, or by a stabilising effect blocking turnover. The long time lag occuring before a significant effect was apparent suggests that, in this case, TH do not act directly on translation or fucosylation, and implies a nuclear mediated effect. Studies currently in progress are intended to answer these questions, and provide information on the specificity of the effect

Neurological development is well coordinated, both temporally and spatially, and the existence of developmental "windows" for the action of exogenous controlling factors such as TH seems plausible. These "windows" may differ between cell types, and result in long term effects by acting at the levels of the system; controlling the establishment of neural circuitry, or the cell; inducing, for example, a cascade of sequential gene expression.

CONCLUSIONS

In recent years, studies from several laboratories have shown that, at least in early rat pregnancy maternal T4 crosses the placenta and is available to the developing fetus. In late pregnancy, maternal T4 transfer is reduced, coincident with the establishment of independent fetal TH economy. These observations challenge the notion of minimal transplacental passage throughout pregnancy and suggest a possible fetal need for maternal TH during critical phases of fetal growth. The high degree of localisation of transferred T4 in fetal brain, relative to the distribution pattern in the mother, suggests a specific role in development of the fetal neural system. Indeed, the fetal brain appear to possess the capability to utilise maternal T4 in early pregnancy, since both T3 and rT3 have been recovered from subcellular organelles, indicating the presence of functional 5'- and 5-deiodinase enzyme systems.

The predominant pattern of feto-placental T4 metabolism changes from 5'-deiodination in early pregnancy to 5-deiodination in late pregnancy. The observation that this switch precedes the establishment of independent fetal thyroid function, coupled with the observations that feto-placental deiodinase activities are predominantly localised within the placenta[21], suggest a degree of fetal control over its own TH environment, mediated at the level of the placenta. Since high levels of rT3 may serve to inhibit local T4 to T3 conversion in fetal tissues, this (hypothetical) signal may serve to protect the fetus from high levels of T3 (derived from both maternal and fetal sources), while concurrently maintaining an adequate iodide supply. Consequently, fetal T4 per se may play an important role in late development.

The identification of a thyroid stimulating function in chorionic gonodotropin also suggests an integrated system for TH economy in the mother, fetus and placenta. Control of maternal hypothalamic-pituitary-thyroid status by a placental factor has been indicated in the literature, since placental carcinomatous conditions, hydatidform mole, hyperplasia and other placental dysfunction has been reported to impinge upon maternal thyroid function.

Use of a rat model has permitted the influence of maternal T4 on fetal brain development to be studied in detail. A wide range of biochemical deficits has been observed in brains from both young and adult progeny of hypothyroxinemic rat dams. Decreases in protein/DNA ratios and DNA concentrations indicate disturbances in cell size and cell loss; in particular, the significant reductions seen in ß-D-galactosidase activity are indicative of neuronal loss. Protein content appears to be reduced, and more specific studies have suggested compromised brain cell function in the progeny of hypothyroxinemic dams. Alterations are also observed in the activities of selected enzymes associated with parameters such as energy metabolism, lysosomal function, synaptic function, myelin synthesis and calcium metabolism. These differences persist up to 7 months of age and, therefore, cannot be corrected by a normal thyroid state in the progeny. These observations accord with the existence of specific phases in (early) brain development, characterised by absolute requirements for maternal T4. Failure to provide an adequate supply of TH during such periods may thus result in irreversible brain damage in postnatal life.

Our findings provide a possible explanation for the behavioural and cognitive deficits observed in children born to euthyroid but severely hypothyroxinemic mothers in iodine deficient regions. It should be emphasised that not all enzymes studied were affected in the animal model and changes were region-specific, analogous to the selective dysfunctions observed in humans.

It may be argued that effects observed in the in vivo rat model may be secondary to placental dysfunction or maternal nutritional disadvantage. These possibilities, although clearly pertinent to observations in humans, have been minimised in our experiments, since the rat dams were euthyroid (T3 levels were in the normal range) and food consumption (including iodine intake) was normal. Although an effect of decreased maternal T4 on placental function per se cannot be entirely ruled out, our experiments with primary cultures of neural cells are supportive of a specific role for TH in cellular proliferation and differentiation of the CNS.

In conclusion, we have presented a basic model for detailed investigation of the influence of maternal TH, especially T4, in early neurogenesis and the consequences for the functional efficiency of the adult brain in iodine deficient regions.

REFERENCES

1. Fisher, D. A., J. K. Dussault, J. Sack and I. J. Chopra, Ontogenesis of hypothalamic-pituitary-thyroid function and metabolism in man, sheep and rat, in: "Recent Progress in Hormone Research", R. O. Greep, ed., Academic Press, New York (1977).
2. Nunez, J., Effects of thyroid hormones during brain differentiation, Mol. Cell. Endocrinol. 37:125 (1984).
3. Ekins, R. P., Hypothesis: The roles of serum thyroxine binding proteins and maternal thyroid hormones in foetal development, Lancet (i):1129 (1985).
4. Sinha, A., Woods R., Ekins R. P., Thyroxine metabolism by maternal and foetal tissues in the rat in early pregnancy, Ann d'Endocrinol 44:9A (1983).
5. Osorio, C. and N. B. Myant, The passage of thyroid hormones from mother to foetus and its

relation to foetal development, Brit. Med. Bull. 16:159 (1960).
6. Robbins, J., and J. E. Rall, Thyroid hormone transport in blood and extravascular fluids, in: "Hormones in Blood", C. H. Gray and V. H. T. James, eds., Academic Press, London (1979).
7. Tait, J. F. and S. Burstein, In vivo studies of steroid dynamics in man, in: "The Hormones", vol. V, Pincus, K. V. Thimann and E. B. Astwood, eds., Academic Press, New York, (1964).
8. Keller, N., U. I. Richardson and F. E. Yates, Protein binding and the biological activity of corticosteroids: in vivo induction of hepatic and pancreatic alanine aminotransferases by corticosteroids in normal and estrogen treated rats, Endocrinology 84:49 (1969).
9. Siiteri, P. K., J. T. Murai, G. L. Hammond, J. A., Nicker, W. J. Raymoure and R. W. Kuhn, The serum transport of steroid hormones, in: "Recent Progress in Hormone Research", R. O. Greep, ed., Academic Press, New York, (1982).
10. Pardridge, W. M., Transport of protein-bound hormones into tissues in vivo, Endocr. Rev. 2:102 (1981).
11. Pardridge, W. M., Transport of protein-bound thyroid and steroid hormones into tissues in vivo: a new hypothesis on the role of hormone binding plasma proteins, in: "Free Hormones in Blood", A. Albertini and R. P. Ekins, eds., Elsevier/North Holland, Amsterdam (1982).
12. Pardridge, W. M., Plasma protein-mediated transport of steroid and thyroid hormones, Am. J. Physiol. 252 (Endocrinol. Metab.15):E157 (1987).
13. Ekins R. P. and P. R. Edwards, Plasma protein-mediated transport of steroid and thyroid hormones: a critique, Ann. N.Y. Acad. Sci. (in press).
14. Ekins, R. P., P. R. Edwards, and B. Newman, The role of binding proteins in hormone delivery, in: "Free Hormones in Blood", A. Albertini and R. P. Ekins, eds., Elsevier/North Holland, Amsterdam (1982).
15. Pardridge, W. M. and E. M. Landaw, Tracer kinetic model of blood-brain barrier transport of plasma protein-bound ligands. Empiric testing of the free hormone hypothesis. J. Clin. Invest. 74:745 (1984).
16. Ekins, R. P., The free hormone concept, in: "Thyroid hormone metabolism", G. Hennemann, ed., Marcel Dekker, New York (1985).
17. Seal, V. S. and R. P. Doe, Corticosteroid-binding globulin: biochemistry, physiology and phylogeny, in: "Steroid dynamics", G. Pincus, T. Nadao and J. F.Tait, eds., Academic Press, New York (1966).
18. Obregon, M. J., J. Mallol, R. Pastor, G. Morreale de Escobar and F. Escobar del Rey, L-Thyroxine and 3,5,3'-Triiodo-L-Thyronine in rat embryos before onset of foetal thyroid function, Endocrinology 114:305 (1984).
19. Pharoah, P. O .D., I. H. Buttfield and B. S. Hetzel, The effect of iodine prophylaxis on the incidence of endemic cretinism, Advances in Exp. Med. and Biol. 30:201 (1972).
20. Morreale de Escobar,G., M. . Obregon and C. R. Ona, Comparison of maternal to fetal transfer of T3 versus T4 in rats, as assessed by T3 levels in fetal tissues. Ann. d'Endocrin. 48:165 (1987).
21. Ekins R. P., Woods R. J., Sinha A. K., Maternal thyroid hormones and development of the fetal brain, in: "Iodine nutrition, thyroxine and brain development", N. Kochpillai and V. Ramalingaswami, eds.,Tata McGraw Hill, New Delhi (1986).
22. Devaskar U. P., S. U. Devaskar, H. Farouk Sadig, V. Chechani, Ontogeny of plasma-free thyroxine and triiodothyronine concentrations during the perinatal period and maternofetal transfer of thyroid hormones in the rabbit, Dev. Pharmacol. Ther. 9:115 (1986).
23. Carayon P., Lefort G. and B. Nisula, Interaction of human chorionic gonadotropin and human luteinizing hormone with human thyroid membranes, Endocrinology 106:1907 (1980).
24. Ballabio M., Sinha A. K. and R. P. Ekins, Thyrotropic activity of crude hCG in FRTL-5 rat thyroid cells, Acta Endocrinologica 116:479 (1987).
25. Hubank M., Ballabio M., Sinha A. K., Gullo D., Bidey S., Bashir A., Ekins R. P., Long-term effects of maternal thyroidectomy on the development of the central nervous system of the progeny in the rat, Ann. d'Endocrinol. 47:48 (1986).
26. Pickard M. R., A. K. Sinha, D. Gullo, N. Patel, M. Hubank, R. P. Ekins, The effect of 3,5,3'-triiodothyronine on leucine uptake and incorporation into protein in cultured neurons and subcellular fractions of rat central nervous system, Endocrinology 121:2018 (1987).

DISCUSSION

<u>BRAVERMAN (Worcester)</u> : I would like to make 4 brief comments : 1) Nelson's study on schizophrenia is irrelevant because the data were questionable; 2) You do not take into account the intersitial space; 3) if TBG enhances transport in man, there is no TBG in any of the rodents studied; 4) Cavalieri and others have clearly shown that addition of proteins decreases the entrance of T4 both in perfused liver systems and in hepatocytes.

<u>EKINS (London)</u> : Of course when you add TBG or a binding protein acutely you obviously reduce the free hormone concentration of the sample to which it is added and that reduces hormone transport into any target cell. But the homeostatic system which exists in man keeps the free concentration constant. The experiment you are referring to is not appropriate. What happens in man is that the total hormone level rises and the free concentration remains roughly constant. Therefore, the coexistence of a more or less constant free concentration of thyroxine and an increased bound hormone concentration enhances hormone transport. Concerning the possible role of the binding proteins in hormone delivery to the fetus, we must assume that there are species differences in the placentas. I am only suggesting that a high level of bound hormone due to high TBG levels assists hormone transport to the fetal-placental unit.

MATURATION OF THYROID HORMONE ACTIONS

Delbert A. Fisher, and Daniel H. Polk

Department of Pediatrics, UCLA School of Medicine
Harbor-UCLA Medical Center, Torrance, CA
and
Department of Pediatrics, Drew University Medical School
Martin Luther King Medical Center, Los Angeles, CA

INTRODUCTION

There is now a large literature to support the view that thyroid hormones have little role in fetal growth and development (1-5). Athyroid human newborns have few if any signs of their chemical hypothyroidism, and length, weight and head circumference at birth are normal for gestation age. Moreover, IQ and neurological function parameters are normal in the vast majority of hypothyroid newborns detected by chemical screening and treated in the early neonatal period (6,7). Selected intrauterine effects of thyroid hormone deficiency have been recognized. Serum TSH concentrations are regularly elevated at birth in hypothyroid infants indicating active negative feedback control of pituitary TSH secretion by thyroid hormone in utero (2-5,8). Also, newborn thermogenesis may be mildly impaired, and the fall in body temperature in the athyroid human infant may be more marked than in a normal, euthyroid neonate (2-4). Thermogenesis in the newborn is largely a function of brown adipose tissue (BAT), and recent evidence in the sheep suggests that norepinephrine-stimulated oxygen consumption in newborn BAT is obtunded by hypothyroidism (9,10).

The above manifestations of the neonatal hypothyroid state are mild and subtle, however, and the classic signs and symptoms of cretinism develop in infants with congenital hypothyroidism during the early months of extrauterine life. These signs and symptoms include the deficiences in thermogenesis and metabolism characteristic of the hypothyroid state in adults. In addition there are profound effects of thyroid hormone deficiency on growth and development. These include deficient statural growth and growth of lean body mass, delayed epiphyseal maturation, deficiencies of brain growth and development, and delayed development/maturation of a variety of organs and tissues (1,11).

ONTOGENESIS OF THYROID HORMONE ACTIONS IN RODENTS

A number of investigators have studied the timing of appearance of thyroid hormone effects in developing rodents. These altricial species (in particular the rat and mouse) are born relatively immature with an undeveloped hypothalamic-pituitary-thyroid system as well as poikilothermia. They have a pattern of thyroid system development analogous to that of humans except that the late ontogenic events occur after birth (12). The ontogenesis of nuclear triiodothyronine (T3) receptors in brain and liver as these relate to serum TSH and serum thyroid hormone levels are demonstrated in figure 1 (12-16).

The ontogenesis of a variety of thyroid actions have been characterized in these altricial species. These include tissue thermogenesis; hepatic enzyme activities; beta adrenergic receptor binding in the heart, lung and brain; body weight gain; carcass and muscle growth; skeletal maturation; skin maturation including eye opening and tooth eruption; brain maturation; pituitary and serum growth hormone concentrations; submandibular gland nerve growth factor (NGF) and epidermal growth factor (EGF) concentrations; EGF levels in skin, eye, kidney and urine; and EGF receptor binding in skin and liver (17-41).

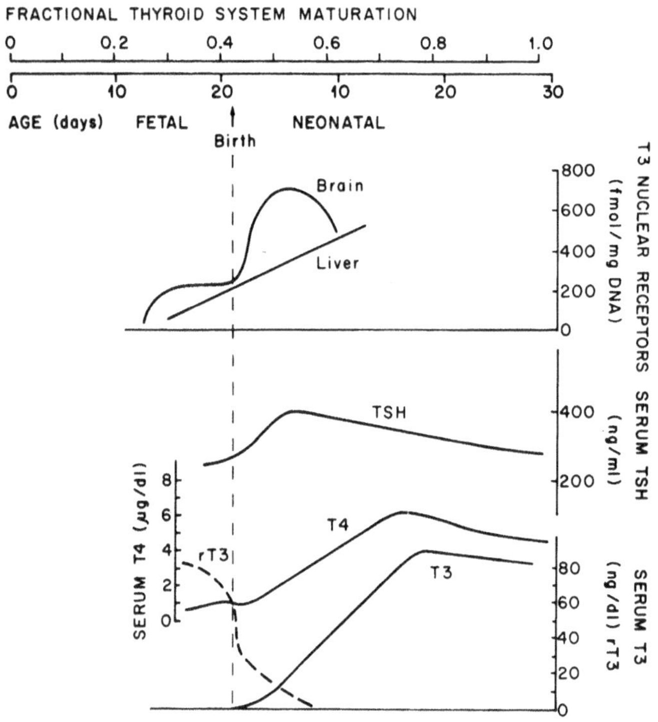

Figure 1. Maturation of thyroid receptors, serum TSH and thyroid hormone secretion in the developing rodent. Fractional thyroid system maturation is shown with the age to allow comparisons among species (see figures 3, 5). Data are derived from references 12-16.

Some of these effects represent thyroid hormone actions on specific gene products probably mediated via thyroid hormone receptor control of gene transcription. These include effects on hepatic enzymes, pituitary growth hormone synthesis, EGF and NGF synthesis in submandibular glands and EGF synthesis in kidneys; and EGF receptor binding in skin and liver (42-45). Other thyroid hormone actions represent complex events the mechanism(s) of which are not yet clear. Such actions include thyroid hormone stimulation of growth, skeletal maturation and brain development. Recent reviews of the role(s) of thyroid hormones in these events are available (1,17-20).

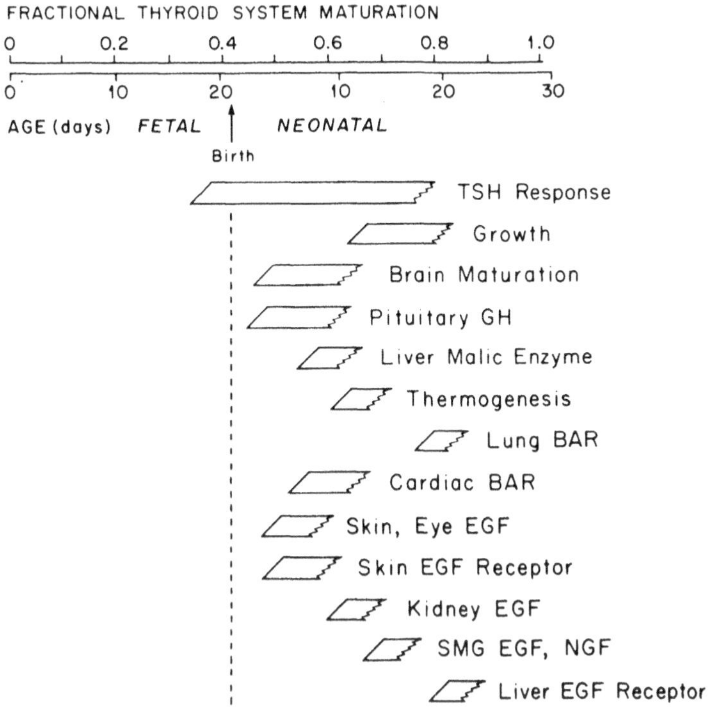

Figure 2. Maturation of thyroid hormone actions in the developing rodent. Fractional thyroid system maturation is shown with age at the top. The approximate time of appearance of an effect of thyroid hormone on the indicated parameters is shown by the beveled left edge of the horizontal bars. See text for details.

The time of appearance of a number of these thyroid effects in the neonatal mouse and rat is summarized in figure 2. All appear in the postnatal period, in general during the second and third postnatal weeks. Thyroid hormone stimulation of pituitary growth hormone (GH) content, of skin and eye EGF concentrations and of skin EGF receptor binding appear during the first few days of life (22,27,31,35,36). Effects on various aspects of brain maturation and liver enzymes have been described between 3 and 10 days of age (17,19,20,26,30,40,41). Thyroid hormone stimulation of thermogenesis and of kidney and urine EGF concentrations appear during the second week (23,34,37). The stimulatory effects of thyroid hormone on lung and heart adrenergic receptors, on submandibular gland NGF and EGF concentrations, and on liver EGF receptor binding appear during the third and fourth weeks of extrauterine life (24,37,38).

As indicated in figure 2, there is a discrepancy of several weeks in the timing of appearance of selected thyroid hormone effects in the developing rodent whereas thyroid hormone nuclear receptors are present in most fetal tissues at birth (13). Nuclear thyroid hormone receptors in lung and heart tissues (not shown in figure 1) show similar ontogenic patterns as in liver (13). Thus, much of the variation in timing of appearance of thyroid hormone actions in the rodents relates to maturation of postreceptor mechanisms rather than the appearance of thyroid nuclear receptors. Other thyroid hormone receptors have been identified including mitochondrial and plasma membrane receptors (46,47). However, the significance of mitochrondrial and plasma membrane receptors is not yet clear. The effects of thyroid hormone described (figure 2) are believed to be mediated via the nuclear receptors.

THYROID HORMONE ACTIONS IN DEVELOPING SHEEP

The sheep, in contrast to the rodent is a precocial species (12). It is born homeothermic and with relatively advanced brain maturation. The patterns of maturation of thyroid nuclear receptors in brain and liver as related to the ontogenesis of plasma TSH and thyroid hormone concentrations in fetal sheep are summarized in figure 3 (12,48,49). As indicated, the pattern of thyroid system ontogenesis is analogous to that of rodents (figure 1) except that most of thyroid maturation occurs in utero in the sheep. Thus, the chronically catheterized sheep fetus has been utilized as a model of human thyroid development.

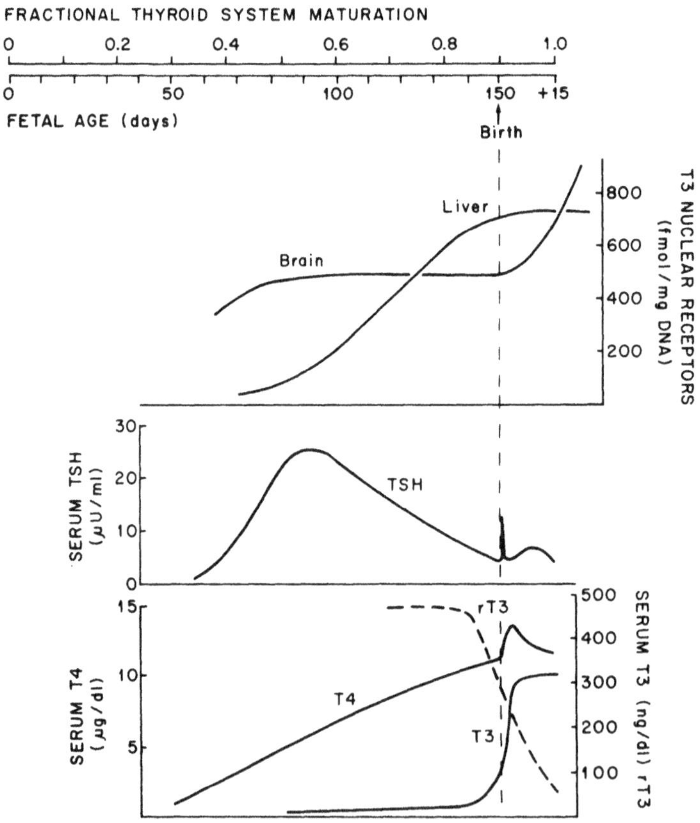

Figure 3. Maturation of thyroid receptors, serum TSH, and thyroid hormone secretion in the fetal and newborn sheep. Fractional thyroid system maturation is shown with the age to allow comparisons among species (see figures 1,5). Data are derived from references 12,48,49.

The timing of appearance of several thyroid hormone effects in the developing sheep is summarized in figure 4. Effects on carcass growth, bone and skin maturation and brain maturation first become detectable in the thyroidectomized fetus at 90-110 days gestation (50-53). The enzyme activities for the monodeiodination of thyroxine (T4) to 3, 5, 3′ triiodothyronine (T3) are thyroid hormone responsive. In adult animals and humans Type 1 enzyme activity in liver increases in response to T4 whereas type II activity in brain tissue decreases. Fetal thyroidectomy at 99-107 days in the sheep significantly increases type II iodothyronine 5′-monodeiodinase activity in fetal brain tissue but does not decrease type III activity in liver (54). In contrast, after fetal thyroidectomy at 129-132 days, both activities are altered; liver type I iodothyronine 5′-monodeiodinase activity decreases and brain type II activity increases (54).

Atrial natriuretic factor (ANF) levels in cardiac tissue are known to be thyroid hormone responsive (55). A significant reduction in ANF levels in fetal cardiac atria was observed when fetal sheep were thyroidectomized at 115 days and measurements conducted at 130 days (56). Decreased cardiac output at birth also has been demonstrated in fetal sheep thyroidectomized at 128-129 days (57). Thyroid hormone effects to increase malic enzyme activity and increase EGF receptor binding in liver do not occur after late fetal thyroidectomy but are observed after thyroidectomy of the 2 month old neonatal

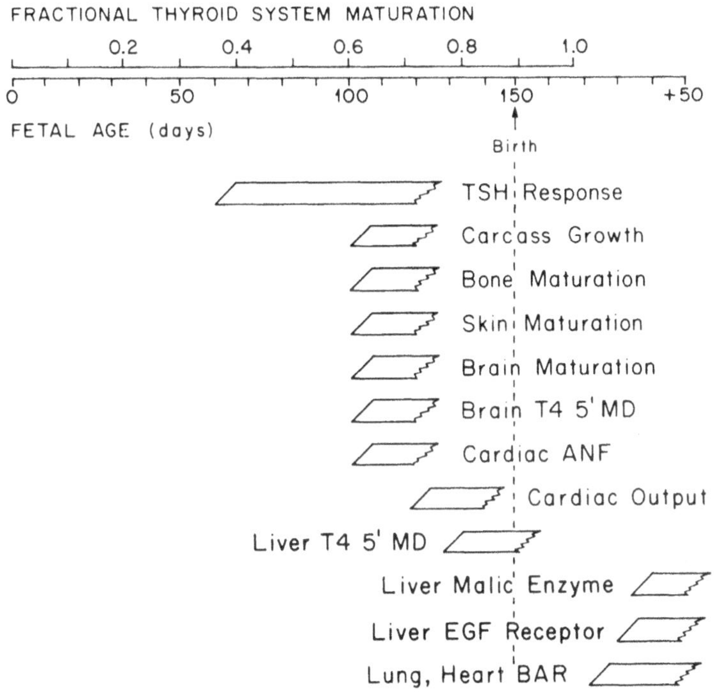

Figure 4. Maturation of thyroid hormone actions in the developing sheep. Fractional thyroid system maturation is shown with age at the top. The approximate time of appearance of an effect of thyroid hormone on the indicated parameters is shown by the beveled left edge of the horizontal bars. See text for details.

lamb (Polk DH and Fisher DA, unpublished data). The effect of thyroidectomy to decrease beta adrenergic receptor (BAR) binding in heart and lung tissue in the sheep was not observed after late fetal thyroidectomy whereas a decrease in BAR receptor binding activity did occur after early neonatal (1-2 day) thyroidectomy (58). Moreover T3 treatment significantly increased BAR binding in thyroidectomized neonates (58).

These results in the developing sheep, although less comprehensive, resemble those in developing rodents (figure 2). In both species relatively early effects of thyroid hormones on various parameters of brain development appear at 0.5 to 0.6 of thyroid system development. Thyroid effects on skin and growth parameters also appear relatively early whereas selected effects on heart, lung, kidney and liver are delayed. Also in the sheep, as in the rodent, the delay in appearance of selected thyroid hormone actions, such as the effects on liver malic enzyme and liver EGF receptor, are not accountable on the basis of delayed appearance of thyroid hormone nuclear receptors (figures 3,4). This applies, as well, to the delayed appearance of lung beta adrenergic receptors (58). Thus, the timing of maturation of thyroid hormone actions is variable in both species and relates both to the ontogenesis of nuclear thyroid hormone receptors as well as maturation of postreceptor response mechanisms.

The TSH response to hypothyroxinemia appears relatively early in both rodent and sheep species (figures 2,4). In adult animals this negative feedback effect of thyroid hormones on pituitary TSH release in mediated via both nuclear thyroid receptors in pituitary thyrotroph cells as well as modulation of hypothalamic TRH synthesis (59). In neonatal rodents and fetal sheep, thyrotropin releasing hormone (TRH) is present in relatively high concentrations in extrahypothalamic tissues, including pancreas, gut and placenta, before TRH appears in significant concentration in hypothalamic tissue (60,61). Moreover, in fetal sheep, TRH in these extraneural sites and in serum is thyroid hormone responsive increasing markedly in thyroidectomized animals (61). These data suggest that at this time of development (0.4 to 0.5 of thyroid system ontogenesis) pituitary and serum TSH is dependent on extrahypothalamic TRH. These extrahypothalamic TRH production sites presumably express nuclear thyroid hormone receptors, and TRH synthesis in these tissues may be thyroid hormone responsive at this early period of development.

The c-erb A gene protein product has been shown to be a high affinity receptor for thyroid hormone (62,63), and novel thyroid hormone receptor genes have been characterized in rat brain and human testis (64,65). These receptors which have a high degree of homology to the c-erb A thyroid receptor protein have been localized to human chromosomes 3 and 17 (62,64). Thus at least two mammalian thyroid receptor genes have been at least partially characterized. The first is expressed predominantly in brain but not in liver (64); the second, expressed in placenta and testis is postulated to be the native (liver) receptor (62,65).

The differing patterns of nuclear thyroid receptor ontogenesis (figures 1,3) in developing brain and liver are consistent with different receptors, and our laboratory has shown different binding affinities for brain and liver receptors in developing sheep (48). The relatively early appearance of the thyroid hormone effects on brain development (figures 2, 4) and late appearance of liver effects is likely attributable to differing ontogenesis of separate receptors. However, the significant lag in selected liver responses, and perhaps brain responses as well, are not accountable on the basis of current information about receptor binding ontogenesis. Whether there is receptor heterogeneity within organs or variation in translational or posttranslational maturation of thyroid nuclear receptors within organs is not known. Heart and lung nuclear receptor binding ontogenesis resemble maturation of liver receptor binding in the rat (13). Thus, there appear to be delays in maturation of postreceptor responses in these tissues as well (lung BAR, figure 2). Thyroid hormone responses appear relatively early in skin (figures 2,4) so that an early, brain-like receptor ontogenesis pattern might be expected. However, the pattern of maturation of thyroid hormone nuclear receptors in skin is not known. To date there is no information regarding the receptor mRNA species in heart, lung or skin of sheep.

THYROID HORMONE ACTIONS IN HUMAN INFANTS

The pattern of development of thyroid hormone secretion in the human fetus and infant is shown in figure 5 (12,66). The human fetus is intermediate between the altricial rat and precocial sheep at birth, being homeothermic but with a relatively immature central nervous system. There is limited information about ontogenesis of thyroid nuclear receptors in humans. Bernal and Pekonen studied 10 human fetuses of 10 to 18 weeks estimated gestation age and obtained after prostaglandin abortion (67). In 5 fetuses there was an increase in brain receptor concentrations from low levels at 10 weeks to values in the 350-500 fmol/mg DNA range by 16 to 18 weeks. Liver, heart and lung receptors were measured in 1 or 2 fetuses each at 16-18 weeks with receptor concentrations in the 200-600 fmol/mg DNA range (67). Gonzales and Ballard assessed thyroid hormone receptor binding in 10 human fetal lungs obtained after dilatation and evacuation abortion between 13 and 19 weeks gestation (68). A single class of high affinity receptors was characterized with an increase in receptor concentration from 200-300 fmol/mg DNA at

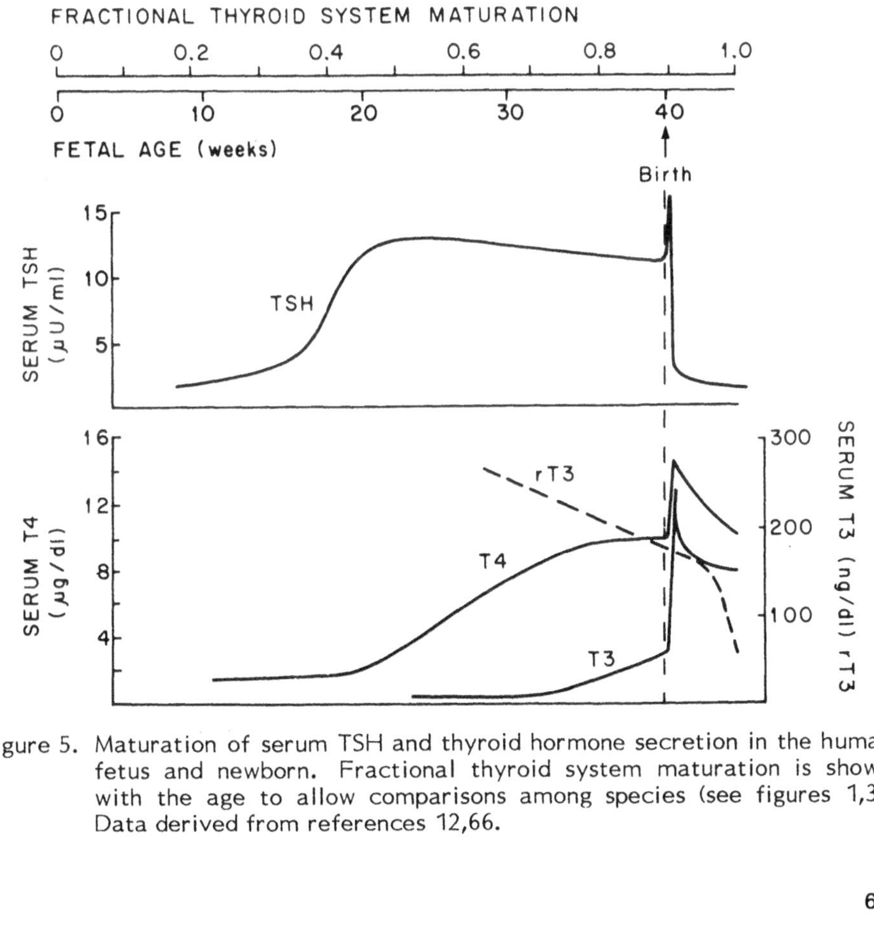

Figure 5. Maturation of serum TSH and thyroid hormone secretion in the human fetus and newborn. Fractional thyroid system maturation is shown with the age to allow comparisons among species (see figures 1,3). Data derived from references 12,66.

12-13 weeks (n=2) to 350-500 fmol/mg DNA at 16-19 weeks (n=8)(68). There are no data in term infants. However, the available information suggests significant concentrations of thyroid hormone nuclear receptors in human fetal brain, liver, heart and lung by midgestation.

The pattern of ontogenesis of thyroid hormone effects in the human fetus and newborn are summarized in figure 6. The TSH response to hypothyroxinemia appears relatively early in development as in the rodent and sheep. This early TSH response may relate to increased extrahypothalamic TRH production in response to hypothyroxinemia; relatively high concentrations of TRH have been reported in the human fetal pancreas during the first half of gestation (69). Also, as indicated earlier athyroid infants detected at birth by neonatal screening have appeared normal at birth and when treated with thyroid hormone in adequate doses before 45 days of age have normal growth and development, normal IQ values and no evidence of brain dysfunction assessed at 6-8 years of age (1-7, 70).

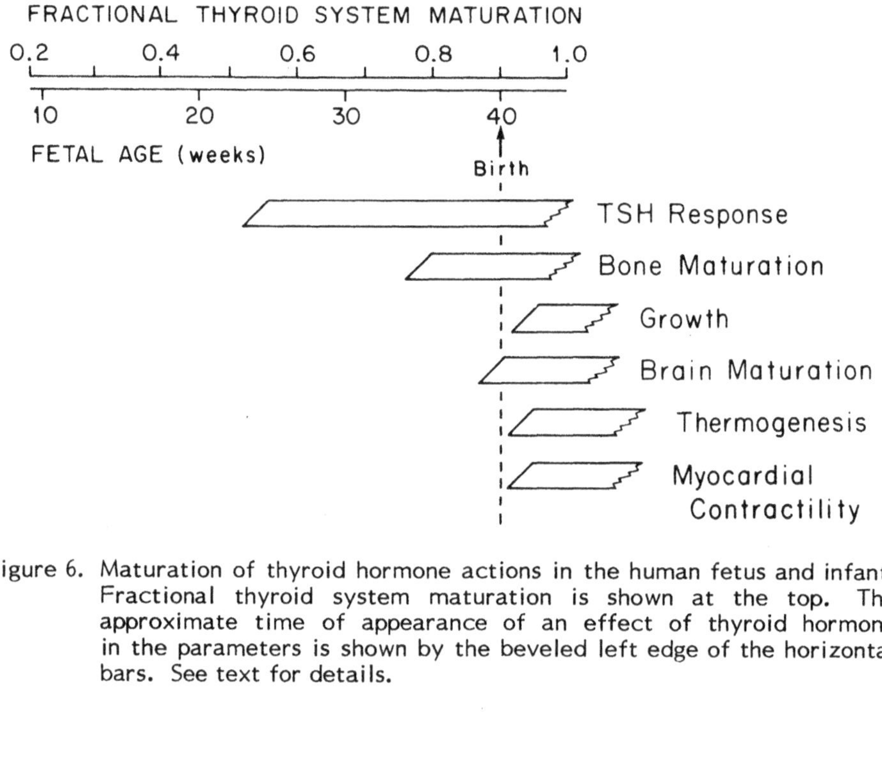

Figure 6. Maturation of thyroid hormone actions in the human fetus and infant. Fractional thyroid system maturation is shown at the top. The approximate time of appearance of an effect of thyroid hormone in the parameters is shown by the beveled left edge of the horizontal bars. See text for details.

Bone maturation is responsive to thyroid hormone in utero in some human fetuses; one third to one half of athyroid human infants are born with a modest retardation of epiphyseal maturation (1-5). Quantitatively this amounts to 2-6 weeks and is quickly corrected with treatment. There also may be some delay in fontanelle closure (2-5). Linear bone growth and weight gain become thyroid responsive soon after birth, and by six weeks of age significant reduction

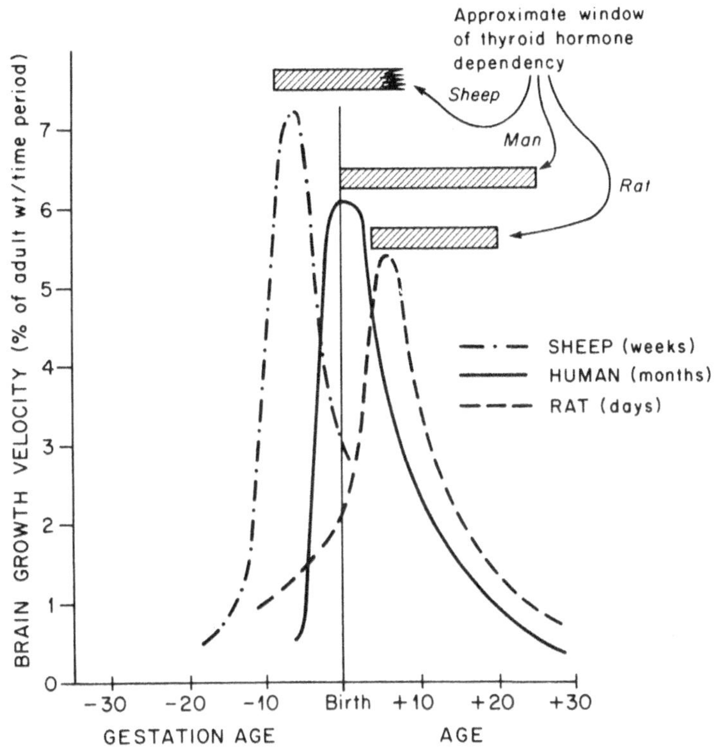

Figure 7. Relative brain maturation in rodents, sheep and humans. Brain growth velocity relative to time of birth is shown for each species. Growth velocity is plotted as % of adult weight per time period (per day in rats, per week in sheep, and per month in human infants). The overhead bars indicate the period of thyroid dependency of brain maturation.

in growth can be detected in athyroid infants (2-3). Athyroid infants generally show no metabolic signs of hypothyroidism at birth, but a few infants manifest prolonged jaundice and/or hypothermia in the neonatal period. Other classic signs and symptoms develop during the first 2-4 months (1-5).

Brain development in infants with congenital hypothyroidism has been studied carefully during the nearly 15 years since neonatal screening for congenital hypothyroidism was introduced. Several hundred infants have been evaluated at 6-8 years of age with Stanford Binet or WISC intelligence tests and with careful neurologic and hearing testing. The mean and range of IQ values in these infants has been similar to simultaneously studied control infants, and to date there has been no evidence of consistent neurological abnormalities and no evidence of learning disorders (6,7,70). In a few children in Toronto and Quebec a correlation of lower range IQ values with higher serum TSH, lower serum T4 values, delayed bone age at birth, and/or delayed treatment, (beyond 45 days) has been reported (69-71). These correlations have been observed only in programs in which the initial dose of replacement thyroxine was relatively low (<10 μg/kg/day) (70-72). Current information suggests that a dose of replacement thyroxine of 10-15 μg/kg/day begun before 45 days of age will assure normal brain maturation in infants with congenital hypothyroidism (70). This indicates that any brain dysfunction due to thyroid hormone deficiency in utero is corrected by early adequate treatment at birth. The time of onset of a thyroid hormone effect on brain development in human infants, therefore, would appear to be the perinatal period (3 to 4 weeks before or after birth).

The relative brain maturation patterns in rodents, sheep and humans is graphically depicted in figure 7. Brain growth velocity is shown in each species plotted as % of adult brain weight per time period (days in rats, weeks in sheep, months in humans). The time of apparent thyroid dependency of brain maturation in the 3 species has been indicated by the horizontal bars for each species. The period of thyroid dependency of developing brain begins in each species at the time of peak brain growth velocity and corresponds temporally to the period of glial proliferation and migration. Thyroid hormones also stimulate neuroblast differentation including neurite outgrowth and neuronal connectivity (20).

The relative somatic growth rates of the three species are shown in figure 8 with the timing of thyroid dependency of growth indicated by the horizontal bars. As indicated the early period of rapid intrauterine growth is not thyroid hormone dependent. The onset of thyroid dependency correlates in each species with the onset of slowing of growth rate. Interestingly, the timing of brain maturation and maximal growth rates in the three species differ relative to the timing of thyroid system ontogenesis, and while the timing of thyroid dependency of these events is relatively similar in the three species this timing does not correlate with the other thyroid system ontogenic events. Thus, important phylogenic determinants appear to condition the maturation of thyroid hormone actions and these determinants are superimposed independently on a similar pattern of hypothalamic-pituitary thyroid system ontogeny in the three species.

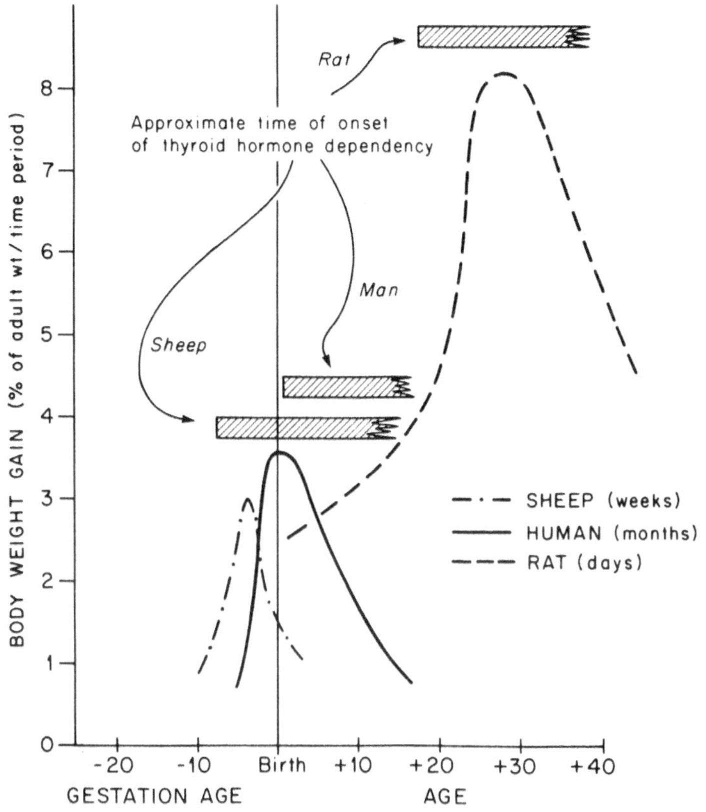

Figure 8. Relative growth rates in developing rodents, sheep and humans. Body weight gain is plotted as % of adult weight per time period (per day in rodents, per week in sheep and per month in humans). The approximate time of onset of an effect of thyroid hormone on body weight gain is shown as the left edge of the horizontal bars. See text for details.

REFERENCES

1. Fisher, D.A., Thyroid hormone effects on growth and development, in: "Pediatric Thyroidology", F. Delange, D.A. Fisher and P. Malvaux, eds, Basel Karger, (1985).
2. J. Letarte, H. Guyda, and J.H. Dussault, Clinical biochemical and radiological features of neonatal hypothyroid infants, in: "Neonatal Screening", J.H. Dussault, G.N. Burrow, eds, Raven Press, New York, p. 225, (1980).
3. D.A. Price, R. Ehrlich, and P.G. Walfish, Congenital hypothyroidism, clinical and laboratory characteristics of infants detected by neonatal screening. Arch Dis Child 56:845 (1981).
4. A. Larsson, J.G. Ljunggren, K. Ekman, A. Nilsson, P. Olin, and G. Bodegard, Screening for congenital hypothyroidism. II Clinical findings in infants with positive screening tests. Acta Paediatr Scand 70: 147, (1981).
5. J. Letarte, and S. LaFranchi, Clinical features of congenital hypothyroidism, in: "Congenital Hypothyroidism", P. Walker and J.H. Dussault, eds, Marcel Dekker, p. 351 (1983).
6. J. Glorieux, J.H. Dussault, J. Morissette, M. Desjardins, J. Letarte, and H. Guyda, Follow up at ages 5 and 7 years on mental development in children with hypothyroidism detected by Quebec screening program. J Pediatr 107:913, (1985).
7. New England Congenital Hypothyroid Collaborative. Neonatal screening: status of patients at 6 years of age. J Pediatr 107:915, (1985).
8. R.B. McCrossin, L.J. Sheffield, and E.F. Robertson, Persisting abnormality in the pituitary-thyroid axis in congenital hypothyroidism, in: "Thyroid Research", Canberra, Australian Academy of Science, Vol. VIII, p. 36, (1980).
9. A.H. Klein, A. Reviczky, and J.F. Padbury, Thyroid hormones augment catecholamine stimulated brown adipose tissue thermogenesis in the ovine. Endocrinology 114:1065 (1984).
10. D.H. Polk, J.F. Padbury, C.C. Callegari, J.P. Newnham, A.L. Reviczky, A.H. Klein, and D.A. Fisher, Effect of fetal thyroidectomy on newborn thermogenesis in lambs. Pediatr Res 21:453, (1987).
11. T.P. Foley, Jr., Sporadic congenital hypothyroidism, in: "Congenital Hypothyroidism", J.H. Dussault and P. Walker, eds, Marcel Dekker, New York, p. 231, (1983).
12. D.A. Fisher, J.H. Dussault, J. Sack, and I.J. Chopra, Ontogenesis of pituitary-thyroid function and metabolism in man, sheep and rat. Rec Prog Horm Res 33: 59, (1977).
13. A. Perez-Castillo, J. Bernal, B. Ferreiro, and T. Pans, The early ontogenesis of thyroid hormone receptors in the rat fetus. Endocrinology 117: 2457, (1985).
14. J.D. Dubois, and J.H. Dussault, Ontogenesis of thyroid function in the neonatal rat. Thyroxine and triiodothyronine production rates. Endocrinology 101: 435, (1977).
15. A.R.C. Harris, S.L. Fang, J. Prosky, L.E. Braverman, and A.G. Vagenakis. Decreased outer ring monodeiodination of thyroxine and reverse triiodothyronine in the fetal and neonatal rat. Endocrinology 103: 2216 (1978).
16. P. Walker, J.D. Dubois, and J.H. Dussault, Free thyroid hormone concentrations during development in the rat, Pediatr Res 14:247, (1980).
17. H.L. Schwartz, Effect of thyroid hormone on growth and development, in: "Molecular Basis of Thyroid Hormone Action", J. Oppenheimer and H. Samuels, eds., Academic Press, New York, p 413, (1983).
18. J. Legrand, Thyroid hormone effects on growth and development, in: "Thyroid Hormone Metabolism", G. Henneman, ed, Marcel Dekker, New York, p 503, (1986).
19. P. Walker, Developmental action of thyroid hormones, in: "Congenital Hypothyroidism", J.H. Dussault and P. Walker, eds, Marcel Dekker, New York, p. 63, (1983).

20. G. Morreale de Escobar, F. Morreale de Escobar, and A.P. Ruiz-Marcos, Thyroid hormone and the developing brain, in: "Congenital Hypothyroidism", J.H. Dussault and P. Walker, eds, Marcel Dekker, New York, p. 85, (1983).
21. A.F. Glassock and C.S. Nicoll, Hormonal control of growth in the infant rat. Endocrinology 109: 176 (1981).
22. F. Herves, G. Morreale de Escobar, and F. Escobar del Rey, Rapid effects of a single small dose of l-thyroxine and triiodo-l-thyronine on growth hormone as studied in the rat by radioimmunoassay. Endocrinology 97:91 (1975).
23. R.E. Steele and D.R. Wekstein, Influence of thyroid hormone on homeothermic development of the rat, Amer J Physiol 222: 1528 (1972).
24. J.A. Whitsett, C. Darovec-Beckerman, J. Pollinger, and J.J. Moore, Jr., Ontogeny of beta adrenergic receptors on the rat lung: effects of hypothyroidism, Pediatr Res 16: 381 (1982).
25. J.A. Whitsett, J. Pollinger, and S. Matz, Beta adrenergic receptors and catecholamine sensitive adenylate cyclase in developing rat ventricular myocardium: effect of thryoid status. Pediatr Res 16: 463 (1982).
26. P. Hahn and S. Hassanali, The effect of 3, 5, 3′-triiodothyronine on phosphoenol-pyruvate carboxykinase, fatty acid synthetase and malic enzyme activity of liver and brown fat of fetal and neonatal rats. Biol Neonate 41:1 (1982).
27. P. Coulombe, J. Ruel, and J.H. Dussault, Effects of neonatal hypo and hyper-thyroidism on pituitary growth hormone content in the rat, Endocrinology 107: 2027 (1980).
28. P. Walker, P. Coulombe, and J.H. Dussault, Effects of triiodothyronine on thyrotropin-releasing hormone-induced thyrotropin release in the neonatal rat. Endocrinology 107: 1731 (1980).
29. P. Walker and J.H. Dussault, Hypothalamic somatostatin and pituitary and serum growth hormone concentrations during postnatal development in rats exposed chronically to propylthiouracil or a low iodine diet, J Devel Physiol 2: 111 (1980).
30. R.M. Smith, A.J. Patel, A.E. Kingsbury, A. Hunt, and R. Balazs, Effect of thyroid state on brain development: beta adrenergic receptors and 5′ nucleotidase activity. Brain Res 198:375 (1980).
31. S.B. Hoath, J. Lakshmanan, S.M. Scott, and D.A. Fisher, Effect of thyroid hormones on epidermal growth factor concentration in neonatal mouse skin. Endocrinology 112: 308 (1983).
32. S.M. Scott, P.H. Chou, and D.A. Fisher, Nerve growth factor concentration in a congenitally hypothyroid mouse model hyt/hyt and its responsivity to thyroxine treatment, J Devel Physiol 5:413 (1982).
33. J. Lakshmanan, U. Beri, J. Perheentupa, A. Grueters, H. Kim, T. Macaso, and D.A. Fisher, Acquisition of submandibular gland nerve growth factor responsiveness to thyroxine in neonatal mice. J Neurosci Res 12: 71 (1984).
34. J. Perheentupa, J. Lakshmanan, and D.A. Fisher, Epidermal growth factor in neonatal mouse urine: maturative effective of thyroxine. Pediatr Res 18: 1080 (1984).
35. J. Lakshmanan, J. Perheentupa, S.B. Hoath, H. Kim, A. Grueters, C. Odell, and D.A. Fisher, Epidermal growth factor in mouse ocular tissue: effect of thyroxine and exogenous EGF. Pediatr Res 19:315 (1985).
36. S.B. Hoath, J. Lakshmanan, and D.A. Fisher, Epidermal growth factor binding to neonatal mouse skin explants and membrane preparations: effect of triiodothyronine. Pediatr Res 19: 277 (1985).
37. J. Lakshmanan, J. Perheentupa, T. Macaso, and D.A. Fisher, Acquisition of urine, kidney and submandibular gland epidermal growth factor responsiveness to thyroxine adminstration to neonatal mice. Acta Endocrinol 109: 511 (1985).
38. J. Alm, S. Scott, and D.A. Fisher, Epidermal growth factor receptor ontogeny in mice with congenital hypothyroidism, J Devel Physiol 8: 377 (1986).

39. P. Walker, Thyroxine increases submandibular gland nerve growth factor and epidermal growth factor concentrations precociously in neonatal mice: evidence for thyroid hormone mediated growth factor synthesis, Pediatr Res 20: 281 (1986)
40. A. Rami, A. Rabie, and A.J. Patel, Thyroid hormone and development of the rat hippocampus: cell acquisition in the dentate gyrus, Neuroscience 19:1207 (1986).
41. A. Rami, A.J. Patel, and A. Rabie, Thyroid hormone and development of the rat hippocampus: morphological alterations in granule and pyramidal cells, Neuroscience 19:1217 (1986).
42. J.H. Oppenheimer, H.L. Schwartz, C.N. Mariash, W.B. Kinlaw, N.C.W. Wong, and H.C. Freake, Advances in our understanding of thyroid hormone action at the cellular level, Endocr Rev 8: 288 (1987).
43. H.H. Samuels, B.M. Forman, Z.O. Horowitz, and Z.S. Ye, Regulation of gene expression by thyroid hormone, J Clin Invest 81: 957 (1988).
44. R.M. Gubits, P.A. Shaw, E.W. Gresik, A. Onetti-Muda, and T. Barka, Epidermal growth factor gene expression is regulated differently in mouse kidney and submandibular gland, Endocrinology 119: 1382 (1986).
45. V.R. Mukku, Regulation of epidermal growth factor receptor levels by thyroid hormones, J Biol Chem 259: 6453 (1984).
46. K. Sterling, G.A. Campbell, and M.A. Brenner, Purification of the mitochondrial triiodothyronine receptor from rat liver, Acta Endocrinol 105:391 (1984).
47. J. Segal and S.H. Ingbar, Specific binding sites for triiodothyronine in the plasma membrane of rat thymocytes, J Clin Invest 70: 919 (1982).
48. D.H. Polk, D. Cheromcha, A. Reviczky, and D.A. Fisher, Nuclear thyroid hormone receptors: ontogeny and thyroid hormone effects in fetal ovine liver and brain, submitted.
49. B. Ferreiro, J. Bernal, and B.J. Potter, Ontogenesis of thyroid hormone receptor in foetal lambs. Acta Endocrinol (Copenh) 116: 205 (1987).
50. G.H. McIntosh, K.I. Baghurst, B.J. Potter, and B.S. Hetzel, Foetal thyroidectomy and brain development in the sheep, Neuropath Appl Neurobiol 5:363 (1979).
51. G.H. McIntosh, B.J. Potter, and B.S. Hetzel, The effects of 98 day fetal thyroidectomy on brain development in the sheep, J Comp Path (1982).
52. B.J. Potter, G.H. McIntosh, M.T. Mano, D.M. Martin, P.F. Rogers, C. Hua, B.G. Cragg, and B.S. Hetzel, The effect of maternal thyroidectomy prior to conception on fetal brain development in sheep. Acta Endocrinol 112: 93 (1986).
53. A. Erenberg, K. Omori, J.H. Menkes, W. Oh, and D.A. Fisher, Growth and development of the thyroidectomized ovine fetus., Pediatr Res 88:783 (1974).
54. D.H. Polk, S.Y. Wu, C. Wright, A.L. Reviczky, and Fisher, D.A., Ontogeny of thyroid hormone effect on tissue 5′-monodeiodinase activity in fetal sheep. Am J Physiol: Endocr Metab 17:E337 (1988).
55. S. Agentin, J. Growin, and M. Nemer, Thyroid hormone stimulates rat pro-natriodilatin mRNA levels in primary cardiocyte cultures, Biochem Biophys Res Comm 146: 1336 (1987).
56. R. Castro, D.H. Polk, R.W. Lam, R.D. Leake, and D.A. Fisher, Atrial natriuretic factor: effect of thyroidectomy on concentration in fetal sheep atria and ventricles, Pediatr Res 23:274A (1988), abstract.
57. J.A. Breall, A.M. Rudolph, and M.A. Heymann, Role of thyroid hormone in postnatal circulatory and metabolic adjustments, J Clin Invest 73:1418 (1984).
58. J.F. Padbury, A.H. Klein, D.H. Polk, R.W. Lam, C. Hobel, and D.A. Fisher, Effect of thyroid status on lung and heart beta-adrenergic receptors in fetal and newborn sheep, Dev Pharmacol Ther 9:44 (1986).
59. T.P. Segerson, J. Kauer, H.C. Wolfe, H. Mobtaker, P. Wu, I.M.D. Jackson, and R.M. Lechan, Thyroid hormone regulates TRH biosynthesis in the paraventricular nucleus of the rat hypothalamus. Science 238: 78 (1987).

60. D. Engler, M.F. Scanlon, and I.M.D. Jackson, Thyrotropin releasing hormone in the systemic circulation of the neonatal rat is derived from the pancreas and other extraneural tissues, J Clin Invest 67:800 (1981).
61. D.H. Polk, A.L. Reviczky, R.W.Lam, and D.A. Fisher, Thyrotropin releasing hormone: effect of thyroid status on tissue concentrations in fetal sheep, Clin Res 36:203A (1988).
62. C. Weinberger, C.C. Thompson, E.S. Ong, R. Lebo, G.H. Gruol, and R.M. Evans, The c-erb-A gene encodes a thyroid hormone receptor, Nature 324: 641 (1986).
63. J. Sap, A. Munoz, K. Damm, Y. Goldberg, J. Ghyseal, A. Leutz, H. Beng, and B. Vennstrom, The c-erb-A protein is a high affinity receptor for thyroid hormone, Nature 324:635 (1986).
64. C.C. Thompson, C. Weinberger, R. Lebo, and R.M. Evans, Identification of a novel thyroid hormone receptor expressed in the mammalian central nervous system, Science 237: 1610 (1987).
65. D. Benbrook and M. Pfahl, A novel thyroid hormone receptor encoded by a cDNA clone from a human testis library, Science 238:788 (1987).
66. D.A. Fisher and A.H. Klein, Thyroid development and disorders of thyroid function in the newborn, New Eng J Med 304:702 (1981).
67. J. Bernal and F. Pekonen, Ontogenesis of nuclear 3, 5, 3′ triiodothyronine receptor in human fetal brain, Endocrinology 114:677 (1984).
68. L.W. Gonzales and P.L. Ballard, Identification and characterization of nuclear 3, 5, 3′ triiodothyronine binding sites in fetal human lung, J Clin Endocr Metab 53:21 (1981).
69. P. Leduque, S. Aratan-Spire, P. Czernichow, and P.M. Dubois, Ontogenesis of thyrotropin releasing hormone in human fetal pancreas, J Clin Invest 78: 1028 (1986).
70. D.A. Fisher and B. Foley, Early treatment of congenital hypothyroidism, Pediatrics, in press.
71. J. Rovet, R. Ehrlich and D. Sorbara, Intellectual outcome in children with fetal hypothyroidism, J Pediatrics 110:700 (1987).
72. J. Glorieux, M. Desjardins, and J.H. Dussault, Useful parameters to predict the eventual outcome of hypothyroid children, Pediatr Res, in press, (1988).

DISCUSSION

SACK (Tel Aviv) : How do you assess that the bone maturation is delayed only two to six weeks ?

FISHER (Torrance) : Because we measure bone age relative to a gestational fetal age. Looking at the picture and comparing it with normal pictures of a given bone maturation we know what gestational age it corresponds to.

SACK (Tel Aviv) : It is very difficult to assess accurately if the delay is only of several weeks.

SINHA (London) : Would you consider the possibility that T4 exerts a direct hormonal role in these early periods of fetal development, rather than a "prohormone" role ?

FISHER (Torrance) : You ask me what is the role of thyroxine in fetal brain maturation. I am suggesting that we should think of the possibility that even though thyroxine is there and clearly crosses the placenta, the quantity crossing varies widely among species. Nonetheless, what gets across could certainly bind to receptors. In addition, there is an active 5'-deiodinase in brain that appears about as early as the T3 receptor appears, and there are a variety of extra-hypothalamic tissues manufacturing TRH which might stimulate total TSH release, so that the system in the fetus is oriented strongly to the protection of the brain : that is, to maintaining the production of T4, its conversion to T3 and the binding of T3 to brain tissue. What effect T3 has during this period of brain maturation is not yet clear. The fact that the hormone is secreted by the fetus or gets across the placenta and that T3 is available at the receptor level does not imply that T3 is having an effect. It's been dramatic to us that thyroid hormone effects in the liver are very clearly delayed many weeks in the ovine species (2 to 3 months after birth for thermogenesis).

Dr. Potter and colleagues in Adelaide have published a nice article in Acta Endocrinologica describing maternal thyroidectomy of the sheep. They looked at the fetus at various times after about 60 days of gestation and found no decrease in fetal T4, no signs of decrease in crown-rump length, no signs of decrease in fetal weight, and no evidence of DNA or protein changes in the brain in this species before about 80 days of gestation : that is, at a time when the fetus is producing little or nothing in the way of thyroid hormone. The transient lag thereafter in brain growth is probably accountable on the basis of the hypothyroid placenta; brain growth is normalized by the time of birth. So, I don't think that very early in embryogenesis thyroid hormone has any role in brain maturation, but proof of this will require further data.

ROVET (Toronto) : Do you have any information on specific brain sites rather than considering the brain as a whole ? Are different brain tissues receptive to thyroxine at different times ?

FISHER (Torrance) : No. Dr. Potter looked at cerebrum and cerebellum and brain stem and found no major differences. There was a later maturation of cerebellum and with the transient lag in brain growth during the last half of gestation, he saw a more prominent effect on cerebellum, but it was not a dramatic difference.

REGULATION OF INTRACELLULAR THYROID HORMONE CONCENTRATIONS IN THE FETUS

María Jesús Obregón, Carmen Ruiz de Oña, Francisco Escobar del Rey and Gabriella Morreale de Escobar

Unidad de Endocrinología Experimental, Instituto de Investigaciones Biomédicas, C.S.I.C., Facultad de Medicina, Arz. Morcillo,4. 28029 Madrid Spain

Thyroid hormones play an important role in development[1], especially in brain development[2,3]. The role of thyroid hormones in development include the stimulation of skeletal maturation, the diferentiation of central nervous system, and they seem to regulate other processes such as the maturation of lung and the process of non-shivering thermogenesis in the neonate. Most of the information available is based on experimental models, mainly in the rat and sheep, due to the obvious difficulties to examine the effects of thyroid hormone deficiency in the human developing brain.

The work done by Jost and his group,established the principle of fetal autonomy for the pituitary-thyroid axis.The best demostration of this autonomy are the babies with congenital hypothyroid that have high TSH values. The ontogenesis of the pituitary-thyroid axis has been extensively studied and reviewed for different species[4], such as man, sheep and rat. In man, very low values of serum TSH, and T4 can be detected by 12 weeks of pregnancy, with a large spurt in serum TSH at 18 weeks, that preceeds the increase in serum T4. In the rat, thyroglobulin can be identified by fluorescence labelling at 17-19 dg. Thyroid hormone levels in fetal plasma have been determined in the rat[4,5] and found to be around 10% of adult levels, while plasma T3 was found to be undetectable. This fact has led to the generalized idea that thyroid hormones are very low during fetal life and therefore of little importance for fetal development. Thyroid hormones in fetal tissues had not been measured due to the lack of radioimmunoassays sensitive enough to acomplish the task, and due to the interferences derived from lipids and particulate material present in the tissular extracts. Some groups have studied the mechanisms that regulate thyroid hormone concentrations, namely 5'- and 5 -deiodinase activities during the fetal and neonatal period[6-10], and the capacity to respond to hypothyroidism during development[11].

In the present communication we review: i) the ontogenesis of thyroid hormone nuclear receptors, ii) the presence of thyroid hormones during fetal life, in brain as well as in other fetal tissues , and iii) the development of mechanisms that regulate T3 levels in the fetus, mainly 5'-D activities in relation to tissular thyroid hormone concentrations. Most work we will present here is done within the period comprising 17-22 dg, e.g. somewhat before onset of fetal thyroid secretion (17.5-18 dg) up to few hours before birth.

FETAL THYROID HORMONES AND T3 NUCLEAR RECEPTOR BEFORE ONSET OF FETAL THYROID FUNCTION

Before onset of fetal thyroid function,thyroid hormones are present in the fetus[12-14], and these hormones are of maternal origin[13], as their levels are undetectable in the progeny of thyroidectomized rats. This point has been treated in more detail by G. Morreale et al (chapter 2, "Transfer of thyroid hormones from the mother to the fetus"), but some points could be of interest here. The first point is that T3/T4 molar ratios are the highest in the embryos (0.21), followed by the placentas (0.09) and maternal plasma (0.02), excluding that our values reflect contaminations from maternal plasma, and suggesting that the placenta could play a role in the production of T3, as well as in the regulation of thyroid hormone levels in the embryo at this early time. This role has been studied by Roti et al[15],and other groups[16,17], that found maximal T4 5-deiodinase activities (inner ring) at 16 dg, and suggested that 5-Deiodination may serve as barrier for a more extensive passage of T4 to T3 from mother to fetus. Kaplan et Shaw[17] have studied 5'D in human and rat placentas, and the maximal 5' Deiodinase activity was found, in the rat, in the basal zone, adjacent to the endometrium, and in humans in the chorionic membrane decidua, but production of T3 (5'D activities)were 10% or less of the total deiodinase activity (5D+5'D). Unfortunately, few reports have studied the placental 5'-deiodinases before onset of fetal thyroid function. The origin of fetal thyroid hormones in this early period is obviously maternal, though the placenta is likely to play a role in regulating the transfer of thyroid hormones. Other authors have obtained similar results with different experimental approaches[14],as regards early transfer of maternal thyroid hormones.

The second point is that T4 and T3 concentrations found in whole embryos before onset of fetal thyroid function are very low, about 0.11 nM T4 and 0.02 nM T3(days 16-17 of gestation). Such low levels, could raise doubts about their importance for fetal development, but it is possible that there are higher concentrations specific tissues. Nonetheless, several defects are observed in the fetuses and progeny of hypothyroid rats including delays in the appearance of ossification centers[18], alterations in carbohydrate metabolism[19], high TSH and T4[20], reduction in DNA content and protein to DNA ratio in brain,[21] as well as other brain enzymes[21] related with myeline metabolism. If these effects are due to lack of thyroid hormones or other indirect effects such as the low GH levels or other growth factors remains to be clarified. Bonet and Herrera[22] have found low pituitary GH in 21-day-old rat fetuses from T mothers.

T3 nuclear receptors (T3 NR) are present in the rat fetus, and have been identified from 13th dg in whole rat embryos[23], and from day 14th in fetal rat brain. Its concentration increase three-fold up to day 17 (Figure 1), when a plateau is reached. After birth, there is a postnatal surge, and adult T3NR levels are reached. In Fig 1 the patterns of T3 NR are also depicted for heart, liver and lung. The number of receptor sites increased progressively in heart and liver,

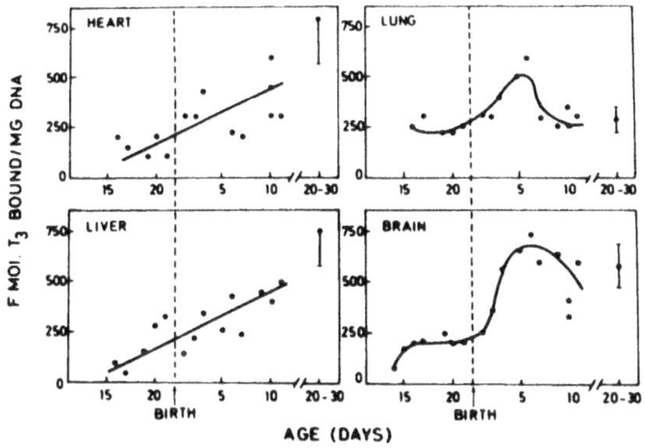

Figure 1. Concentrations of nuclear T3 receptors in several organs from fetal and neonatal rats (From Perez-Castillo et al[23], Endocrinology 117: 2457, 1985; with the permission of the publishers).

while characteristic patterns were observed in lung and brain These patterns of development are presently been explored using either specific monoclonal Ab[24], or the c-DNA for the c-erb-A[25], which encodes a protein described as the putative nuclear T3 receptor[26,27].

In addition,T3 NR has been described in brain of 10-weeks-old human embryos[28], its concentrations increased 10-fold by week 16th, at which time T3 NR was present also in liver, lung and heart[28]. T3 and T4 were also present in human fetal brain, T3 been detected from the earlier age (10 weeks), while in the other organs studied, T3 was absent, despite the presence of T4. This implies: 1) That in humans, although serum T3 is undetectable before the 30th week of pregnancy, T3 could be produced locally in the fetal brain by brain 5'Deiodinases, which are still not active in fetal liver and lung. As will be explained below, that is what seems to happen in the rat. 2) The presence of T3 NR, T3 and T4 in fetal human brain is coincident with the phase of active neuroblast multiplication[29] Recently[30], ontogenesis of T3 NR has been studied in human fetal brain, in relation with its T3 content, with essentially the same results. In the same symposium, the presence of T3 and T4 was reported, as well as 5'D activities in human fetal brain[31] from 11 weeks to 25 weeks of gestation. Both, 5'D activity, and thyroid hormone levels were increasing from the 11 up to 22th weeks of gestation.

THYROID HORMONES AND 5'DEIODINASE ACTIVITIES IN FETAL TISSUES AFTER ONSET OF FETAL THYROID FUNCTION

We will refer now to our studies done in the rat. We have examined if the mechanisms that maintain T3 levels are already developed in fetal brain, as well as in other tissues. For this, we have measured T4 and T3 concentrations in fetal tissues from 17 to 22 dg (dg). We have developed specific and sensitive RIAs for thyroid hormones, based on methods described by Weeke and Orskow[32], with modifications[33,34]. Plasma and tissue samples were extensively purified, as described[13], to determine very small amounts of T3 and T4.

5'D activities were measured in fetal brain, lung and BAT as described[35], using either ^{125}I-T4 or ^{125}I-rT3 (2 nM),20 mM DTT, in the absence (5'D-Total) or presence (5'D-II) of 1 mM PTU for 1 hour. Lung 5'D-I = (5'D-Total) - (5'D-II). When using ^{125}I-T4, 1μM T3 was added to prevent inner ring deiodination.

Fetal samples (brain, plasma, thyroids, lung, BAT, liver and carcass) were obtained at different dg, from pregnant rats (C or thyroidectomized (T) on different treatments : 0.02% MMI was given in the drinking water to produce fetal as well as

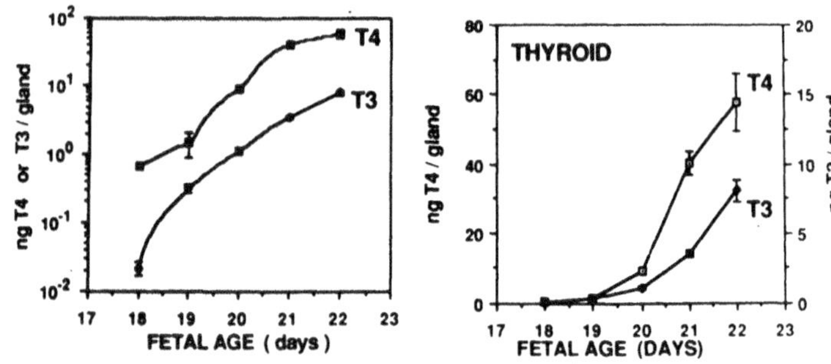

Figure 2. Concentrations of T4 and T3 in pronase digests of fetal thyroids between 18 and 22 dg. The left hand panel shows the data on logarithmic scale. Unpublished data.

maternal hypothyroidism, T4 or T3 were infused into the MMI-treated mothers. For details see references[13,36].

Thyroid Hormones in the Fetal Thyroid

Thyroglobulin can be identified in the fetal thyroid at 17 dg, while fetal thyroid secretion starts at 17.5-18 dg. We have measured total T4 and T3 contents in fetal thyroids between 18 and 22 dg, after digestion with Pronase. Figure 2 shows the results for C rats, using logaritmic or linear representations for comparison. T4 increased from 0.66±0.07 (SEM) ng T4 / gland at 18 dg to 57.7±8.2 at 22 dg, while T3 increased from 0.021±0.005 to 8.07±0.78 ng T3/gland in the same period. The increases were about 100-fold for T4 and 400-fold for T3. The ratio of thyroidal T4 to T3 decreased from 30 on day 18, to 5 on day 19, and was around 7 on 20-22 dg. The thyroidal T4/T3 ratios found in adult rats are in the same range[37].

When the mother is hypothyroid, T3 and T4 thyroidal contents are the same as in fetuses from C dams up to day 20, but during the last two dg thyroidal T4 and T3 are lower becoming 24% and 28%, respectively of C values (as discussed in chapter 2). This finding could suggest than the thyroid of the fetuses from T mothers is stimulated to compensate for the poor maternal contribution. This would agree with higher TSH found in those fetuses[20,22].

Thyroid hormones in Fetal Plasma

T4 and T3 were measured in fetal plasma and tissues between 18 and 22 dg. During this period the concentrations of T4 in fetal plasma are very low, and only extraction of large amounts of serum (0.5 ml) allow this determination. Fetal plasma T4 increased 9-fold, from 0.053±0.01 at 18 dg to 0.455±0.07 µgT4/dL at 22 dg (See Figure 3 middle pannel). This values are 8-10 times lower than the ones found in adult rats. In contrast with this increase in plasma T4 and with the marked increase in thyroidal T3 content, plasma T3 increased less than 2-fold during the same period (Figure 3, right pannel). In the adult animal, hepatic T3 production has been considered as the main source of circulating T3. Hepatic 5'D activities has been described to be low during the fetal life[6,10,38], and this could be an explanation of the very low T3 concentrations in fetal plasma, unless plasma T3 is rapidly sequestered by fetal tissues.

T4, T3, and 5'-Deiodinase in Fetal Brain and other Tissues

T4 concentrations increased 9-fold in most fetal tissues between 17 and 22 dg, in parallel with the increase in plasma T4 (See Fig 3, 4 and 5 for brain T4, lung T4 and liver T4). The brain/plasma T4 ratio is 0.24, four times higher than the

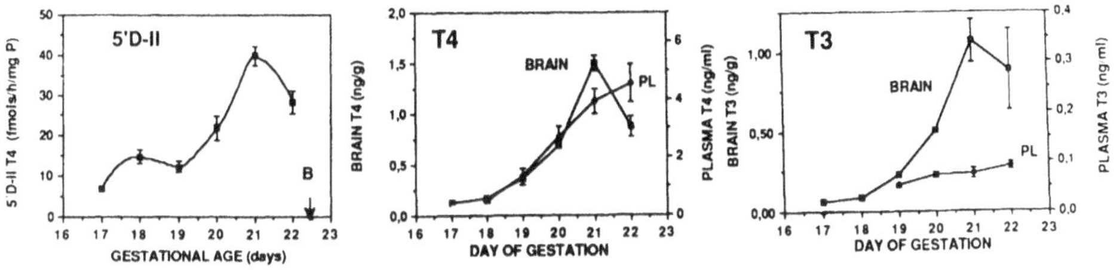

Figure 3. T4 5'D-II activities (left panel), T4 and T3 concentrations (middle and rigth hand panel) in fetal brain and plasma between 17 and 22 days of gestation, using normal pregnant rats.

values found in adult brain (0.06), suggesting that T4 uptake in the fetal brain is increased. Fetal brain 5'D-II activities and T3 concentrations are also shown in Figure 3. A 6-fold increase was observed in brain 5'D-II activities (days 17-21), mainly between 19 and 21 dg, while T3 concentrations increased 18 times between 17 and 21 dg, 6 times more than expected from plasma T3 levels. These data, together with those about transfer of maternal T4 and T3 (See below and Chapter 2), strongly suggest that T4 is the only source of T3 in the fetal brain, and that plasma T3 does not contribute to the maintainance of T3 concentrations in the fetal brain. No major differences were found in brain T4, T3 or 5'D-II in fetuses from T dams as compared with C dams.

T4 and T3 concentrations increased about 7-8 times in fetal carcass during the same period. Carcass is the total body minus brain, liver, lung, BAT and thyroid. T4 concentrations in fetal carcass increased from 0.22±0.02 to 1.75±0.28 ng T4/g between 18 and 22 dg, and T3 from 0.051±0.01 to 0.41±0.03 ng T3/g. The carcass to plasma T4 ratio was around 0.4 . We do not know if T4 and T3 values in carcass are homogeneous or are localized in certain areas, e.g. brown fat pads (BAT), randomly distributed, skin, or other compartments.

Data concerning the fetal lung are shown in Figure 4. As already described[39], lung 5'D is 90% inhibited by PTU (defined as type I). As shown in Figure 4, lung 5'D-I increases more

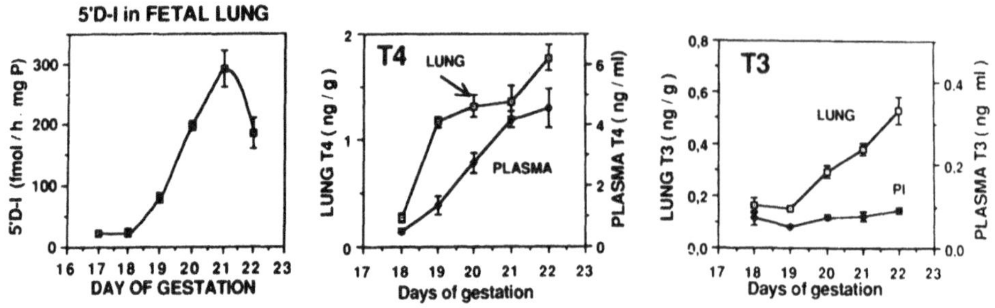

Figure 4. 5'D-I activities (left-hand panel) and T4 and T3 concentrations (middle and right-hand panel) in fetal lung and plasma between 17 and 22 days of gestation, from C pregnant rats. rT3 was used as substrate in 5'D assay.

than 10 times between 18 and 21 dg. This increases are in agreement with results from Segall-Blank[40] that demostrate conversion of T4 to T3 in cultures of fetal lung cells. Lung T4 increases 7 times. Although the increase in lung 5'D-I activity is greater than brain 5'D-II, lung T3 increases 3-fold (6-fold increases were found in some experiments). In contrast, and despite a 6-fold increase in liver T4, (Figure 5) liver T3 parallels the small increase observed in plasma T3. Hepatic 5'D has been described as very low at 14 and 19 dg in the rat[6], (5-10% of adult levels). McCann et al found[39] very low 5'D-II and I in the rat brain and lung, respectively, at 17 dg, increasing on the day of birth. Present data agree with these two timings.

In summary, the fetus and newborn have been considered asthyroid hormone deficient, and especially T3 deficient[4], due to low rate of production in nonthyroidal tissues (as suggested by hepatic data). But this statement reflects only T4 or T3 levels in fetal plasma. Fetal tissues are not thyroid hormone deficient and reach 50-100% of adult levels at 21 dg, only 3 days after onset on fetal thyroid secretion. Tissular T4 increases in parallel with plasma T4 with high tissue/plasma ratios, and T3 is regulated by 5'D in different tissues. Brain 5'D-II plays an important role in the regulation of T3 production in fetal brain, independently of serum T3 levels.

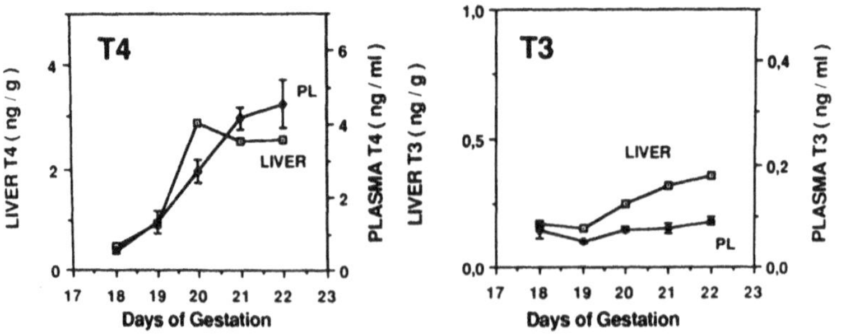

Figure 5. T4 and T3 concentrations in fetal liver and plasma between 18 and 22 days of gestation, from C pregnant rats.

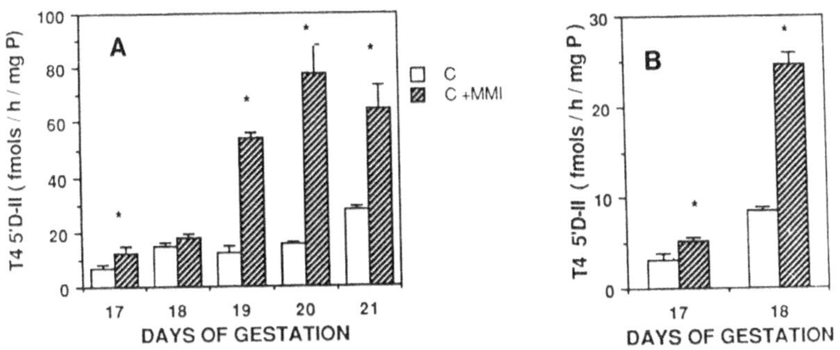

Figure 6.A.T4 5'D-II activities in fetal brain from C and MMI-treated dams, between 17 and 22 days of gestation. Treatment with MMI started on 14 dg.
6 B. T4 5'D-II activities in fetal brain from C and MMI-treated dams, on 17 and 18 dg. Treatment with MMI started on 12 dg.
*, indicate $p < 0.05$.

Responses to hypothyroidism and regulation by maternal thyroid hormones

One of our aims was to study the response of the fetal brain to hypothyroidism. Treatment of the mothers with MMI, results in very low T3 and T4 in the fetal thyroids[36], and reduced T4 and T3 in plasma and all fetal tissues[36], brain included. MMI treatment increased T4 5'D-II activitites in fetal brain from the earlier time studied, e.g. 17 dg. The responses are depicted in Figure 6.A. All increases were significant, except on day 18. When the rats were started on MMI treatment 2 days earlier (at 12 dg)(Figure 6, panel B) two to three-fold increases in brain 5'D-II were observed at 17 and 18 dg, respectively, demostrating that if hypothyroidism is severe enough, the fetal brain is able to respond to this very early in gestation. Kaplan and Yaskoski[11] described four-fold increases in brain T4 5'D activities in hypothyroid pups on the 1st day of life, suggesting that during fetal life mechanisms to increase 5'D should already be active. When 5'D-I was measured in hypothyroid fetal lungs, a small decrease in lung 5'D-I value was observed on days 20 and 21 of gestation which was statistically significant only on day 20.

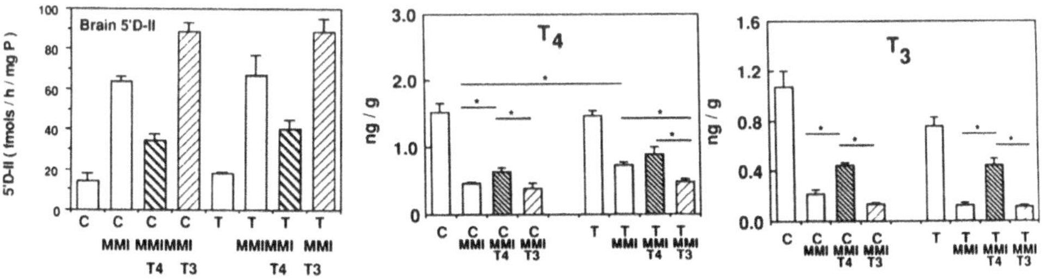

Figure 7. Brain T4 5'D-II activities (left-hand panel), brain T4 and T3 concentrations in 21-day-old fetuses, from C and T mothers. Treatment of the mother is indicated under each column. (*, indicates $p < 0.05$).

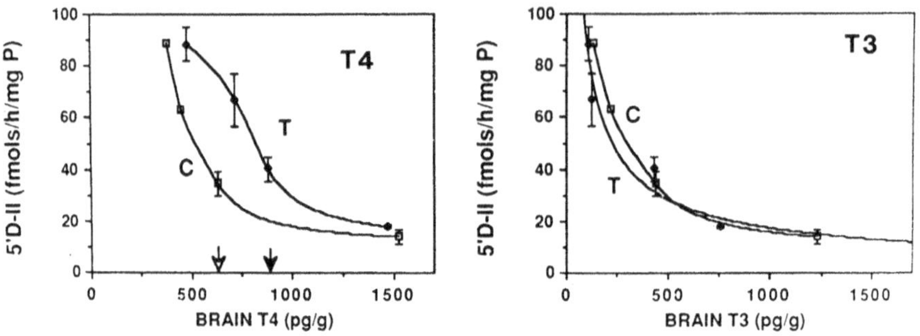

Figure 8. Correlation between brain T4 and T3 concentrations and brain 5'-Deiodinase (type II) activities, in the same groups as indicated in Figure 7. Values obtained in fetuses from C and T mothers were plotted separatedly, each point indicates groups with different treatments. The values are the same as those used in Figure 7.

We have shown that infusion of the mothers with physiological doses of T4 increased thyroid hormones in most fetal tissues[36] (See Chapter 2). Though maternal transfer of T4 in the rat was described to be less than 23%[4], even such small amounts of T4 may produce significant levels of T3 in some fetal tissues. In Figure 7, the results for fetal brain are shown using C and T mothers. Infusion of the rats with T4 increases T4 and T3 levels in the fetal brain, while infusion with T3 does not increase T3 in the fetal brain, despite increased T3 levels in fetal plasma, lung and other tissues. In the right-hand panel, the effects on brain 5'D-II are shown. The stimulation caused by MMI, was partially prevented by T4 treatment, while T3 caused further increases in brain 5'D-II. The relationship between 5'D-II and thyroid hormone concentrations is represented in Figure 8. Inverse relationships are found that are identical for 5'D-II versus T3 in fetuses from C and T mothers, but not for T4 versus 5'D-II. 5'D-II is more sensitive to small variations in brain T4 concentrations in fetuses from T mothers. These results suggest a different set point of regulation of 5'D-II in the brain of fetuses from T mothers. The greater sensitivity could be developed as a result of hypothyroidism suffered before onset of fetal thyroid function.

The Fetal Brown Adipose Tissue (BAT)

BAT has been considered as a tissue important for hibernators and its importance has been recognized during the neonatal period[41]. In man it has been identified in newborn babies and in adults bearing pheochromocytomas[42]. Very dramatic changes occur in BAT during adrenergic stimulation[43], in situations such as exposure to cold. In the rat, BAT has been described to have a 5'D-II[44] similar in its kinetic characteristics to the one described in brain, pituitary, and placenta. BAT 5'D-II is regulated in a very singular and characteristic way. It

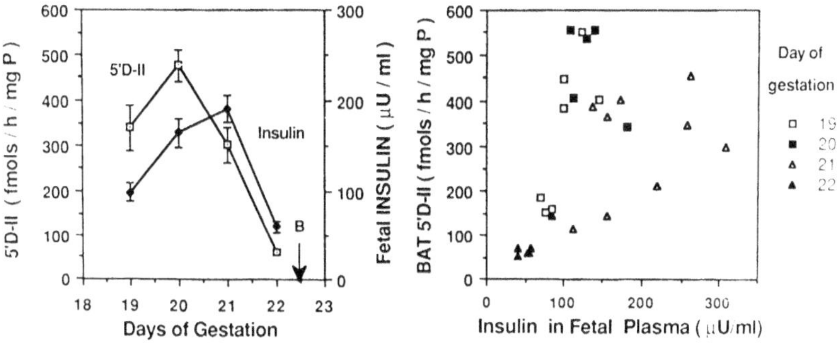

Figure 9. T4 5'D- II activities in fetal BAT between 19 and 22 days of gestation, and insulin values in fetal plasma on the same days (closed circles).

On the right-hand panel, a correlation (r= 0.44, $p < 0.05$) using individual values for plasma insulin and BAT 5'D-II activities, at different days of gestation (represented by different symbols).

increases in hypothyroidism[44,45] and under adrenergic stimulation[46], and it has been described as a potential source of local and systemic T3[47] in situations such as cold exposure. It is also stimulated by insulin and glucagon[48]. GH[45] has been proposed to exert inhibitory actions on 5'D-II. T3 NR has been described in BAT with a high degree of saturation[49], 70% of the nuclear T3 coming from local deiodination. In addition, high BAT 5'D-II activities have been described on day 20 of gestation[38].

We have studied BAT 5'D-II activities and thyroid hormones in rat fetuses. BAT 5'D-II activities are shown in Figure 9. Maximal values are observed on days 19 and 20, with a sharp decrease from days 21 to 22, hours before birth. Figure 9 also shows, insulin levels in fetal plasma. Changes in insulin levels correlate well with BAT 5'D-II activities, although the patterns are not superimposable. In the right panel we show a correlation of individual values obtained for insulin and

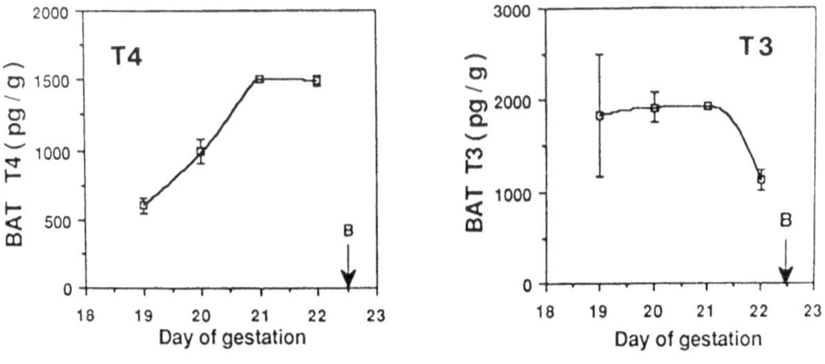

Figure 10. T4 and T3 concentrations in fetal BAT, between 19 and 22 days of gestation. B, indicates birth.

5'D-II at different fetal ages. When plasma catecholamine levels were measured at the same ages, their values increased continuosly up to the hour of birth (Drs Goya and Pascual-Leone, personal communication), and did not correlate with the pattern observed for BAT 5'D-II. Thyroid hormone concentrations are shown in Figure 10. BAT T4 increases steadily in parallel with Pl T4, with a high BAT to Pl T4 ratio of 0.38, that is almost identical to the one we have found in carcass. T3 concentrations are the highest from all fetal tissues studied and are in agreement with high 5'D-II activities. This is specially relevant considering the very premature stage of development of BAT.

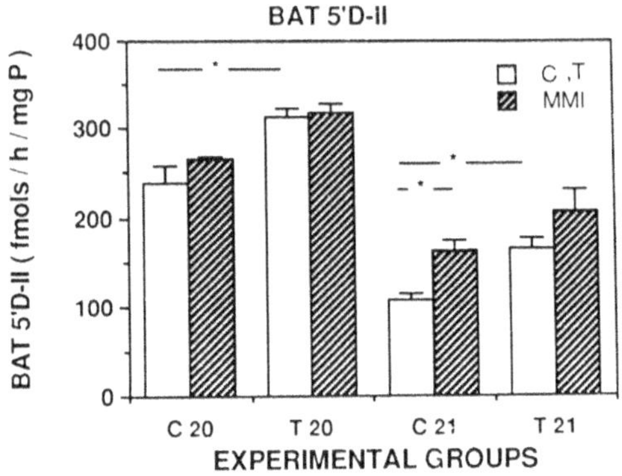

Figure 11. BAT 5'D-II activities in 20 and 21-day-old fetuses from C and T dams, treated or not with MMI. (*,indicate $p<0.05$)

In the hypothyroid rat fetus, MMI treatment leads to a decrease in BAT thyroid hormone concentrations (<25% of C values). Despite the low T4 and T3 concentrations, BAT 5'D-II was not increased by MMI treatment, (Figure 11), except at day 21, in fetuses from C+MMI rats. We do not know if activities are already maximal at 20 dg, or if, as observed in adult rats[45], 5'D-II increases only after long-term hypothyroidism. The same pattern of regulation is observed in thyroid hormone deficiency caused by low iodine diets (LID), that will be explained further on. In Figure 12, we compare BAT 5'D-II (days 20 & 21) in both situations: MMI treatment and LID. It is observed that LID leads to a more severe hypothyroidism at 21 dg, at least in terms of 5'D-II responses. BAT T3 was also lower.

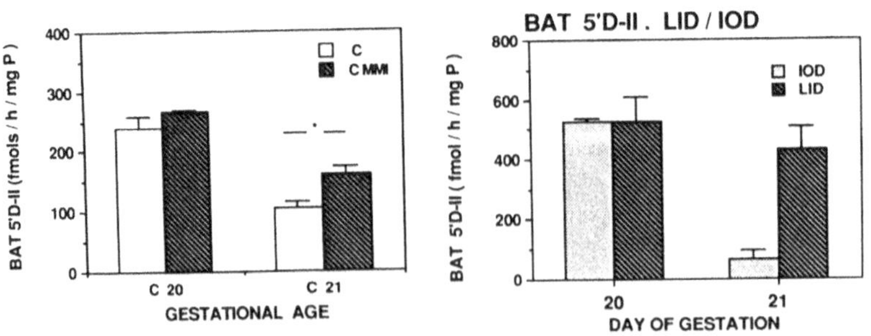

Figure 12. Comparison of BAT 5'D-II activities, in fetus from dams treated with MMI or fed LID or IOD diets. Fetal ages are 20 and 21 days of gestation.

In Figure 11 it is also shown that maternal hypothyroidism increased BAT 5'D-II, though increases were small. Thyroid hormones were similar in fetuses from C or T dams.The BAT weight, and relative weight was lower in the fetuses from T dams, as compared with C dams. In the former, we also found some impairement in the responses of BAT 5'D-II to maternal thyroid hormones, that in fetuses from C dams were similar to the ones observed for brain 5'D-II (described in Figure 7).

From the present data, it is evident that 5'D-II regulation in fetal BAT cannot be explained in simple terms. A multirregulation that includes insulin, thyroid hormones and possibly GH, might participate or overlap in the regulation of its activity. Further experiments are needed to know if the high thyroid hormone concentrations present in the fetal BAT are relevant for the development of the tissue. As plasma T3 is very low, the role of BAT in the maintainance of circulating T3 does not seem to be very important during fetal life.

<u>Responses to iodine deficiency</u>

The effects of thyroid hormone deficiency on the CNS are most dramatic in areas of endemic goiter where neurological cretinism is found.

We have previously studied the effects of LID during gestation in the rat[50]. LID fetuses had goiter and thyroidal iodine was 5% of their controls, fed LID+KI (IOD). Before onset of fetal thyroid function, the concentrations of T4 and T3 in embryos resemble those of maternal plasma: low T4 and normal T3 levels. At 17 dg, T3 levels were decreased to 50% of IOD levels. At 21 dg, T4 levels were very low in plasma, carcass, liver and brain, while T3 were reduced to 40-50% in the same organs[50].

We will present here a longitudinal study we have recently performed, which includes the last period of gestation (days 18-21) using pregnant rats on a LID, as described[50]. We have measured thyroid hormone concentrations and 5'D in fetal brain, lung and BAT. Results are shown in Figure 13 for fetal brain. Except on day 18, fetal brain T4 and T3 are very low in the LID groups, (12% of IOD values on day 21). They decrease with age, despite very high 5'D-II activities in the fetal brain, that are not able to mantain brain T3 levels, possibly due to the lack of substrate: T4. Figure 14 shows results for fetal lung. T3 and T4 are also very low in LID fetuses,

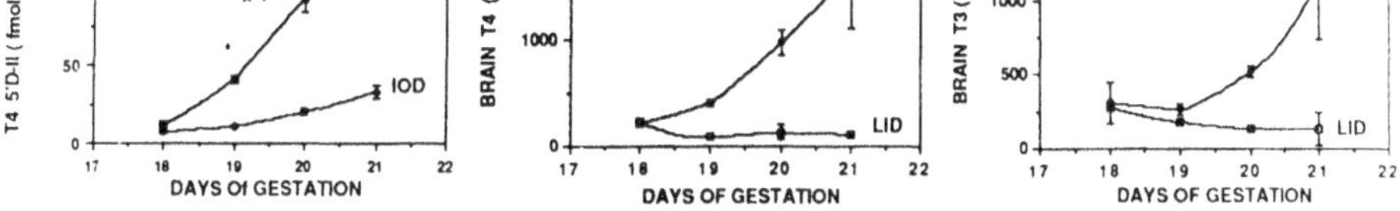

Figure 13. 5'D-II activities, T4 and T3 concentrations in fetal brain, between 18 and 21 days of gestation, from pregnant rats, fed LID or IOD diets (*, indicate $p < 0.05$).

and there is a decrease in lung 5'D-I activity at days 20 and 21 of gestation.

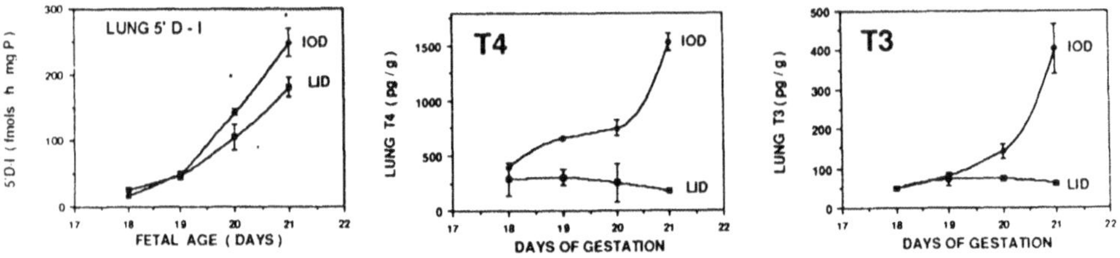

Figure 14. 5'D-I activities, T4 and T3 concentrations in fetal lung, in the same groups as in Figure 13.

Results obtained for BAT has been previously shown in Figure 12. The severity of thyroid hormone deficiency observed in the three organs studied is greater than in MMI-treated fetuses.

CONCLUSIONS

We have presented evidence that thyroid hormones reach the fetal compartment before onset of fetal thyroid function, and T3 NR are present during this early period, both in humans and the rat. With the onset of fetal thyroid function, thyroidal T4 and T3 increase exponentially, and there is a progresive increase of plasma T4 concentrations, accompanied by a parallel increase of T4 concentrations in all the fetal tissues studied. These reach 50% of T4 adult levels in brain and BAT, 25% in lung, while only 15% of T4 adult levels are found in plasma and less than 10% in liver. The Tissue to plasma T4 ratios are higher than in the adult.

In parallel, some tissues develop the capacity for production of T3, e.g. specific 5'-deiodinases for each tissue with different ontogenic patterns. 5'D-II in brain has a characteristic pattern, with a 2-fold increase over adult levels, and produces locally most of the T3 present in the fetal brain. Brain 5'D-II responds to hypothyroidism from the earliest age tested, and can be modulated by maternal thyroid hormones. In lung, 5'D-I increases 10-fold. BAT has very high 5'D-II activities from the earliest age tested, that are in good agreement with the high T3 concentrations found in BAT. The T3 concentrations reached are 15% of adult levels in the fetal plasma and liver, 35% in lung, and 50-100% in brain and BAT.

In summary, during fetal life T3 NR, T3 and T4 are present in the fetal tissues, at concentrations near adult levels in brain and BAT. The mechanisms that regulate T3 production are also active. These facts lead to the conclusion that the fetus is not T3 deficient, although plasma T3 concentrations might indicate the opposite. In our opinion, these are all strong suggetions for a role of T3 in fetal development. This is supported by the following findings: in BAT, one of us has shown that the specific mRNA of thermogenin, expressed during fetal life in BAT by day 21, decreases to 30% of normal values when the fetus is hypothyroid[51]. This decrease to 30% of C values is also observed during the first days of life. The effects on lung maturation caused by thyroid hormones had been studied by Hitckcock[52]. In brain, most studies are done in the neonatal rat, but Smith et al. were able to show changes due to hypothyroidism (decrease in ß-adrenergic receptors) on day 21 of fetal life[53].

REFERENCES

1. Legrand, J., Thyroid hormone effects on growth and development, in "Thyroid Hormone Metabolism", G. Hennemann, Marcel Dekker, New York & Basel, (1986), p 503.

2. Morreale de Escobar, G., A. Ruiz-Marcos and F. Escobar del Rey, Thyroid hormones and the developing brain, in: "Congenital Hypothyroidism", J. Dussault and P. Walker, eds., Marcel Dekker, New York (1983), p 85.

3. Rosman, N.P. and M. J. Malone, Brain myelination in experimental hypothyroidism: Morphological and biochemical observations, in: "Thyroid Hormones and the Developing Brain", G. D. Grave, ed., Raven Press, New York (19) p 169.

4. Fisher, D.A.,J. H. Dussault, J. Sack, I. J. Chopra, Ontogenesis of hypothalamic-pituitary-thyroid function and metabolism in man, sheep and rat, Rec Prog. Horm. Res.33: 59 (1977).

5. Wrutniak C. and G. Cabello, Changes in the concentration of thyroxine in the plasma of rat fetuses during late gestation: influence of ligation of the maternal uterine vein and artery, J. Endocrinol (London) 99: 233 (1983)

6. Harris, A. R .C., S. L. Fang, J. Prosky, L. E. Braverman and A. G. Vagenakis, Decreased outer ring monodeiodination of thyroxine and reverse triiodothyronine in the fetal and neonatal rat, Endocrinology 103: 2216 (1978)

7. Cheron, R. G., M. M. Kaplan and P. R. Larsen, Divergent changes of thyroxine-5'-monodeiodination in rat pituitary and liver during maturation, Endocrinology 106: 1405 (1980)

8. El-Zaheri, M. M., L. E. Braverman and A. G. Vagenakis, Enhanced conversion of thyroxine to triiodothyronine by the neonatal rat pituitary, Endocrinology 106: 1735 (1980)

9. Kaplan, M. M. and K. A. Yaskoski, Maturational patterns of iodothyronine phenolic and tyrosyl ring deiodinase activities in rat cerebrum, cerebelum and hypothalamus, J. Clin. Invest. 67: 1208 (1981).

10. Suzuki, Y., K. Kita, Y. Uchigata, I. Takata and T. Sato, Maturation of renal and hepatic monodeiodination of thyroxine to triiodothyronine and post-natal changes of serum thyroid hormones in young rats, Acta Endocrinologica 99: 540 (1982).

11. Kaplan, M. M. and K. A. Yaskoski, Effects of congenital hypothyroidism and partial and complete food deprivation on phenolic and tyrosyl ring iodothyronine deiodination in rat brain, Endocrinology 110: 761 (1982).

12. Obregon, M. J., J. Mallol, R. Pastor, G. Morreale de Escobar and F. Escobar del Rey, L-Thyroxine and 3,5,3'-triiodothyronine in rat embryos before onset of fetal thyroid function, Endocrinology 114: 305 (1984).

13. Morreale de Escobar, G., R. Pastor, M. J. Obregón and F. Escobar del Rey, Effects of maternal hypothyroidism on the weight and thyroid hormone content of rat embryonic tissues, before and after onset of fetal thyroid function, Endocrinology 117: 1890 (1985).

14. Woods, R. J., A. K. Sinha and R. P. Ekins, Uptake and metabolism of thyroid hormones by the rat foetus in early pregnancy, Clinical Science 67: 359 (1984).

15. Roti, E., L. E. Braverman, S-L. Fang, S. Alex and C. H. Emerson, Ontogenesis of placental inner ring thyroxine deiodinase and amniotic fluid 3,3',5'-triiodothyronine concentration in the rat, Endocrinology 111: 959 (1982).
16. Banovac, K., Lj. Bzik, D. Tislaric and M. Sekso, Conversion of thyroxine to triiodothyronine and reverse triiodothyronine in human placenta and fetal membranes, Hormone Res. 12: 253 (1980).
17. Kaplan, M.M. and E. A. Shaw, Type II iodothyronine 5'-deiodination by human and rat placenta in vitro, J. Clin. Endocrinol. Met. 59: 253 (1984).
18. Weiss, R. M. and C. R. Noback, The effects of thyroxine and thiouracil on the time of appearance of ossification centers of rat fetuses, Endocrinology 45: 389 (1949).
19. Porterfield, S. P. and C. E. Hendrich, The effects of growth hormone treatment on thyroid-deficient pregnant rats on maternal and fetal carbohydrate metabolism, Endocrinology 99: 786 (1976)
20. Porterfield, S. P. and C. E. Hendrich, Alterations of serum thyroxine, triiodothyronine, and thyrotropin in the progeny of hypothyroid rats, Endocrinology 108: 1060 (1981)
21. Hubank, M., M. Ballabio, A. Sinha, D. Gullo, S. Bidey, A. Bashir and R Ekins, Long term effects of maternal thyroidectomy on the development of the central nervous system of the progeny in the rat, Annales d'Endocrinologie 47: 48, (Abstract 79) (1986).
22. Bonet, B. and E. Herrera, Different response to maternal hypothyroidism during the first and second half of gestation in the rat, Endocrinology 122: 450 (1988)
23. Pérez-Castillo, A., J. Bernal, B. Ferreiro and T. Pans, The early ontogenesis of thyroid hormone receptor in the rat fetus, Endocrinology 117: 2457 (1985).
24. Luo,M. and J.H. Dussault, Immunocytochemical mapping of nuclear T3 receptors using a monoclonal antibody in the developing and adult brain, in:"Iodine in Neurodevelopment", R. Delong and J. Robbins, eds., Plenum Press, New York, (1988).
25. Thompson, C. C., C. Weinberger, R. Lebo and R. M. Evans, Identification of a novel thyroid hormone receptor expressed in the mammalian central nervous system,Science 237:1610(1987)
26. Sap, J., A. Muñoz, K. Damm, Y. Goldberg, J. Ghysdeal, A. Leutz, H. Beug and B. Vennstrom, The c-erb-A protein is a high affinity receptor for thyroid hormone. Nature 324: 635 (1986).
27. Weinberger, C., C. C. Thompson, E. S. Ong, R. Lebo, D. J. Grunol and R. M. Evans, The c-erb-A gene encodes a thyroid hormone receptor, Nature 324: 641 (1986).
28. Bernal, J. and F. Pekonen, Ontogenesis of the nuclear 3,5,3'-triiodothyronine receptor in the human fetal brain, Endocrinology 114: 677 (1984).
29. Dobbing, J. and J. Sands, Timing of neuroblast multiplication in developing human brain, Nature 226: 639 (1970).
30. Su, H. L., P. Ling, R. K. Yang and H.C. Chao, Ontogenesis of neclear T3 receptor in human fetal brain, in:"Iodine in Neurodevelopment", R. Delong and J. Robbins, eds., Plenum Press, New York, (1988).
31. Karmarkar, M.G., D. Prabakaran, M. M. Godbole and M. M. S. Ahuja, Thyroid hormone contents and 5 & 5' monodeiodinase activites in developping human cerebral cortex, in:"Iodine in Neurodevelopment", R. Delong and J. Robbins, eds., Plenum Press, New York, (1988).

32. Weeke, J. and H. Orskov, Ultrasensitive radioimmunoassay for direct determination of free triiodothyronine concentration in serum, Scand. J. Clin. Lab. Invest. 35: 237 (1975)

33. Obregón, M.J., G. Morreale de Escobar and F, Escobar del Rey, Concentrations of triiodo-L-thyronine in the plasma and tissues of normal rats as determined by radioimmunoassay: Comparison with results obtained by an isotopic equilibrium technique, Endocrinology 103: 2154 (1978)

34. Obregón, M.J., J. Mallol, F. Escobar del Rey and G. Morreale de Escobar, Presence of L-thyroxine and 3,5,3'-triiodo-L-thyronine in tissues from thyroidectomized rats, Endocrinology 109: 908 (1981)

35. Visser, T.J., J. L. Leonard, M. M. Kaplan and P. R. Larsen, Kinetic evidence suggesting two mechanisms for iodothyronine 5'-deiodination in rat cerebral cortex, Proc. Natl. Acad. Sci.79: 5080 (1982).

36. Morreale de Escobar, G., M. J. Obregón, C. Ruiz de Oña and F. Escobar del Rey, Transfer of thyroxine from the mother to the rat fetus near term: effects on brain 3,5,3'-triiodothyronine deficiency, Endocrinology 122: 1521 (1988).

37. Abrams, G.M. and P.R. Larsen, Triiodothyronine and thyroxine in the serum and thyroid glands of iodine-deficient rats, J Clin Invest 52: 2522 (1973).

38. Iglesias, R., J.A. Fernandez, T. Mampel, M.J. Obregón and F. Villarroya, Iodothyronine 5'-deiodinase activity in rat brown adipose tissue during development, B. B. A. 923: 233 (1987).

39. McCann, U.D., E.A. Shaw and M.M. Kaplan, Iodothyronine deiodination reaction types in several rat tissues: Effects of age, thyroid status, and glucocorticoid treatment, Endocrinology 114: 1513 (1984).

40. Segall-Blank, M., W.H. Douglas, R.L. Sanders and K.R. Hitchcock, Thyroxine metabolism in cultured cells derived from fetal rat lung, Pediatric Res. 17: 596 (1983).

41. Nedergaard, J., E. Connolly and B. Cannon, Brown adipose tissue in the mammalian neonate, in: "Brown adipose tissue", P. Trayhurn and D.G. Nicholls, eds, Edward Arnold, London, (1986), p 152.

42.Lean, M.E.J. and P.T. James, Brown adipose tissue in man, in: "Brown adipose tissue", P. Trayhurn and D.G. Nicholls, eds, Edward Arnold, London, (1986), p 339.

43.Himms-Hagen, J., Brown adipose tissue and cold-acclimatation, in: "Brown adipose tissue", P. Trayhurn and D.G. Nicholls, eds, Edward Arnold, London, (1986), p 214.

44.Leonard, J. L., S. A. Mellen and P. R. Larsen, Thyroxine 5'-deiodinase activity in brown adipose tissue, Endocrinology 112: 1153 (1983).

45. Silva, J. E. and P. R. Larsen, Interrelationships among thyroxine, growth hormone, and the sympathetic nervous system in the regulation of 5'-iodothyronine deiodinase in rat brown adipose tissue, J. Clin. Invest. 77: 1214 (1986).

46. Silva, J. E. and P. R. Larsen, Adrenergic activation of triiodothyronine production in brown adipose tissue, Nature 305: 712 (1983).

47. Silva, J. E. and P. R. Larsen, Potential of brown adipose tissue type II thyroxine 5'-deiodinase as a local and systemic source of triiodothyronine in rats, J. Clin. Invest. 76: 2296 (1985).

48. Silva, J.E. and P. R. Larsen, Hormonal regulation of iodothyronine 5'-deiodinase in rat brown adipose tissue, Am. J. Physiol. 251: E639 (1986).

49.Bianco, A.C. and J. E. Silva, Nuclear 3,5,3'-triiodothyronine (T3) in brown adipose tissue: Receptor occupancy and sources of T3 as determined by in vivo techniques, Endocrinology 120: 55 (1987)

50. Escobar del Rey, F., R. Pastor, J. Mallol, and G. Morreale de Escobar, Effects of maternal iodine deficiency on the L-thyroxine and 3,5,3'-triiodothyronine contents of rat embryonic tissues before and after the onset of fetal thyroid function, Endocrinology 118: 1259 (1986).

51. Obregon, M. J., R. Pitamber, A. Jacobsson, J. Nedergaard and B. Cannon, Euthyroid status is essential for the perinatal increase in thermogenin mRNA in brown adipose tissue of rat pups, B. B .R. C. 148: 9 (1987).

52. Hitchcock, K. R., Hormones and the lung. I. Thyroid hormones and glucocorticoids in lung development. Anat Rec 194: 15 (1979)

53. Smith, R. M. Thyroid Hormones and Brain Development, in: "Fetal Brain Disorders" B. S. Hetzel, R. M. Smith, eds., Elsevier North Holland Publishing Co, Amsterdam (1981), p 149.

ACTION OF THYROID HORMONES ON BRAIN DEVELOPMENT

Jean H. Dussault

Unité d'ontogénèse et génétique moléculaire
CHUL
Ste-Foy, Québec

INTRODUCTION

Relations between the central nervous system (CNS) and thyroid hormones have only been recognized in the nineteen century. The first detailed description of cretinism are found in the publication of the Royal Commission of the King of Sardaigna in 1848.[1] Cretinism is described as "a person with abnormal corporal development and mental retardation. They are born in regions were many such cases are found and endemic goiter is prevalent".

The most striking effects of thyroid hormones are observed during maturation of the CNS. The absence of thyroid hormones during this period produces multiple morphological and biochemical alterations and in human leads to irreversible mental retardation. Efforts to understand the mechanism of the hormonal effect have been hampered by the enormous complexity of the developing brain. Although sparse data exists for human it is known that the period between the end of the first trimester of gestation and 6 months after birth is the period of active neurogenesis and the most active phase of the brain growth spurt. The brain is particularly vulnerable to various insults during this period.[2-4] However most of our knowledge on the effects of thyroid hormones in the brain comes from experiments on animals, particularly the rat. Grossly the newborn rat can be compared to the human fetus during the second trimester of gestation and the 6-10 day old rat to the newborn infant.[5]

In this chapter we will try to summarize the latest knowledge on the role of the thyroid hormones on the developing brain. The reader is referred to other publications for broader review on specific aspects of the subject.[6-9]

THYROID HORMONE RECEPTORS

The mechanism of action of thyroid hormone is though to be initiated by its binding to a specific nuclear receptor.[10] The demonstration of local conversion of T_4 to T_3 in the rat brain with the result that the T_3 receptor is highly saturated under basal conditions suggests a primary role for the receptor in the maturation of the CNS.[11]

In the human fetus, Bernal & Pekonen[12] have been able to demonstrate a high affinity binding site with the binding specificity of the nuclear T_3 receptor which concentration is very low at 10 weeks of gestation but increases tenfold by week 16. Recent evidence from Su et al[13] obtained from fetus of 12 to 36th week of gestation confirms these preliminary data. The affinity is similar in all the regions of the brain studied (2,54 ± 0,63 $10^{-10}M^{-}$) and the maximum binding capacity (MBC) increases ten fold during that interval in the cerebrum whereas the increase is only twofold in the cerebellum. There is a correlation between the T_3 content and the MBC with an average occupancy of 28%. (Table I)

In the rat forebrain the T_3 receptors are present in comparatively high concentration with a peak at nine days and a rapid decrease thereafter. The same phenomena can be absent in the cerebellum but with much lower concentrations.[14-16] In the adult brain[17] we have demonstrated that the distribution of the receptors varies with the regions studied. The highest concentrations were found in the amygdala, the hypocampus and the cortex while the lowest were in the brain stem and the cerebellum: there was a ninefold difference between the lowest and highest capacities observed.

Using a specific monoclonal antibody developed against the nuclear T_3 receptor (NT3R)[18] we were able to demonstrate specific immunostaining in different region of the developing and adult rat brain.[19] In the 16 days old embryo only weakly NT3R immunostaining neurons were detected in the globus pallidus, amygdala, thalamus and hypothalamic area. In the 18 day old embryo the staining increased in both density and intensity but still no staining was found in the cerebral cortex areas. Thereafter the staining increased in the above mentioned region particularly in the cerebral cortex.

After birth the highest density of NT3R staining cells was found in 6 day old rats. The highest density of stained neurons was seen in the cerebral cortex particularly in the cingulate and superior frontal gyrus and some nuclei of the hypothalamus. In the cerebellum only a few weakly

TABLE I

ONTOGENESIS OF NUCLEAR T_3 RECEPTOR
IN HUMAN FETAL BRAIN
H L SU et AL

SOURCE : 23 cerebra 12-36 week
12 cerebella 18-36 week

AFFINITY : SIMILAR IN ALL REGIONS
$2.54 \pm 0.63 \times 10^{10}\ M^{-1}$

MBC
CEREBRA : 104 → 649 f mol/mg DNA
CERELLA : 2 FOLD INCREASE

T_3 CONTENTS
INCREASES WITH DEVELOPMENT
CORRELATION WITH MBC

OCCUPANCY
28%: STABLE

stained Purkinje's cells were found. In the 14 day old rat the density and intensity of NT3R immunoreactivity were slightly decreased except in the hypocampus and cerebellar cortex.

To pursue our investigation of the ontogenesis of the receptor we have used separate primary cultures from rat embryo (18 days) of astrocytes, neurons, and olygodendrocytes as models.

Receptors in neuronal cultures show a maximum MBC after 12 days of culture whereas the peak for the astrocytes occurs at 21 days. The receptors of the two cell lines have the same affinity for T3 but the capacity is threefold higher in the neurons.[20] In the oligodendrocytes, binding studies show that the maximal binding capacity and affinity were similar to that obtained in the same experimental condition in mixed neuronal cultures at 7 days.

Using the NT3R antibody and double staining with specific antibodies to the various cell lives to identify them we pursued further our studies. (Table II) In the astrocytes from 2 days old rat, 47% of the cell are positive for the NT3R antibody after 3 days in culture and increase to 75% after 15 days. 40% of protoplasmic astrocytes were stained whereas 80% of the fibrillar astrocytes were positive.[21] In the oligodendrocytes our studies revealed that 80% of the morphologically immature cells were positive whereas only 40% of the mature were stained.[22] These results are in accordance with our previous results using scatchard analysis with an increase of the MBC with time in culture.

After triiodothyronine binds to its specific receptor in the different cell populations of the central nervous system it initiates various responses and at different time during the maturation of the system (critical period).

ANATOMICAL AND HISTOLOGICAL ALTERATIONS

Ruiz Marcos and coworkers recently studied the effect of hypothyroidism (thyroidectomy) on the number of spines per apical shaft in pyramid cells of the rat visual and auditory cortex.[23,24] Groups of rats thyroidectomized at 10 days of age were treated with thyroxine starting at different times after the operation. Pyramidal cells of thyroidectomized rat treated from 12 days of age developed normally. When treatment was delayed until 30 days of age, these cells were severely damaged as

TABLE II

IN VITRO ONTOGENESIS OF THE NT3R

	Neurons	Astrocytes	Oligodendrocytes
MBC (ngT3/mg DNA)	0.30	0.18	.30
Immunostaining and cell types	+	75% Proto 40% Fibrillar 80%	Immature 80% Mature 40%
Days in culture	9	21	7

assessed by the distribution of apical shaft spines. More precise studies revealed that when treatment was started at 15 days, the distribution of spines found at 60 days of age was intermediate between the control and the hypothyroid animals; when treatment was begun at 20 days, the damage was irreversible. Thus a critical period appears to occur between day 10 and day 20 when the sole opportunity for maturation for the cells has been lost.

Cerebellum

Legrand and coworkers[25] have studied the effects of hypothyroidism on the development of the rat cerebellum and on the maturation of different "circuits" leading to the establishment of the adult "wiring" pattern of the brain. The only efferent neurons of the cerebellar cortex are the Purkinje's cells which are formed first before birth in the rat. Although the number of Purkinje's cells is not decreased by hypothyroidism, their maturation is permanently affected, as reflected by their arborization and number of dentritic spines. After a period of mitotic activity, cells from the external germinal layer migrate toward the internal granular layer, differentiate in the process and establish contacts with afferent mossy fibers and with Purkinje's cells to yield the normal "wiring" pattern. Hypothyroidism delays the desappearance of the external germinal layer and decreases the number and density of synoptic contacts with the already defective Purkinje's cells, resulting in a permanent impairment of neuronal connectivity. The critical period appears to occur before day 21 after birth, since treatment given after that date is no larger effective in preventing cellular damage.

BIOCHEMICAL ALTERATIONS

Examples of biochemical effects of thyroid hormones in the developing rat brain are given in Table III. This list is not exhaustive and most of the biochemical effects of hypothyroidism become irreversible if replacement therapy is delayed after the critical period of development which in rats

TABLE III

Examples of effects of thyroid hormones in various biochemical parameters in the developing brain:*

Energy metabolism :	oxygen consumption glucose + ketone body metabolism
Polyamines	content metabolism
Microtubules	Tubulin Microtubule-Associated protein
Myelin	Composition Metabolism
Neurotransmitters and neuropeptides	content metabolism receptors

* For the effect of thyroid hormones on these various parameters and exact references the reader is referred to the review by Dussault et al ref: 9

usually spans the first 10-14 days after birth: unfortunately many of these effects can only be detected 2-3 weeks after birth when they no longer can be reversed.

These numerous functions of the brain affected by thyroid hormones do so with a very long lag time in many cases. These observations strongly suggest that many of these late effects are secondary to one or several actions of thyroid hormones at an earlier stage of development. To unravel the mechanism of action of thyroid hormones in the brain it became imperative to determine which are their primary target cells. We have addressed ourselves to this problem by establishing separate cultures of neurons, astrocytes and oligodendrocytes. These "in vitro" systems should allow us to determine whether for example all cell types within their family are sensitive to thyroid hormones and whether same hormonal actions require interactions between different cell types.

ELECTROPHYSIOLOGICAL ALTERATIONS

The particular susceptibility of the developing peripheral auditory system to thyroid hormone deprivation is well known and can serve as a model for alteration detected by electrical activity. The onset of evoqued cochlear electrical activity (which is postnatal in the rat) is delayed by hypothyroidism and is returned to normal by thyroid hormone administration.[26,27] In contrast relatively few abnormalities are noted in hypothyroid animals when hypothyroidism is induced in adulthood which again indicated that these is a "critical period" of peripheral auditory system development that depends on thyroid hormone.[28,29] We have recently determined this period of susceptibility using auditory brain stem response (ABR). To determine the "critical period" groups of rat were treated with propylthiouracil (PTU) for 10 days periods beginning at different age and ABR evaluated at 120 days. PTU treatment significantly increases wave I latency when administered from 3 days before delivery through 6 days of age, but was without permanent effect when administered for 10 days starting 10 days after birth. T_4 administration prevented permanent abnormalities when given during the first 10 days of life. These data suggest that the period of greatest vulnerability to thyroid hormone depletion extends from at least 3 days before delivery through between 5 and 10 days of age.

In the human[30] in a group of 34 congenitally hypothyroid children between ages 5 and 12 under thyroid therapy since the third week of life, significant abnormalities were observed in their auditory brain stem response compared to a control group. 20% showed prolonged wave latencies and 30% had shortened interpeak latencies.

SUMMARY AND CONCLUSIONS

A short "critical period" during which thyroid hormones are essential for normal maturation of the CNS is evident at anatomical, biochemical and neurophysiological levels. In the last decade we have made progress toward understanding the mechanism of action of thyroid hormones due in part ot numerous studies of the ontogenesis and distribution of the nuclear

T3 receptor. The availability of a nuclear T3 receptor antibody and a cDNA probe[31] for the receptor should help us to further identify region and cell types influenced by thyroid hormones both by immunocytochemical methods and in situ hybridization. These studies might indicate when the molecular events that control the growth and maturation of the brain are initiated.

However much further research in this area is needed to comprehend further the relation between thyroid hormones and brain development such as the importance of the severity of the thyroid hormone depletion and its occurrence during the "critical period" which appears to vary depending on the neurophysiological function involved.

REFERENCES

1. Rapport de la commission créée par sa majesté le Roi de Sardaigne pour étudier le crétinisme. Imprimerie Royale, Turin (1848).
2. J. Dobbing, Undernutrition and the developing brain, Am. J. Dis. Child. 120:411 (1970).
3. J. Dobbing, The later growth of the brain and its vulnerability, Pediatr. 53:2 (1977).
4. J. Dobbing and J.L. Smart, Vulnerability of the developing brain and behaviour, Br. Med. Bull. 30:164 (1974).
5. P.G. Croskerry, G.K. Smith, B.J. Shepard and K.B. Freeman, Perinatal brain DNA in the normal and growth hormone treated rat, Brain Res. 52:413 (1973).
6. D.H. Ford, and E.B. Cramer, Developing nervous system in relation to thyroid hormones, in: "Thyroid Hormones and Brain Development," ed. G.D. Grave, Raven, New York, p. 1 (1977).
7. J. Legrand, Hormones thyroïdiennes et maturation du système nerveux, J. Physiol. 78:603 (1982-1983).
8. G. Morreale de Escobar, A. Ruiz-Marcos and F. Escobar del Rey, Thyroid hormone and the developing brain, in: "Congenital Hypothyroidism," ed. J. H. Dussault, P. Walker, Dekker, New York, p. 85 (1983).
9. S. G. Amur, G. Shanker and R. A. Pieringer, Regulation of myelin basic protein (arginine) methyltransferase by thyroid hormone in myelinogenic cultures of cells dissociated from embryonic mouse brain, J. Neurochem. 43:494 (1984).
10. J.H. Oppenheimer, Thyroid hormone action at the cellular level, Science 203:971 (1979).
11. M.M. Kaplan, U.D. McCann, K.A. Yaskoski, P.R. Larsen and J.L. Leonard, Anatomical distribution of phenolic and tyrosyl ring iodothyronine deiodinase in the nervous system of normal and hypothyroid rats. Endocrinol. 109:397 (1981).
12. J. Bernal and F. Pekonen, Ontogenesis of the nuclear T3 receptor in the human fetal brain, Endocrinol. 114:677 (1984).
13. J.H. Su, P. Lung, R.K. Yang and H.C. Chao, Ontogenesis of nuclear T3 receptor in human fetal brain. Abstract Iodine and the brain, Bethesda, Maryland (1988).
14. P. Coulombe, R. Ruel, J.H. Dussault, Récepteurs nucléaires de la T3 dans le cerveau et le cervelet du rat au cours du développement, Union Med. Can. 110:658 (1981).

15. N.L. Eberhardt, T. Valcana and P.S. Timiras, Triiodothyronine nuclear receptors: An in vitro comparison of the binding of triiodothyronine to nuclei of adult rat liver, cerebral hemispher, and anterior pituitary, Endocrinol 102:556 (1978).
16. H.L. Schwartz and J.H. Oppenheimer, Nuclear triiodothyronine receptor sites in the brain: Probable identity with hepatic receptors and regional distribution, Endocrinol. 103:267 (1978).
17. J. Ruel, R. Faure and J.H. Dussault, Regional distribution of nuclear T3 receptors in rat brain an evidence for preferential localization in neurons. J. Endocrinol. Invest. 8:343 (1985).
18. M. Luo, R. Faure, J. Ruel and J.H. Dussault, A monoclonal antibody to the rat nuclear T3 receptor: production and characterization, Endocrinol. in press (1988).
19. M. Luo and J.H. Dussault, Immunocytochemical mapping of nuclear T3 receptors using a monoclonal antibody in the developing and adult rat brain. Abstract The Endocrine Society, New Orleans, Louisiana (1988).
20. M. Luo, R. Faure and J.H. Dussault, Ontogenesis of nuclear T3 receptor in primary cultured astrocytes and neurons, Brain Res. 381:275 (1986).
21. J. Puymirat, L. Min and J.H. Dussault, Immunocytochemical localization of L-triiodothyronine nuclear receptors (TNR) in astroglial cell cultures Abstract American Society for Neurochemistry, New Orleans, Louisiana (1988).
22. F. Besnard, M. Luo, J. Puymirat, C. Fressinaud, G. Labourdette, J.H. Dussault and L.L. Sarliève, Immunocytochemical localization of 3, 5, 3'-Triiodothyronino (T3) receptors in pure rat oligodendrocytes producing myelin-related membranes. Abstract Huitième séminaire hivernal des neurosciences européennes, Tigens, France (1988).
23. A. Ruiz-Marcos, A. Salas, F. Sanchez-Toscano, F. Morreale de Escobar and G. Morreale de Escobar, Effect of neonatal and adult-onset hypothyroidism on pyramidal cells of the rat auditory cortex, Dev. Brain Res. 9:205 (1983).
24. A. Ruiz-Marcos, F. Sanchez-Toscano, F. Escobar del Rey and G. Morreale de Escobar, Severe hypothyroidism and the maturation of the rat cerebral cortex, Brain Res. 162:315 (1979).
25. J. Legrand, Morphogenetic action of thyroid hormones, Trends Neurosci. 2:234 (1979).
26. W.L. Meyeroff, Hypothyroidism and the ear: Electrophysiological morphological and chemical considerations, Laryngoscope 89:1(1979).
27. A. Uziel, A. Rabie and M. Marot, The effect of hypothyroidism on the onset of cochlear potentials in developming rats. Brain Res. 182:172 (1980).
28. R. Hébert, J.M. Langlois and J.H. Dussault, Permanent defects in rat peripheral auditory function following perinatal hypothyroidism: Determination of a critical period, Dev. Brain Res. 23:161(1985).
29. F.N. Ritter, The effects of hypothyroidism upon the ear, nose and throat, Laryngoscope 77:1427 (1967).
30. R. Hébert, E. Laureau, M. Vanasse, J.E. Richard and J. Glorieux et al, Auditory brainstem response (ABR) audiometry in congenitally hypothyroid children under early replacement therapy, Pediatr. Res. 20:570 (1986).

31. Catherine C. Thompson, Cary Weinberger, Roger Lebo and Ronald M. Evans, Identification of a Novel Thyroid Hormone Receptor Expressed in the Mammalian Central Nervous System. Science 237: 1610 (1987).

DISCUSSION

FISHER (Torrance) : Jean, you have done a beautiful job in summarizing the complexity of what is going on in the brain with regard to thyroid function. I would like to focus on the heterogeneity of the thyroid receptor. We do not know how heterogeneous it might be in the brain. We have obtained data on thyroid hormone analog binding (TRIAC, T3, T4, reverse T3) to liver and brain nuclei from fetal and neonatal sheep. The results indicate clearcut differences in receptor binding affinities between both tissues (10 fold higher in the brain compared to liver), as well as differences in the timing of receptor ontogenesis. In addition, there is the recent fascinating work of Weinberg et al. and Sap et al., indicating that the c-ERB oncogene is a homolog of the T3 receptor. In these studies, the authors showed that the rat brain T3 receptor which they characterized doesn't seem to be present in liver - or if so in low cencentration - suggesting that the liver and brain receptors may be different gene products. We do not know whether the receptor in brain is a single receptor species. It is possible that different receptor types are being expressed in various segments of the brain at different times.

DUSSAULT (Quebec) : I agree with you. Also one of the recently described T3 receptor probes was isolated from testes which shouldn't have a T3 receptor. Testis has been considered to be non responsive to thyroid hormone.

DE NAYER (Brussels) : I think that possible differences between detection by T3 binding (capacity) and detection with antibody probes could explain some variance observed in the data.

PHYSIOPATHOLOGY - PANEL DISCUSSION

Moderator : P. De Nayer

BRAVERMAN (Worcester) : Dr. Morreale, in your 1988 article in Endocrinology, I noticed that the levels of maternal T4 during the perfusion were about 50 % higher than in the control dams. I wonder if that may account for some of the transport that you have seen.

MORREALE (Madrid) : Yes indeed. In that paper, the mothers had been on MMI only from day 16 onwards and had a T4 in plasma which was higher (although lower than in non pregnant female rats) than in other pregnant rats. But T3 in some of their tissues was normal. It depends on what you consider to be too high and whether you are looking at tissue or plasma. In later experiments, we started treating 2 days earlier, and with a T4 daily dose of 2.4 µg, the maternal T4 level was the same as that of controls. Also, T3 in the liver was normal. The 2.4 µg/day dose is known to restore a normal liver -GPD level in hypothyroid adult rats. This dose normalized the mothers, but the fetal plasma T4 was still low and TSH high. In spite of this, fetal brain had achieved normal T3 levels which indicated that the brain was capable of making very much out of rather little T4.

BRAVERMAN (Worcester) : I agree with that but even so there is a gradient between maternal T4 and fetal T4 so that the placenta represents a partial barrier.

MORREALE (Madrid) : I agree, it is not a free transfer. There is certainly a gradient. But the presence of a gradient doesn't mean that there is no transfer. The amount coming over might be very important to protect the brain of a hypothyroid fetus.

DE NAYER (Brussels) : Do you have any experience with thyroid hormone analogs such as DIMIT which crosses the placenta more readily ?

MORREALE (Madrid) : No. All our work has been based on natural hormone concentrations determined by radioimmunoassay. Apart from that, we were very worried by reports from studies performed on the rat fetus with DIMIT showing gross fetal abnormalities.

DE NAYER (Brussels) : There have been many reports about placental transport, but I have been looking unsuccessfully through the literature for the presence of T3 receptors in the placenta. Of course Weinberger isolated his T3 receptor c-DNA probe from human placenta. Has anyone in the audience information on receptors in the placenta ?

EMERSON (Worcester) : I believe I recall a paper in which S.Y. Cheng reported T3 receptors in the placenta, and my recollection is that the binding capacity was lower than in most other tissues.

BRAVERMAN (Worcester) : We have some data on the rat uterus indicating that there are T3 receptors.

DE NAYER (Brussels) : Concerning the topic of hormone conversion, could Dr. Obregon or Dr. Morreale tell us more about the contribution in the fetal rat of brown adipose tissue to T3 production as compared to liver and kidney, for example ?

MORREALE (Madrid) : We think that in the fetal stage brown adipose tissue T3 is not contributing much to circulating T3. We always find very low T3 levels in plasma during fetal development. Apparently, the T3 which is produced inside brown adipose tissue cells is important for the brown adipose tissue itself. It would seem that in the fetal stage, as probably in adults, the liver is (in normal conditions) the main source of circulating T3. Since the fetal T4 deiodinase in liver is very low, we find very low levels of circulating T3. But we don't know why the fetal thyroid is not secreting much T3 because it is present in the gland.
However, in the rat there is a high activity of brain T4 deiodinase at a very initial stage of development and this provides high brain concentrations of T3 from T4. The experiments infusing T4 to the mothers show that T3 does not get into the fetal brain; all the T3 you find there comes from T4. The brain deiodinase is already responsive to thyroid hormone deficiency and the receptors are also there. All these mechanisms "do their best" to protect the brain from hormone deficiency. Our problem is that we have not been able yet to delineate an adequate end-point to evaluate these protective mechanisms. Similarly, for many reasons, it has been claimed tha the adult rat brain is not sensitive to (the lack of) thyroid hormones. We rather believe we have not looked at the right end-point or at the critical stage, as Dr Dussault has pointed out.

DELANGE (Brussels) : I have a pratical question related to the function of the placenta and the content of reverse T3 in amniotic fluid. Since we are now convinced that the largest part of reverse T3 in amniotic fluid is actually coming from the placenta itself, could this placental rT3 have some value for the prenatal diagnosis of CH ?

BRAVERMAN (Worcester) : I think it is of no or little use. Is there any data on the sensitive TSH assay in amniotic fluid ?

FISHER (Torrance) : I am asked relatively frequently whether some approach to prenatal diagnosis (through amniocentesis) might be useful. There is still an occasional pregnant woman given radioiodide inadvertently for Graves' disease. My response has been that we need not do this. The risk of the procedure probably is higher than whatever small risk exists for fetal hypothyroidism. In addition, there is no reliable assay for the diagnosis of prenatal hypothyroidism. Preliminary data from I. Kourides suggest that a sensitive TSH assay could be applied to this purpose. However we have no normal data. Moreover if we diagnose a hypothyroid fetus, how are we going to treat it ? DIMIT is not a proven method of treatment. Making mothers hyperthyroid with thyroxine in order to get enough hormone across the placenta to satisfy the fetus carries considerable risk. Thus, there is no reliable approach to prenatal diagnosis via amniocentesis and no method of treating fetal hypothyroidism if we diagnose it.

I would like also to make a comment about thermogenin which Dr. E. Silva has reported recently to be thyroxine-responsive. We are looking in the sheep at brown adipose tissue thermogenin, but we have no definitive data at the moment. The oxygen consumption of brown adipose tissue increases progressively in the fetal sheep. Peak activity corresponds with term gestation in that species. In the hypothyroid newborn sheep, there is marked hypothermia and a very significant reduction in norepinephrine-stimulated oxygen consumption by brown adipose tissue. The newborn sheep like a newborn human is dependent on brown adipose tissue for newborn thermogenesis and the deficiency of thermogenesis in the hypothyroid newborn almost surely is due to thermogenin deficiency. The rat newborn is intriguing and different : it doesn't produce enough heat to maintain body temperature. Is there a deficiency in brown adipose tissue thermogenesis in those animals ? Why in fact can't they maintain their body heat ? Perhaps Dr. Obregon can comment about this.

OBREGON (Madrid) : In the rat, newborn pups are able to maintain temperature but only for short periods of time. The mRNA for thermogenin is present less than 24 hours after birth and the protein is detected already after 2 days. When newborn pups are exposed to cold temperatures (28), an increase in thermogenin mRNA is observed for at least 6 hours.

FISHER (Torrance) : As I indicated, it is believed that brown adipose tissue is critical for extra-uterine survival and heat production in the newborn sheep. The data is not so clear for humans. The transition from brown adipose tissue thermogenesis to other tissue thermogenesis and the mechanisms of thyroid mediation of (the other) tissue thermogenesis is not yet clear. In the rat, which is a poikilothermic species, I haven't thought that brown adipose tissue was very important, but rather that rats were dependent on the development of thyroid dependent (other) tissue thermogenesis progressively with time over 2 or 3 weeks after birth. Your data obviously suggest otherwise.

MORREALE (Madrid) : I wish to comment about a previous point. It is true from the experience of all groups doing neonatal thyroid screening, that if you start treatment early enough you can prevent severe mental retardation. However there also are data which indicate that all these babies do not have a completely normal central nervous system. Dr. Dussault this morning described the abnormal evoked potentials of the auditory stem and there are data from Wolter in Brussels showing that some of these children have minimal brain damage. If I understood correctly, the data from the screening program in Quebec suggest that the children who had very low T4 values at birth or evidence of intra-uterine hypothyroidism as assessed by skeletal growth are also the ones who had the poorest prognosis : they have IQ values in the normal range but often lower than for other children. The main goal of the screening program is therefore achieved, but it would be worthwhile to obtain a more complete normalization of CNS development. Everybody states that if you want to get T4 across the human placenta you have to give high doses to the mother. I really wonder if the evidence for that is so hard. In our work using T3 (even with 50 μg of T3) there was a little increase of T3 in fetal blood, but no decrease in TSH. I think this is an important point : if the rat data are relevant, then fetal TSH should be considered as a very poor indicator of placental passage of T3, perhaps because TSH is not suppressed in the fetus by T3. Hence, if we could measure T3 directly in the brain of the hypothyroid neonate we could see that some thyroid hormone had crossed the placenta. I agree that our work done with T4 was, indeed, with very large doses, and T4 was found in the cord blood. One should perhaps aim at maintaining pregnant women with high normal T4 levels during the later stages of pregnancy. This would be of no harm to them, and it might be good for the brain of the possibly hypothyroid fetus.

FISHER (Torrance) : The other point that we have not addressed is the premature infant. Premature infants born under 30 weeks of gestation have thyroxine levels fairly routinely under 6 μg/dl without TSH elevation and with low free thyroxine levels. These infants respond nicely to TRH with an increase in TSH and with an increase in T4. Thus they have diagnostic criteria for hypothalamic hypothyroidism. These premature newborns mature in the extra-uterine environment autonomously and normalize their serum T4 over a several week period. They have been studied carefully and have no evident later brain damage. Presumably, they reflect the hypothalamic and thyroid status of the intra-uterine normal human fetus which also manifests hypothalamic hypothyroidism and has no ill effects therefrom.

ROVET (Toronto) : I would like to refer to Dr. Fisher's comment. The kind of deficits noted in premature infants are actually very similar to what is found in the children with congenital hypothyroidism. Also, there is a growing literature from autopsy data of people with severe dyslexia which shows second trimester neurological cell migration deficits in the left cerebral hemisphere. It is interesting that of the cases autopsied, several have had mothers with hypothyroidism. This type of dyslexia is very different from what we have seen in the children with CH who have less severe deficits. They can learn to read; they just are a little behind. This would suggest that there are 2 different kinds of dyslexia : one acquired in the second trimester and one developing during the third trimester or during the neonatal period.

SACK (Tel Aviv) : The best way to treat the fetus with intra-uterine hypothyroidism is to inject thyroxine intra-amniotically. This can be done once a week. To inject 250 µg of l-T4 is safe. We have done this in several hundred pregnancies but the data have never been published : we conducted a trial of such treatment to prevent RDS but there was no statistically significant result. In any event, we could show that all fetuses absorbed T4 from the amniotic fluid. Their cord blood T4 levels were higher than control infants, but not the T3 values. These results, as suggested by Dr. Braverman, indicate that the fetus and the placenta know "how to deal" with T4 and that high values of T4 are not followed by high levels of T3.

PART 2. ETIOLOGY

TRANSIENT HYPOTHYROIDISM IN INFANTS DUE TO MATERNAL TSH-RECEPTOR BLOCKING ANTIBODIES-CHARACTER OF THEIR IgG AND LONG-TERM PROGNOSIS OF THEIR INFANTS

Nobuo Matsuura[1], Junji Konishi[2], Kenji Fujieda[1], Kenji Yuri[1], Shohei Harada[1], Kanji Kasagi[2], Jasuhiro Iida[2], Machiko Endo[3], and Yutaka Yamada[3]

Department of Pediatrics, Hokkaido University School of Medicine, Sapporo,[1] and Department of Nuclear Medicine, University of Kyoto, Kyoto,[2] and Pediatric Clinic Hakodate Chuo Hospital, Hakodate,[3] Japan

INTRODUCTION

Transient neonatal hypothyroidism in infants born to mothers with chronic thyroidism, first reported in 1960 by Sutherland et al.[1], has been established as an important diseases entity in the newborn period. Transplacental transfer of TSH-receptor blocking antibodies has been defined as the cause of the disease.[2] Although similar cases have been reported since then,[3][4] the clinical features, as well as the prognosis of each case, varied greatly. That is, most of infants developed normally; however, some infants were physically, mentally or emotionally retarded in spite of intensive treatment during the newborn period,[1] and their condition was quite different from that found in congenital hypothyroidism. It is not clear whether the difference was due to the nature of TSH-receptor blocking antibodies in the mother or to the severity of thyroid suppression during the fetal and newborn periods. It was suggested that the IgG in the mother's serum not only inhibited TSH binding to its receptor, but also blocked TSH-stimulated cAMP response, TSH-stimulated iodine uptake, organification and ^{3}H-thymidine incorporation into DNA in cultured thyroid cells: that is, it blocked post-receptor processes.[4] The purposes of this paper are twofold: 1) To study more precise mechanism and to evaluate the possible involvement of GTP-binding protein of the inhibition of adenylate cyclase activities using porcine thyroid cells. 2) To establish a basis for the long-term prognosis of these patients, particularly of two of the sibling sets, whose mothers were once hypothyroid and were euthyroid during pregnancy.

MATERIAL AND METHODS

1) Thyroid cell dispersion and culture

Porcine thyroid tissue obtained at a slaughter house was throughly washed with Hanks salt solution depleted of Ca^{2+} and Mg^{2+} (Hanks CMF). The thyroid cells obtained by digestion with collagenase (1mg/ml) were suspended in Krebs-Ringer HEPES Buffer. Aliquots of cell suspension containing 2mM 3-isobutyl-1-methyxanthine (IBMX) and test drugs were incubated for 30 minutes at 37°C, after which incubation was stopped by adding HCL (final concentration of 0.2N) and the cells were disrupted by heating at 100°C for 60 seconds. Intra- as well as extra-cellular cAMP was determined by RIA using a commercially available Kit (Yamasa Co. Tokyo Japan). Aliquots of porcine thyroid cells were disrupted and protein was determined according to the method of Lowry, et al. using bovine serum albumin as the standard. Generation of cAMP was expressed as p mol/mg protein/30 minutes.

2) TSH-binding inhibitor immunoglobulin (TBII) was assayed by radioreceptor assay of TSH using a commercially available kit.[5] Thyroid-stimulation blocking antibodies (TSBAb) were assayed by measuring basal and TSH-induced cAMP accumulation in the presence and the absence of test IgG using cultured thyroid adenoma cells or porcine thyroid cells as targets.[6] Serum T_4, T_3, and TSH were determined by RIA using commercially available kits as previously described.[2] IgG was prepared by protein-A sepharose column.

3) Patients.

Eighteen mothers with chronic thyroiditis and twenty-four of their offspring with transient hypothyroidism were registered in our country until now. In this paper, we particularly report on two of the sibling sets.
Case M-1 (M.K.)[2]: Mother of cases 1 and 2 had been in good health until fifteen years of age, when facial puffiness, intolerance to cold, and constipation developed. Hypothyroidism was diagnosed when she was 17 years of age. However, she never noticed swelling of the anterior area of the neck. The ^{131}I-thyroidal uptake (24 hr.) was 0.4%, and the thyroidal scan revealed no notable accumulation of ^{131}I around the neck. She was treated with 90 mg per day of desiccated thyroid. When she was pregnant with case 1, she decided to reduce the dose to 30 mg per day, for fear of side effects on the fetus. Therefore, she was in a hypothyroid state during the pregnancy. However, when she was pregnant with case 2 she continued to take 140 mg per day of desiccated thyroid regularly and was euthyroid throughout the pregnancy.[2]
Case M-2 (O.M.): Mother of cases 3 and 4 was referred to our hospital because her second child (case 3) was found to have elevated TSH levels at newborn screening. When she visited our hospital, the thyroid function of her second child was returned to normal levels; however, the mother's thyroid function was decreased; that is, T_4, T_3, and TSH levels at 35 days postpartum were 2.5 ug/dl, 66 ng/dl, 64.2uU/ml, respectively and when the test was repeated at 50 days postpartum, the respective levels were 4.4 ug/ml, 93 ng/dl, and 113 uU/ml. Both antithyroglobulin and antimicrosomal antibodies were positive, with titers of 10^2 and 40^2, respectively. The basal TSH level was 162.9 uU/ml and exaggerated response (max. 370.1 uU/ml) was observed after TRH administration. The ^{123}I-thyroid uptakes at 3 and 24 hours were 2.4 and 1.2%, respectively, and the thyroid scan revealed no notable accumulation of ^{123}I around the neck. She was treated with 125 ug per day of L-thyroxine. When she was pregnant with case 4, she continued to take L-thyroxine regularly and was euthyroid throughout her pregnancy.

RESULTS

1. Effects of TSH and prostaglandins on porcine thyroid adenylate cyclase activities.

Addition of TSH (bTSH) at a concentration of 0.5 mU/ml caused a significant increase in cAMP, which reached maximal levels with 10 mU/ml. Addition of PGE_1, PGE_2, PGA_2 in concentrations of 10^{-7} to 10^{-6}M caused a significant increase in cAMP; however, PGB_2 and PGI_2 did not show any effects. (data not shown).

2. Effects of α-adrenergic antagonists and agonists on porcine thyroid adenylate cyclase activities.

TSH-induced cAMP accumulation was significantly inhibited dose-dependently by the addition of epinephrine, norepinephrine, or the α_2-agonist clonidine in concentrations of 10^{-8}M to 10^{-5}M; however, the α_1-agonists methoxamine and phenylephrine caused only a weak inhibition. Epinephrine-induced inhibition in cAMP accumulation was released by the addition of the α_2-antagonist yohimbin dose-dependently; however, the nonspecific α-antagonists phenoxybenzamine and phentolamine caused only 50-60% release and neither the α_1-antagonist prazosin or the β-antagonist, propranolol showed any effects (Fig. 1).

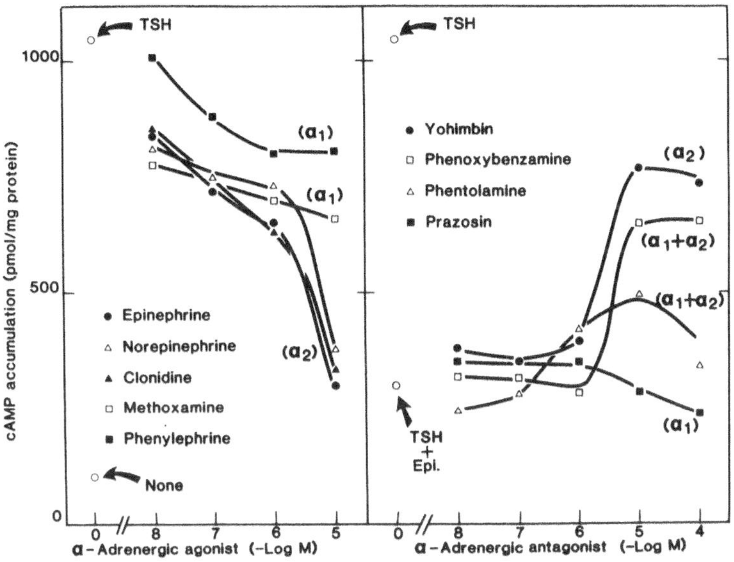

Fig. 1. Effects of α-adrenergic agonists and antagonists on TSH-induced cAMP accumulation

From these experiments, it is concluded that epinephrine inhibited TSH-induced cAMP accumulation through α_2-adrenergic receptors.

3. Effects of Islet-activating protein (IAP) on epinephrine-induced inhibition of adenylate cyclase activities.

To clarify the mechanisms involved in epinephrine-induced adenylate-cyclase suppression, we used IAP, a pertussis toxin, which catalyzed ADP-ribosylation of a GTP-binding protein (Ni) and caused the

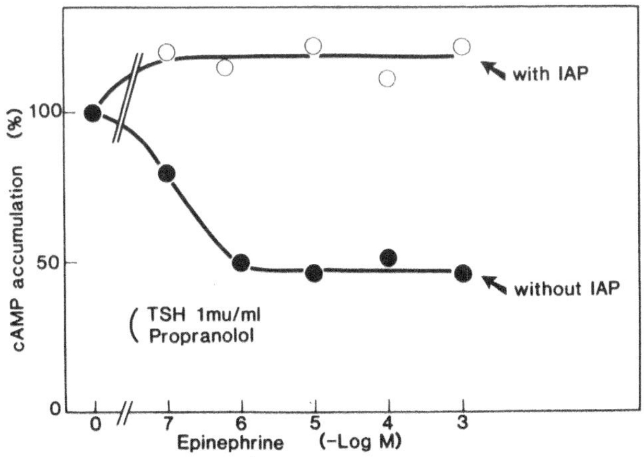

Fig. 2. Effect of IAP and epinephrine on TSH-induced cAMP accumulation

failure of Ni to mediate receptor-linked inhibition of adenylate cyclase activity. Porcine thyroid cells were incubated for twenty hours with IAP (300ng/ml). Epinephrine-induced adenylate cyclase inhibition was completely reversed by IAP treatment (Fig. 2). From the above experiments, it is concluded that epinephrine inhibited adenylate cyclase activity throughly GTP-binding protein (Ni).

4. Effects of IAP on TSH, and PGE_1-stimulated adenylate cyclase activity.

TSH- and PGE_1-induced cAMP accumulations were increased 40 and 60%, respectively, by prior exposure of porcine thyroid cells to IAP. This evidence suggested that Ni might be partly responsible for the stimulation of adenylate cyclase activity. (data not shown)

5. Effects of IgG on the adenylate cyclase system in patients with atrophic autoimmune thyroiditis who possessed TSH-receptor blocking antibodies (M-IgG).

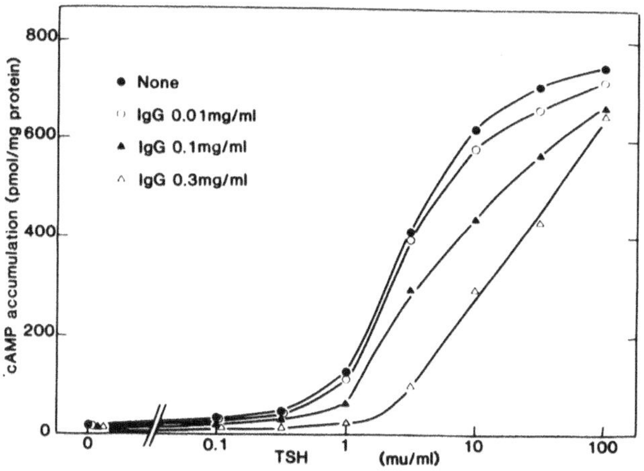

Fig. 3. Dose-dependent inhibition of TSH-induced cAMP accumulation by M-IgG

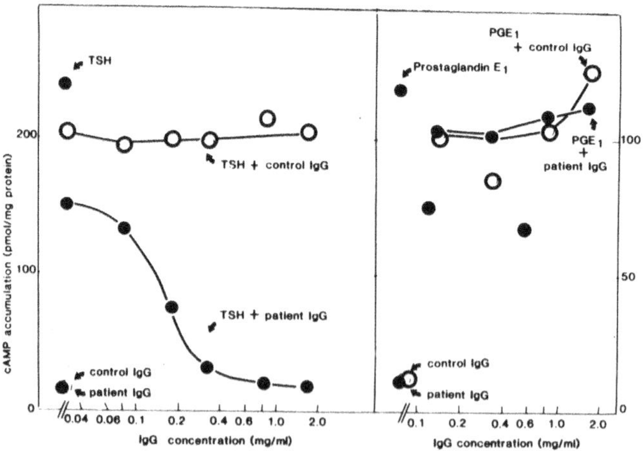

Fig. 4. Effects of M-IgG and control IgG on PGE_1- and forskoline-stimulated cAMP accumulation

TSH-induced cAMP accumulation was inhibited dose-dependently with the addition of 0.01 to 0.3 mg/ml M-IgG (Fig. 3). TSH-induced cAM accumulation was completely inhibited by M-IgG; however, neither PGE_1-nor forskoline-induced cAMP accumulation was influenced by the addition of M-IgG (Fig. 4).

6. Effects of IAP on M-IgG-induced inhibition of adenylate cyclase activity.

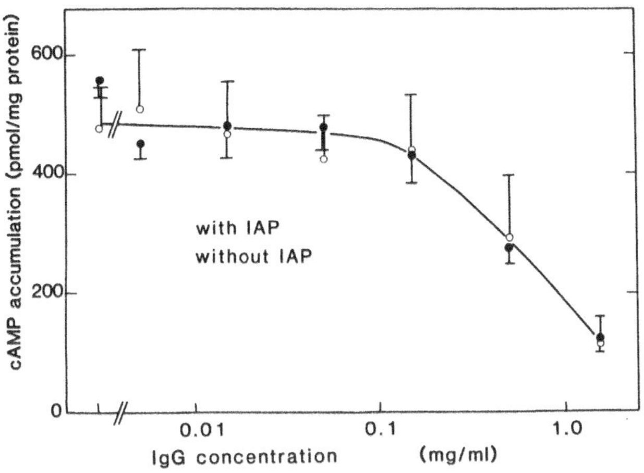

Fig. 5. Effects of M-IgG and IAP on TSH-induced cAMP accumulation

TSH-induced cAMP accumulation was inhibited dose-dependently by the addition of M-IgG that was not influenced by prior exposure of cells to IAP (Fig. 5). From the above experiments, it was concluded that M-IgG inhibited TSH action on the receptor levels competitively and was not influenced by GTP-binding protein (Ni).

7. TSH-receptor antibodies in the mothers who give birth to hypothyroid infants and infants with various thyroid dysfunctions.

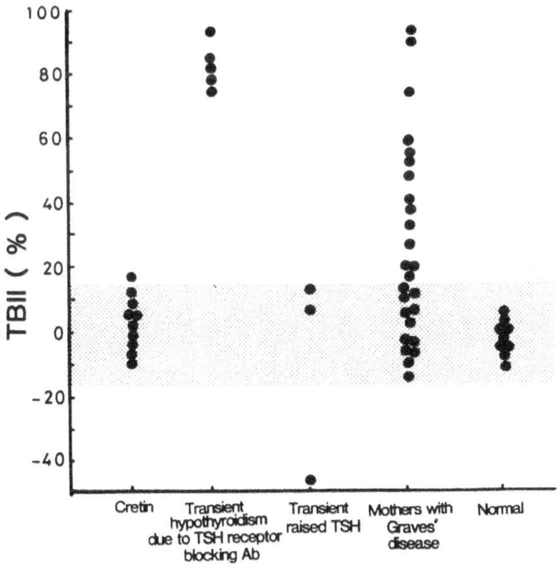

Fig. 6. TBII activities in mothers of infants with various thyroid dysfunction

TBII activities in the mothers who gave birth to infants with various thyroid dysfunctions are shown in Fig. 6. Potent TBII activities were observed in the mothers with Graves' disease and chronic thyroiditis. None of the mothers who gave birth to hypothyroid infants showed positive TBII activities. One mother whose TBII was extremely low was found to have anti-TSH antibodies.

Table 1. Clinical data of four infants (two sibling sets) with transient hypothyroidism

Case number infant	Case number mother	sex	Gestational age (wk)	Birth weight (gm)	Birth length (cm)	Epiphysis of dist. femur (mm)	Start of therapy (day)	Max, L-T4 dose (ug/kg/day)	End of therapy (month)
1	M-1	M	41	3083	49.0	0 x 0	14	6.1*	17.0
2	M-1	M	39	2600	45.0	0 x 0	4	5.0*	12.0
3	M-2	F	41	2970	48.0	8 x 6	none	-	-
4	M-2	M	41	2860	48.3	2 x 2	13	10.0	8.0

* desiccated thyroid (mg/kg/day)

Case No.	age (days)	Thyroid function at diagnosis T4 (ug/dl)	T3 (ng/dl)	TSH (uU/ml)	TBII (%)	Age of walking alone (month)	IQ (age when studied) (years)
1	3	0.4	27	150	N.D.	18	70 (11 1/2)
2	0	2.0	N.D.	320	71.6	12	103 (9 3/12)
3*	35	10.3	231	2.7	41.2	12	76 (5)
4	13	0.9	47	1768	86.1	12	100 (1) (D.Q.)

*TSH levels at 5 days old was 57.1 uU/ml.

Table 2. Potency of TBII activities in mothers and outcome of their infants

Cases No mother	infants	Thyroid function in infants TSH(μU/ml)	T_4(μg/dl)	Epiphysis of dist.femur(mm)	dilution of mother's serum (50% inhibition)	b-TSH eq. (mU/ml)
M-1	case 1	150<	0.4	0 x 0	1:320	384
"	case 2	320<	2.0	0 x 0	1:320	384
M-2	case 3	57.1	11.8	8 x 6	1: 8	9.4
"	case 4	1768	0.9	2 x 2	1: 70	84
M-3	case 5	3.4	10.9	ND	1: 10	12
M-4	case 6	31.6	18.2	ND	1: 30	36
"	case 7	25.3	10.5	ND	1: 28	33.6

ND: not determined

Table 3. Thyroid function of mothers with chronic thyroiditis at delivery

Case number mother	infant	Thyroid function at derivery T4 (ug/dl)	T3 (ng/dl)	TSH (uU/ml)	MCHA	TGHA	TBII (%)	TSBAb (% inhibition)	Treatment during pregnancy
M-1	1	5.3	135	45.0	160^2	160^2	97.0	100	30mg/day*1
"	2	16.2	200	N.D.	160^2	160^2	89.3	100	120mg/day*1
M-2	3	3.3	66	64.2	40^2	10^2	83.3	91.7	none
"	4	14.8	138	9.6	80^2	10^2	81.7	N.D.	125ug/day*2

*1 Desiccated thyroid *2 L-thyroxine N.d.: not done.

8. Potency of TBII activities in mothers with atrophic autoimmune thyroiditis and outcome in their offspring.

Clinical data on four infants (two sibling sets) with transient hypothyroidism and thyroid function before treatment are shown in Table 1. Potent TBII as well as TSBAb, was observed in the sera of both mothers and infants. (Table 3). However, the effects of dilutions of each mother's serum with normal pool serum were quite different. The fifty-percent inhibition of labeled TSH binding to its receptor in M-1 came at approximately 320-fold dilution, and that of M-2 in pregnant-case 4 was 70-fold, while in pregnant case 3 the dilution was only 8-fold. In another mother (M-3 and M-4) who gave birth to a normal infant (case 5) or infants with transient raised TSH. (case 6,7), the 50 % inhibition occurred only 10 to 30-fold (Fig. 7). The thyroid function and size of the epiphysis of distal femur of their offspring are well correlated with potency of TBII activities in the mother. From the above findings, the dilution test on mother's sera are essential for the evaluation for the prognosis of their offspring.

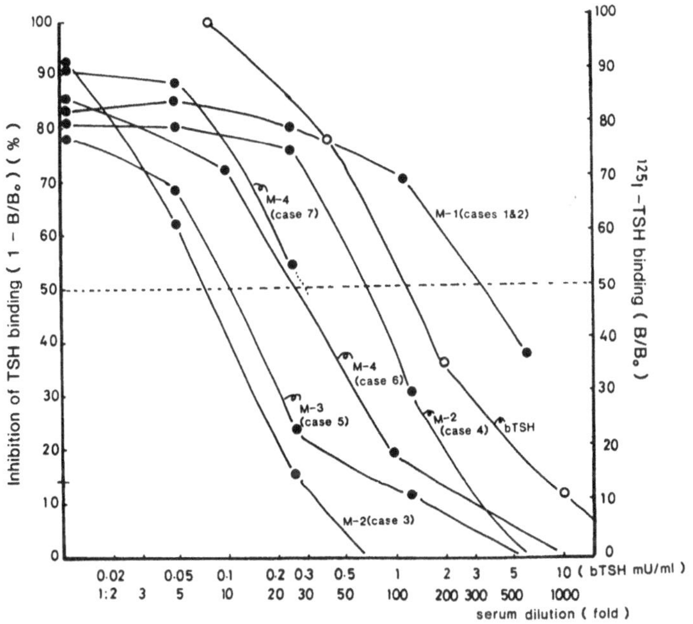

Fig. 7. Dilution curve of the serum of mothers with chronic thyroiditis and bTSH

9. Long term prognosis of two of the sibling sets.

The results of their mother's thyroid function tests are shown in Table 3. The mother of case 3 was not treated at all and the mother of case 1 decided to reduce her dose of thyroid medication to one third when she noticed her pregnancy; therefore, both mothers were in a hypothyroid state for the major part of the pregnancy period. In contrast, both mothers continued taking thyroid medication when they were pregnant with case 2 and case 4. Height and IQ in case 1 at the age of eleven were 129.4 cm (-2.0 S.D.) and 70, while those of case 2 at the age of ten were 135.5 cm (Mean) and 103, respectively; that is, physical and mental development were retarded in case 1 compared with case 2. Height and IQ of case 3 at the age of 4 5/12 were 96.2 cm (-1.8 S.D.) and 76, respectively. While this is too young to reach a conclusion, physical and mental development were comparable to those at the age of one year in case 4.

10. Mental development in cases of transient hypothyroidism due to TSH-receptor blocking antibodies (nationwide studies)

IQ or DQ was evaluated in 17 of 23 living cases with transient hypothyroidism. Five of the 17 cases had IQs below 80. The mothers of all five cases were either not treated at all or reduced their dose of thyroid medication during pregnancy; therefore, they were in a hypothyroid state for the major part of the pregnancy period. Only one mother among the eighteen cases whose IQ was more than 80, continued taking 100 ug a day of L-thyroxine during the pregnancy period; however, the thyroid function at delivery was slightly low. (Fig. 8) The distriburion of IQ or DQ in patients with transient hypothyroidism due to maternal TSH-receptor blocking antibodies are compared to that with the congenital hypothyroidism detected by neonatal screening in Japan. Although absolute number are quite difference between two groups, the incidence of cases below IQ 80 are increased in patients with transient hypothyroidism (Fig. 9).

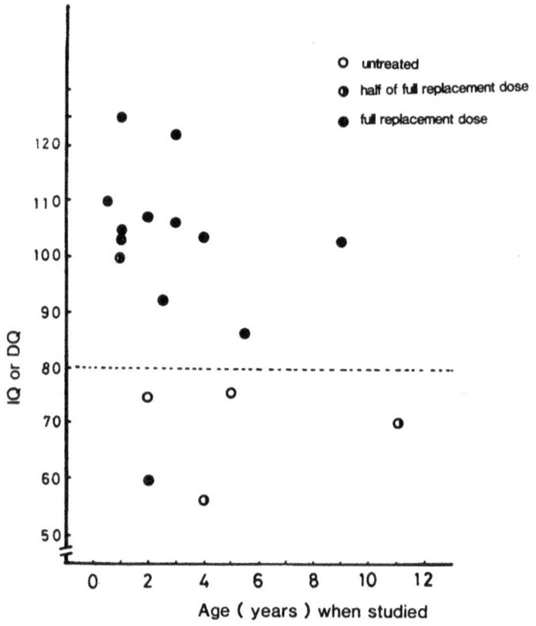

Fig. 8. IQ or DQ of seventeen infants with transient hypothyroidism and status of thyroid medication during pregnancy

DISCUSSION

Transient neonatal hypothyroidism in infants born to mothers with chronic thyroiditis was confirmed as an important disease entity in the newborn period. The presence of TSH-receptor blocking antibodies, which were detected by radioreceptor assay of TSH and thyroid-stimulation blocking activities (TSBAb) had been considered the pathogenesis in this condition.[2)3)4)] To clarify the mechanisms of thyroid suppression and thyroid atrophy of this condition, we studied the adenylate cyclase system using porcine thyroid cells. We used IAP, a pertussis toxin, as a unique probe to study mechanisms involved in receptor-mediated signal transduction in porcine thyroid cells. TSH, PGE_1 and E_2 stimulated adenylate cyclase activity dose-dependently, while epinephrine inhibited TSH-induced cAMP accumulation through α_2-adrenergic receptors. This inhibition was reversed by prior exposure of cells to IAP as a result of the failure of Ni to mediate receptor-linked inhibition of adenylate cyclase activity brought about by the toxin-catalyzed ADP-ribosylation of GTP-binding α-subunits (41K) of Ni. IgG of the mothers with chronic thyroiditis (M-IgG) inhibited TSH-stimulated cAMP accumulation; however, failure to inhibit PGE_1 and horskolin induced cAMP stimulation. Moreover, dose-dependent inhibition of TSH-induced cAMP accumulation by M-IgG was not influenced by prior exposure of cells to IAP. From these experiments, it is concluded that M-IgG inhibited porcine thyroid membrane adenylate cyclase activity on TSH-receptor levels, and did not influence PGE receptors or GTP-binding α-subunits of Ni.

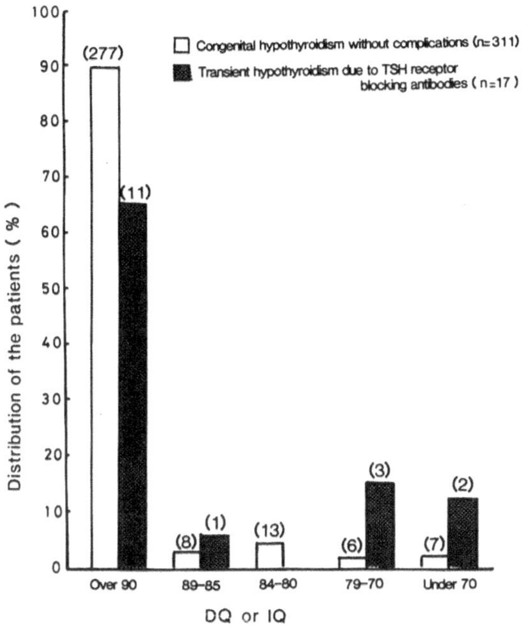

Fig. 9. Distribution of IQ or DQ in patients with congenital hypothyroidism and transient hypothyroidism

Proliferation or differentiation of cells is initiated by the combined addition of two or more growth factors in the medium and a GTP-binding protein was found to be involved in cell proliferation.[7] Phosphatidylinositol hydrolysis via phospholipase C (PI turnover) followed by Ca^{2+} mobilization and C-kinase activation is also involved in cell proliferation. Research on the effects of M-IgG upon PI-turnover, cAMP-stimulated iodine-uptake, and ^3H-thymidine uptake is now underway.

We reported here that the physical and mental development of two sibling sets were quite difference from each other. Cases 1 and 3 were physically and mentally retarded while cases 2 and 4 developed normally. The mothers in cases 1 and 3 were hypothyroid during the whole pregnancy period, while they were euthyroid when they were pregnant with cases 2 and 4. TBII activity was compatible when mother M-1 was pregnant with cases 1 and 2, while it was more potent when mother M-2 was pregnant with case 4 than with case 3. A nationwide survey also revealed that there was phychomotor and mental retardation (IQ less than 80) in five of 23 living cases (21.7%), which is much higher than that of congenital hypothyroidism (2.1% in a 1984 survey of Japan). The mechanisms of psychomotor and mental development delay are not clear; but at least four mechanisms might be speculated. First, the delayed development might depend on the severity of thyroid suppression during the fetal period as well as the degree of hypothyroidism during the newborn period.[8] The second consideration is the period of hypothyroidism and the age at the start of treatment. The nature of the TSH-receptor blocking antibodies, and finally, the degree of maternal hypothyroidism during pregnancy might also be important factors. There were no apparent differences in age at the start of medication, dose of medication, thyroid function before treatment. As shown in this studies, the potency of TBII activities were well correlated with thyroid function in newborn period. Although the dilution test of mother's serum was not performed in all cases, it was speculated that the potency of TSH-receptor blocking activities was not markedly different in the mothers of the two groups. Only the thyroid function status of mothers during pregnancy was quite different. it is too premature to conclude that the hypothyroidism during pregnancy causes psychomotor and mental developmental delay in infants; however, the evidence presented in this report suggests that the role of thyroid function in early embryogenesis be given serious reconsideration [9]. We wish to thank Profs. Michio Ui and Toshihiko Murayama of the Department of Physiological Chemistry, Faculty of Pharmaceutical Science, Hokkaido University, for their kind gift of IAP and advice and comments concerning this experiment.

ACKNOWLEDGEMENT

The study was supported in part by grant-in-aid for scientific research 62480223 from the Ministry of Education, Science and Culture, Japan.

REFERENCES

1) Sutherland, J.M., Esselborn, V.M., Burket, R.L., et al. 1960: Familial nongoitrous cretinism apparently due to maternal antithyroid antibody. Report of a family. N. Engl. J. Med., 263: 336.
2) Matsuura, N., Yamada, Y., Nohara, Y., et al; 1980: Familial neonatal transient hypothyroidism due to maternal TSH-binding inhibitor immunoglobulins. N. Engl. J. Med., 303: 738.
3) Iseki, M., Shimizu, M., Oikawa, T., et al. 1983: Sequential serum measurements of thyrotropin binding inhibitor immunoglobulin G in transient familial neonatal hypothyroidism. J. Clin. Endocrinol. Metab., 57: 384.
4) Takaku, N., Mori, T., Koizumi, Y., et al, 1984: Transient neonatal hypothyroidism due to maternal immunoglobulins that inhibit thyrotropin-binding and post-receptor processes. J. Clin. Endocrinol. Metab., 59: 142.
5) Southgate, K., Cregh, F., Teece, M., et al., 1984: A receptor assay for the measurement of TSH receptor antibodies in unextracted serum. Clin. Endocrinol, 20: 539.
6) Konishi, J., Iida, Y., Endo, K., et al, 1983: Inhibition of thyrotropin-induced adenosine 3',5'-monophosphate increase by immunoglobulins from patients with primary myxedema. J. Clin. Endocrinol. Metab., 57: 544.
7) Murayama, T., Ui, M., 1987: Possible involvement of a GTP-binding protein, the substrate of islet-activating protein, in receptor-mediated signaling responsible for cell proliferation. J. Biol. chem., 262: 12463.
8) Rovet, J., Ehrich, R., Sorbara, D. 1987: intellectual outcome in children with fetal hypothyroidism. J. Pediatr., 110: 700.
9) Ekins, R., 1985: Roles of serum thyroxine-binding proteins and maternal thyroid hormones in fetal developement. Lancet, I: 1129.

DISCUSSION

ORGIAZZI (Lyon) : The effects of blocking antibodies on thyroid function are indeed complex. My question is : in your experiments, how can you be sure the IAP does not modify the blocking effects of the antibodies ? Can you be sure that in the reaction mixture the antibody was not bound to any of the components, since immune complexes could alter the effects of IAP ? With a similar blocking antibody, despite the fact that the antibody would not stimulate cAMP production, we were able to obtain a complete desensitization of thyroid cells (preincubated with blocking antibodies) to subsequent stimulation by TSH or TSAb. Moreover, we demonstrated that the blocking effect was inhibited by coincubation with cycloheximide, which inhibits protein synthesis. In other words, the blocking antibody behaved as if it would stimulate protein synthesis in thyroid cells.

MATSUURA (Sappora) : We did not check for this possibility. I thank you for the comment.

PINCHERA (Pisa) : You mentioned in your slides that the mothers were untreated. They had chronic atrophic thyroiditis and must have been hypothyroid during their pregnancy. Is this correct ?

MATSUURA (Sapporo) : Yes.

PINCHERA (Pisa) : Did you observe cases of congenital hypothyroidism from mothers who were treated before pregnancy ?

MATSUURA (Sapporo) : Most patients were treated before pregnancy.

PINCHERA (Pisa) : In our own cases blocking antibodies are present only in the patients with atrophic thyroiditis who had not been treated. In those who were treated for a sufficient period of time, the antibodies were not present. Therefore, we would anticipate that these patients would be less likely to deliver a congenitally hypothyroid newborn.

BROWN (Worcester) : Looking at thyroid stimulating antibodies as opposed to blocking antibodies, an important question is whether the different effects of the antibodies are mediated by the same or by separate antibody populations. One antibody blocks binding and another blocks stimulation and/or growth. Have you had the opportunity to estimate the half-maximum level of blocking activity and have you looked at the inhibition of thyroid growth ?

MATSUURA (Sapporo) : No. We have only done binding assays.

DREXHAGE (Amsterdam) : I wonder how long these antibodies actually last in the babies ? In Graves' disease, it is sometimes 3-5 weeks.

MATSUURA (Sapporo) : The antibodies were detected until 3 to 5 months of age. This was also true in some patients with neonatal Graves' Disease.

SPORADIC AND ENDEMIC CONGENITAL HYPOTHYROIDISM: EVIDENCE FOR AUTOSENSITIZATION

Steven C. Boyages , Jan Willem Lens, Reindert D. van der Gaag, Glen F. Maberly [2], Creswell J.Eastman [2], and Hemmo A. Drexhage

Laboratory for Clinical Immunology, Department of Pathology, Free University Hospital, Amsterdam, The Netherlands. Endocrine Unit, Department of Medicine, Westmead Hospital, Sydney, Australia [2]

INTRODUCTION

Although aetiologically distinct sporadic and endemic congenital hypothyroidism (cretinism) share a common historical ancestry. Until the early 17th century endemic cretinism remained largely a travellers' curiosity and was attributed to inbreeding, isolation, or various climactic factors (1). Felix Platter of Basle (1602) is credited with the first accurate medical description. The thyroid gland was discovered by Thomas Wharton in 1656 but its physiological role remained unclear until the late 19th century. With improving surgical and antiseptic techniques, total thyroidectomy was increasingly advocated for the treatment of goitre. As expected "cachexia strumaprivia" was reported and its clinical similarity to endemic cretinism noted. It had always been believed that there was a common cause for endemic goitre and cretinism, as both conditions were found side by side, but it was not suspected that there was a causal connection (2).

At about the same time Thomas Curling in 1850 reported two cases of cretinism where he noted: "a defective brain or cretinism combined with an entire absence of the thyroid"(3). Fagge (1871) described a further four cases and designated this condition as "sporadic cretinism" (4). He drew attention to the similarities and differences between the two types. These observations were extended to the adult by Gull in 1873 (" a cretinoid state supervening in adult life in women") and subsequently termed "myxœdema" by Ord in 1878. In 1893 Osler described 60 cases of sporadic cretinism in North America and suggested this was the more prevalent form of congenital hypothyroidism (5).

McCarrison in 1908 classified endemic cretinism into neurological and myxedematous types (6). We now know that iodine deficiency is the major aetiological factor and the neurological type is the predominant clinical form found in all endemias (7). The majority of these affected patients are euthyroid or mildly hypothyroid and goitrous. Myxedematous cretinism is analogous in appearance to sporadic congenital hypothyroidism and is best described from Zaire, Nepal and certain parts of China. Rather than goitre myxedematous cretins have thyroid atrophy and fibrosis (8). It is increasingly believed that neurological abnormalities may be present in myxedematous cretins and the clinical differences between the two types are explained by the effects of ongoing thyroid hormone deficiency (9). The factors that account for this geographical and clinical diversity are unknown.

Sporadic congenital hypothyroidism occurs in about 1 in 4000 live births in the West but its aetiology remains obscure (10). Early diagnosis by neonatal screening and treatment with thyroxine have largely obviated antecedent intellectual disability (11). Autoimmune thyroid processes have been suggested as a possible pathophysiological mechanism for this condition but conventional thyroid antibodies such as anti-microsomal antibody are usually negative. We have previously shown among 34 mothers of infants with CHT, 15 had immunoglobulins blocking thyroid growth (TGBI) induced by thyroid-stimulating hormone (TSH). Fifty percent of affected neonates also had TGBI detected in serum. TGBI persisted in 3 of 7 mothers tested 3 years after delivery (12).

In 1961 Balfour et al. demonstrated autoantibodies to a "second colloid antigen" (CA2) in a variety of autoimmune disorders, including congenital hypothyroidism: CA2 antibodies were detected in the serum of 4 of 11 mothers of athyreotic cretins (13). The specificity and clinical relevance of these autoantibodies has not yet been elucidated but they are found in Graves' disease, atrophic and Hashimoto's thyroiditis, and also in subacute thyroiditis (De Quervain's thyroiditis). In the latter they are considered to be the predominant autoantibody accompanying this disorder.

We sought to examine in this study the presence of TGBI and CA2-Ab in sporadic CHT and whether these autoantibodies persist in mother or child several years (3-5 yrs) after diagnosis. The latter would indicate evidence for autosensitization rather than simple transplacental transfer of maternal immunoglobulins in the pathogenesis of CHT. In addition, the presence of thyroid autoantibodies in endemic cretins was examined not only because of the clinical similarities between sporadic and endemic CHT but also because of thyroid atrophy in severely hypothyroid cretins.

PATIENTS

Group 1- Sporadic Congenital Hypothyroidism

Sera of mothers of CHT infants, detected by neonatal screening programme using filter bloodspot thyroxine and TSH, were taken at the time of diagnosis (approximately 3-4 weeks after delivery). 50 cases came from the Quebec Network of Genetic Medicine, Canada and 14 cases from the Amsterdam - CHT screening programme. The hypothyroid children were classified into 3 groups: those with thyroid dysgenesis (n=54), those with goitre (n=7) and those with transient hyperthyrotropinaemia (n=3). In 12 of the Dutch cases infant serum was also tested.

We sought to determine the persistence of thyroid antibodies by testing sera collected from 15 pairs of mothers and infants with sporadic CHT at follow-up (detected by the Dutch CHT screening programme, 1980-1982). The period after diagnosis was between 3-5 years and no serum was available from the time of diagnosis. All children were on thyroxine replacement therapy at the time of testing.

Sera of 12 healthy mothers giving birth to a healthy baby and of 18 healthy laboratory staff members served as negative controls. Positive control sera for second colloid antigen antibodies was obtained from 5 patients with subacute thyroiditis.

Group 2-Endemic Congenital Hypothyroidism

Patients were a sample of 69 endemic cretins, 37 males and 32 females, randomly selected by the health authorities of Qinghai Province, and representative of the cretin population of the region. Independent of Australian workers, patients were divided into neurological (n = 15), myxedematous (n = 25) and mixed groups (n = 29) according to clinical criteria alone. The mean age for females was 25.9 years (SD =10.4) , range 7 to 44 years, and the mean age for males was 27.1 years (SD =12.9), range 4 to 52 years. These differences were not statistically significant ($P > 0.10$). Qinghai Province is situated in central Western China on the other side of the Himalayas where McCarrison's original studies (6) were conducted. It is a cold, high altitude (1,600 to 5,000 metres) region where the population of 4 million comprises Han Chinese, Hui, Tibetans and numerous other minority groups. The endemia in Qinghai is located around the Huong He (yellow) river, mainly in the eastern agricultural region of the provine. Goitre and cretinism have long been recognised as endemic and estimates suggest that 50% of the population live in iodine deficient areas. The water iodine concentration is low and measures 5 - 20 μ g/l. The prevalence of cretinism is estimated to be as high as 10% in some villages and iodised salt prophylaxis was introduced in 1978.

METHODS

Immunoglobulin (Ig) preparation

Immunoglobulins (IgG) from patient serum were prepared by separation using a Protein A-sepharose column coupled to a UV path spectrophotometer. Glycine-HCl buffer (pH= 2.8) was used to elute IgG from the column and the resultant preparation was extensively dialysed against phosphate buffered saline (pH = 7.4).

Thyroid-growth blocking immunoglobulins (TGBI)

TGBI were determined by measuring DNA synthesis in cultured guinea pig thyroid explants by Feuglen cytophotometry as previously described in detail (12). The majority assays were performed in triplicate at a fixed amount of Ig (125 μg IgG/ml culture fluid). The maternal sera from the Dutch CHT-screening program (n=14) were tested in dose response over a concentration range of 30 to 250 μg IgG/ml culture fluid. From the Qinghai endemic group, 12 sera were tested for TGBI in dose-response over a concentration range of 15 to 100 μg IgG/ ml culture fluid.

Second Colloid Antigen-Antibodies (CA2-Ab)

Second colloid antigen-antibodies (CA2-Ab) were determined by indirect immuno-fluorescence on frozen human thyroid tissue obtained from surgery as previously described (13). Briefly, cryostat sections (5 μm) of these thyroids were fixed in 100% methanol (60 minutes at 25C°) followed by a post-fixation of 10 seconds in 70% methanol. Patient and control sera were tested in dilutions of 1:10 to 1:80 in phosphate bufferred saline (PBS), pH 7.4. Sections were incubated with diluted sera, washed twice in PBS, incubated with fluorescein conjugated calf anti-serum to human total immunoglobulins (Kallestad, Austin, Texas; dilution 1:120), and thereafter two washes in PBS. Sections were mounted in aquamount (BDH Chemicals Ltd, Poole, England) and kept in the dark at 4° C until evaluation.

Other thyroid autoantibody tests

Antithyroglobulin (TgAb) and antimicrosomal (MAb) antibodies were determined by a commercial haemagglutination assay (Wellcome, Breckenham, UK). Thyrotrophin binding inhibitory immunoglobulins (TBII) were measured with a commercial radioreceptor assay (Trak Henning, Berlin).

RESULTS

Controls

We had previously shown that TGBI were absent in 0 of 18 healthy individuals (12). CA2-Ab also showed a low prevalence in those with a normal pregnancy outcome and healthy non-pregnant controls, 1 of 12 (8%) and 2 of 18 (11%) cases, respectively. By contrast, all 5 patients with subacute thyroiditis were strongly positive for CA2-Ab and in 3 of these patients, CA2-Ab were present in the absence of concomitant TgAb and McAb.

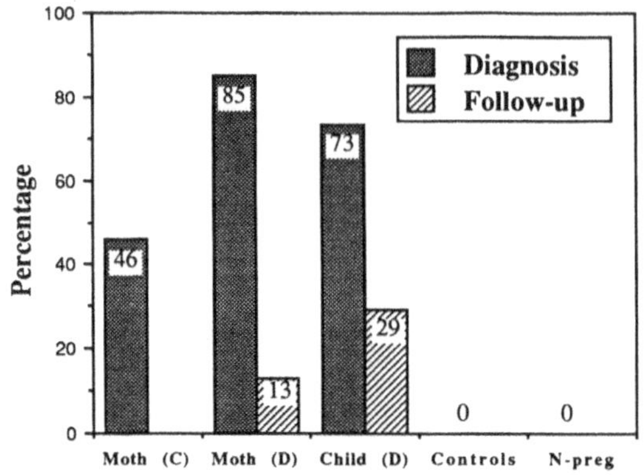

Figure 1. TGBI in mothers (C = Canadian, D = Dutch) and infants with sporadic CHT at diagnosis and at follow-up (3-5 yrs).

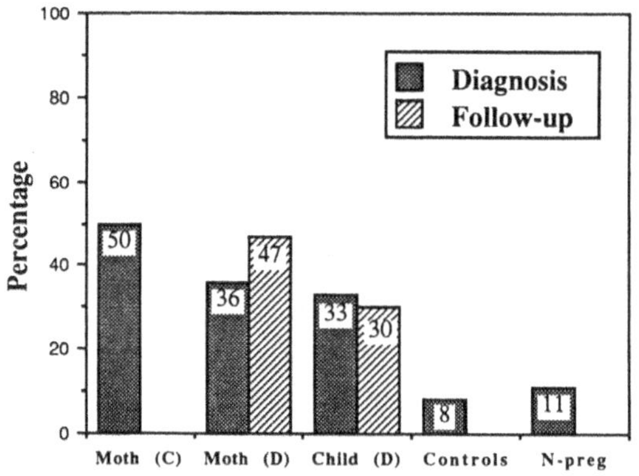

Figure 2. CA2-Ab in mothers (C = Canadian, D = Dutch) and infants at time of diagnosis and at follow-up (3-5 yrs)

Thyroid antibodies in maternal and neonatal serum of sporadic CHT at birth

TGBI. In a previous study of the Quebec programme, TGBI were detected in 15 of 34 (44%) mothers who gave birth to a child with sporadic CHT (12). In the present study, TGBI were present in a further 8 of 16 (50%) mothers of the Canadian series when tested at a single concentration of IgG (125 μg IgG/ml). Therefore a total of 46% of mothers were positive for TGBI from the Quebec programme. By comparison testing in dose response, a more sensitive method, revealed 11 of 13 (85%) Dutch cases were positive for TGBI. In the corresponding sporadic CHT neonate of these mothers TGBI were positive in 8 of 11 (73%) cases (Figure 1).

CA2-Ab. 47% of maternal sera of CHT cases were positive for CA2-Ab, 25 of 50 (50%) from Quebec and a lower incidence in Dutch mothers, 5 of 14 (36%). In the latter group where corresponding neonatal serum was available, 4 of 12 (33%) CHT infants were positive for CA2-Ab. A further 2 children in the Dutch series had equivocal results (figure 2).

McAb,TgAb and TBII. McAb and TgAb were detected in 8 of 65 (12%) and 2 of 65 (3%) mothers respectively. These thyroid antibodies were absent in 0 of 12 (0%) Dutch CHT infants. TBII activity was not detected in mothers or children from either series.

Thyroid antibodies in maternal and infant serum of sporadic CHT 3 - 5 yrs after diagnosis

TGBI. Of the 15 Dutch mother-child pairs investigated during follow-up 3 to 5 years after birth, serum was available to test thyroid-reactive autoantibodies. All these children had dysgenesis (agenesis n=7; ectopic n=6; eutopic hypoplastic n=2). In 2 of 15 (13%) mothers of CHT infants TGBI were present 3 to 5 years after diagnosis. In CHT infants 4 of 14 (29%) were positive for TGBI (figure 1).

CA2-Ab, McAb and TgAb. Of the 15 maternal sera tested for CA2-Ab, 7 (47%) were positive. Two of the mothers were also positive for thyroid McAb. In the children, 4 of 13 (30%) were positive for CA2-Ab. No McAb or TgAb were detected in the infant group (figure 2).

Endemic Congenital Hypothyroidism (endemic cretinism)

Thyroid atrophy was found in the severely hypothyroid group. The latter corresponds to classical myxedematous cretinism (Figure 3). Goitre was detected in euthyroid predominantly neurological cretins. In all 69 cretins McAb and TgAb were negative. CA2-Ab were positive in 12 of 42 (29%) and 7 of 26 (27%) euthyroid and hypothyroid cretins respectively. This incidence is similar to sporadic CHT infants. Thyroid growth blocking immunoglobulins (TGBI) were assessed in 12 patients, 4 euthyroid ("neurological") and 8 hypothyroid ("myxedematous"). The latter had thyroid atrophy on US. TGBI were positive in 7 of 8 (88%) hypothyroid cretins which was significantly greater than 0 of 4 (0%) euthyroid goitrous cretins (Figure 4).

DISCUSSION

We report in this study a close link between the presence of thyroid autoantibodies inhibiting thyroid growth and congenital hypothyroidism. These autoantibodies were found in both sporadic and endemic types of congenital hypothyroidism and particularly in the latter correlate with the nature of disease expression. Although CA2-Ab and TGBI were prevalent in mothers of sporadic CHT infants, the persistence of these antibodies well after the abatement of maternal transfer of immunoglobulins suggest that a primary immune process may be triggered in the affected child. Similarly, these immunoglobulins are found in older endemic cretins.

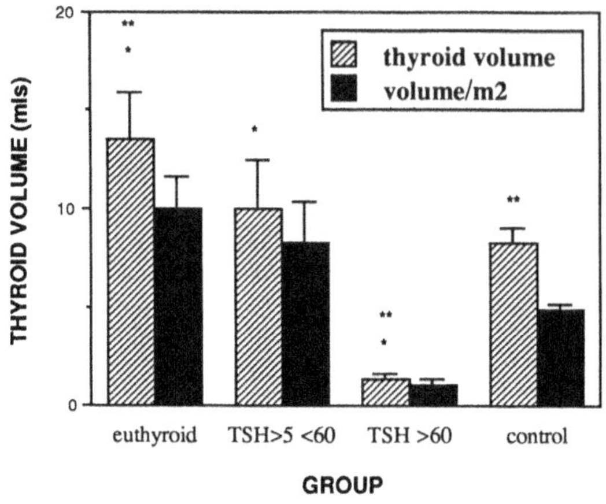

Figure 3. Thyroid volume determined by sonography (7.5 MHz transducer) in control (iodine sufficient) and cretin populations related to thyroid function. The hatched histograms represent raw values whilst closed histograms represent thyroid volume corrected for body mass index. Those patients with severe thyroid homone deficiency ie TSH > 60 mIU/l demonstrated thyroid atrophy.

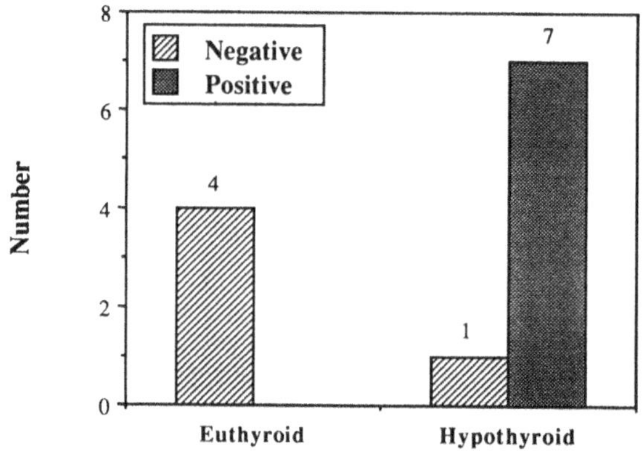

Figure 4. TGBI in euthyroid and hypothyroid endemic cretins from Qinghai Province, China.

In the majority of cases of sporadic CHT the cause is unknown (10). A few cases are related to documented intrauterine inflammatory disease and iatrogenic causes. Genetic factors do not appear to be operating but some studies have suggested a greater incidence of thyroid disease in the families of infants with sporadic CHT (14). This observation has raised the possibility of autoimmune thyroid disease as a cause of CHT. Indeed, transplacental transfer of maternal immunoglobulins blocking TSH induced cyclic AMP stimulation in neonates with transient CHT has been documented (15). An association of thymic hypoplasia and hypothyroidism in infancy is also reported (16). We have previously reported the presence of TGBI in 44% of mothers of CHT infants which persisted after delivery (12). The present study confirms and extends these prior observations and shows the persistence of TGBI and CA2-Ab not only in the mothers but also in the affected child. These findings indicate that transplacental transfer of maternal IgG may not be the sole pathogenetic mechanism of CHT and imply a primary immune phenomena in both mother and child. In the latter hypothyroidism and thyroid dysgenesis ensues, whilst in the mother the presence of these autoantibodies is not usually associated with thyroid dysfunction.

The possible pathogenetic role of thyroid autoantibodies in CHT is unknown. It is well accepted that, while autoantibodies can sometimes be regarded as good markers of the the autoimmune process, damage to thyroid tissue is brought about by other immune effector mechanisms. Monoclonal McAb studies have demonstrated only a few such antibodies are by themselves cytotoxic to thyroid cells (17). We earlier hypothesised that the effect of TGBI occurring in primary myxedema was to prevent the regeneration of immune-destroyed thyrocytes (18). Clinical observations have also revealed that McAb, TgAb and TGBI usually have little effect on the developing fetal thyroid gland. These findings point to the role of other aetiological factors (eg viral, enviromental) for the expression of disease activity. This is best exemplified by endemic congenital hypothyroidism.

We have previously shown that endemic cretins from Qinghai Province, whether classified as typical "neurological" or "myxedematous" display a similar pattern of neurological disability (19). This consists of severe intellectual disability, deafness and a peculiar spasticity which is predominantly proximal and not consistent with prior descriptions of "spastic diplegia". In addition, gait disturbance is almost universal and contributed to by joint deformity and laxity as well as by spastic rigidity. The neurological sequelae are thought to reflect the combined effects of maternal and fetal hypothyroxinaemia on the developing fetal brain (9). Maternal hypothyroxinaemia as evidenced by animal studies is thought to play a crucial role prior to the development of the fetal thyroid gland (20). These mechanisms, at least in the Qinghai endemia, are common to all types of cretinism. In this endemia, clinical differences were explained by the severity and duration of continuing hypothyroidism. The present study presents preliminary evidence that autoimmune thyroid growth inhibition may explain why some endemic cretins have thyroid atrophy and remain severely hypothyroid. The factors that precipitate this autoimmune process are unknown but may be related to the severity of iodine deficiency, genetic predisposition, or some other geographic based element. Selenium deficiency, which is also endemic in Qinghai Province, has recently been suggested as a contributing factor (21).

A schema of a preliminary hypothesis for the pathophysiology of sporadic and endemic CHT linking the autoimmne phenomenna and other concepts discussed above is shown in Figure 5 and 6. Sporadic CHT is primarily a late onset intrauterine fetal disorder. By contrast endemic CHT results from a combination of two differing pathophysiological processes. The first may be occurring in all types of cretins and is related to the degree of maternal hypothyroxinaemia. The second occurs less commonly and appears to result in persisting hypothyroidism and thyroid atrophy.

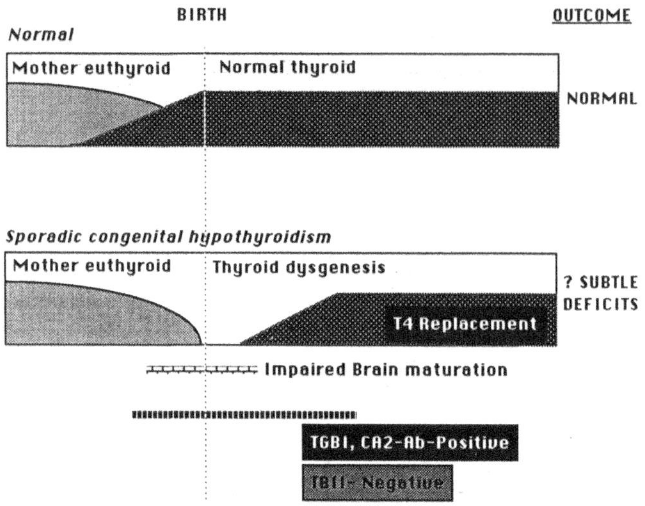

Figure 5. Schematic working hypothesis for the pathogenesis of sporadic CHT. Maternal stores and placental transfer of thyroxine are normal whereas the fetal thyroid gland fails to develop. Thyroid antibodies in the fetus are presumed to appear early in gestation but persist several years after delivery. Neurological outcome is usually normal in treated patients but subtle deficits may be found in some.

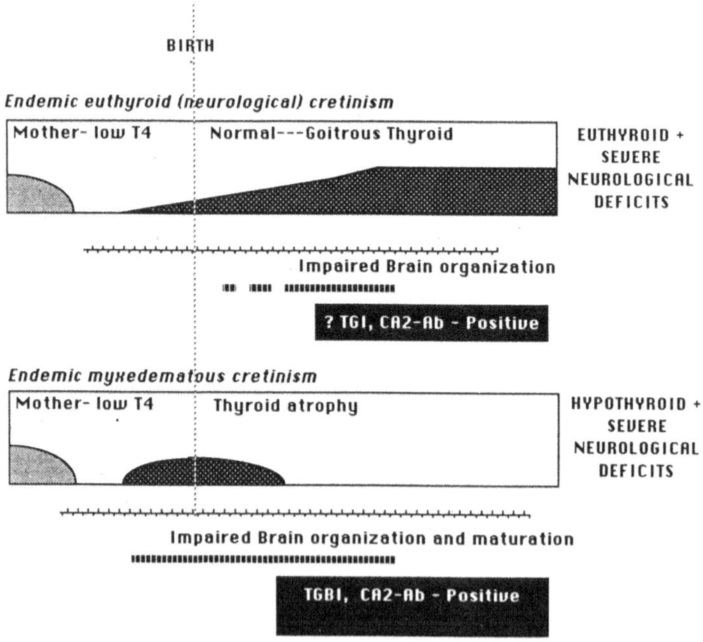

Figure 6. Schematic working hypothesis for the pathogenesis of endemic CHT. In both euthyroid and hypothyroid cretins maternal hypothyroxinaemia is present. The latter results in severe neurological disability in both types of cretins. In the majority of cretins (ie neurological) the fetal thyroid gland develops and after birth, normal or near normal thyroid function is achieved. In myxedematous cretins, thyroid atrophy is found and autoimmune processes may contribute.

REFERENCES

1. Cranefield PF. The discovery of cretinism. Bull. Hist. Med. 1962; 36: 489-511.

2. Merke F. History and iconography of endemic goitre and cretinism. Berne: Hans Huber, 1984

3. Curling TB. Two cases of absence of the thyroid body, and symmetrical swellings of tissue at the sides of the neck, connected with defective cerebral development. Med. Chirug. Tr. 1850; 33:303-306.

4. Fagge CH. On sporadic cretinism in America. Royal Med. Chirug. Socl. (London) 1871; 54: 156-159.

5. Osler W. Sporadic cretinism in America. Trans Congress Am Phys Surg. 1897; 4: 169-206.

6. McCarrison R. Observations on endemic cretinism in the Chitral and Gilgit valleys. Lancet 1908; ii: 1275-1280.

7. Stanbury JB. The iodine deficiency disorders: introduction and general aspects. In: Hetzel BS, Dunn JT, Stanbury JB, eds. The prevention and control of iodine deficiency disorders, Amsterdam: Elsevier, 1987; 35-47.

8. Boyages SC, Collins JK, Maberly GF, Morris JG, Jupp JJ, Eastman CJ. Congenital iodine deficiency disorders (endemic cretinism): history and description. Aust NZ J Develop Disab 1987; 13 : 3-11.

9. Delong R. Neurological involvement in iodine deficiency disorders. In: Hetzel BS, Dunn JT, Stanbury JB, eds. The prevention and control of iodine deficiency disorders, Amsterdam: Elsevier, 1987; 49-63.

10. Foley TP. Sporadic congenital hypothyroidism. In: Dussault JH, Walker P, eds. Congenital hypothyroidism. New York: Marcel Dekker, 1983; 231-259.

11. Glorieux J, LaVecchio FA. Psychological and neurological development in congenital hypothyroidism. In: Dussault JH, Walker P, eds. Congenital hypothyroidism. New York: Marcel Dekker, 1983; 411-430.

12. Van der Gaag RD, Drexhage HA, Dussault JH. Role of maternal immunoglobulins blocking TSH-induced thyroid growth in sporadic forms of congenital hypothyroidism. Lancet 1985; i: 246-250.

13. Balfour BM, Doniach D, Roitt IM, Coughman KG. Fluorescent antibody studies in human thyroiditis: autoantibodies to an antigen of the thyroid distinct from thyroglobulin. Br J Exp Path 1961; 4:307-316.

14. Childs B, Gartner LI. Etiologic factors in sporadic cretinism. An analysis of ninety cases. Ann Human Genetics 1955; 19 : 90-96.

15. Matsuura N, Yamada Y, Nohara Y et al. Familial neonatal transient hypothyroidism due to maternal TSH-binding inhibitor immunoglobulins. N Engl J Med 1980; 303: 738-741.

16. Hong R, Gatti R, Rathburn JC, Good RA. Thymic hypoplasia and thyroid dysfunction. N Engl J Med 1970; 111: 51-53.

17. Banga JP, Mirakian R, Hammond L, et al. Characterization of monoclonal antibodies directed towards the microsomal/microvillar thyroid autoantigen recognised by Hashimoto's thyroiditis. Clin Exp Immunol 1986; 64: 544-554.

18. Drexhage HA, Bottazzo GF, Bitensky L, Chayen J, Doniach D. Thyroid growth blocking antibodies in primary myxœdema. Nature 1981; 289: 594-

19. Boyages SC, J-P Halpern, Maberly GF et al. A comparative study of neurological and myxedematous endemic cretinism in Western China. J Clin Endocrinol Metab (submitted for publication)

20. Morreale de Escobar G, Pastor R, Obregon MJ, Escobar del Rey F. Effects of maternal hypothyroidism on the weight and thyroid hormone content of rat embryonic tissues, before and after onset of fetal thyroid function. Endocrinology 1985; 117:1890

21. Goyens P, Golstein J, Nsombola B, Vis H, Dumont JE. Selenium deficiency as a possible factor in the pathogenesis of myxedematous cretinism. Acta Endocrinol 1987; 114: 497-502.

DISCUSSION

LAZARUS (Cardiff) : Can I ask you about your Chinese patients in whom you found the colloid antigen. Was that vefore you gave iodine prophylaxis and if so, did you repeat measurements after giving iodine prophylaxis ?

BOYAGES (Amsterdam) : It was before.

PINCHERA (Pisa) : Did you find that these antibodies were still present in the congenital hypothyroid endemic cretins even years after birth, implying therefore that they were not passively transferred ?

BOYAGES (Amsterdam) : Yes, that is correct.

PINCHERA (Pisa) : Your patients were positive for thyroid growth blocking antibodies or for the second colloid antigen antibodies. Were they negative in terms of common antibodies to the thyroid ?

BOYAGES (Amsterdam) : Except for some correlation with TBII, antimicrosomal and antithyroglobulin antibodies were negative.

THYROID GROWTH BLOCKING ANTIBODIES AND CONGENITAL HYPOTHYROIDISM

Jean H. Dussault

Centre de Recherche du Réseau
de Médecine Génétique du Québec
CHUL
Ste-Foy, Québec

INTRODUCTION

In 1960 the simultaneous demonstration of thyroid autoantibodies in both mother and newborn sera in many cases of congenital hypothyroidism led to the hypothesis that maternal autoantibodies could pass through the placenta and induce fetal thyroid disease.[1] In a recent study we demonstrated that 14% of normal pregnant women had positive titers for antimicrosomal antibodies with a positive correlation with the titer found in the newborn which had normal thyroid function. On the other hand only one newborn out of 104 with congenital hypothyroidism had positive titer.[2] These results suggested that antimicrosomal antibodies did not play an important role in the pathogenesis of congenital hypothyroidism.

With the demonstration of the existence of a class of antibodies that bind to the TSH receptor and inhibit cyclic AMP (cAMP) stimulation by TSH (TBII or thyrotropin binding inhibitory immunoglobulins)[3] the syndrome of familial transient neonatal hypothyroidism has been identified in infants of mothers with Hashimoto's thyroiditis. In this syndrome there appears to be a transplacental passage of maternal TS1 block (antibodies blocking TSH induced stimulation of the TSH receptor).[4] Thus the interest in the possible role of thyroid antibodies with blocking activity against the TSH receptors.

Drexhage et al have recently exploited the difference between adult hypothyroid patients with goiter or thyroid atrophy, to demonstrate the presence of immunoglobulins that block the growth promoting effect of TSH in the sera of patients with thyroid atrophy.[5] They named this immunoglobulin TG1 block. This observation led Doniach to the hypothesis that these blocking antibodies could be the cause of congenital hypothyroidism.[6] This hypothesis was later confirmed by the report of Matsura et al[7] where transient neonatal hypothyroidism was caused by an immunoglobulin which inhibited TSH binding to its receptor (TBII) and its action on cAMP production (TS1-block). Furthermore Van Der Gaag et al confirmed the results obtained in the adult with their study in newborn with congenital hypthyroidism and their mothers were 50% of the population was TG1-block positive.[8]

The Quebec experience

We have looked at HLA class II antigens in the mothers of hypothyroid newborns compared to our normal population. A greater incidence of HLA DR3 was found in our hypothyroid population with a relative risk of 3.[9] This association might indicate a possible role of genetically transmitted factors involved in the autoimmune response in congenital hypothyroidism. In order to elucidate the mechanism of action of these immunoglobulins on growth and development of thyroid cells we then used as a model the FTRL-5 cell line from rat thyroid.[10]

G-6PD

The first protocol chosen was the effect of the immunoglobulins on the enzyme C-6PD involved in cellular growth versus the effect of TSH. Although the number of subjects studied was limited we could not find any difference between the mothers of hypothyroid newborns and normal pregnant women in blocking the stimulating effect of TSH. Furthermore the same immunoglobulins did not influence the stimulating effect of TSH on the production of cAMP in the FRTL-5 cell line.

I^{131} UPTAKE

On the other hand, using a specific parameter of thyroid function in this cell line, the I^{131} uptake as described by Weiss et al[11] we were able to demonstrate a statistical difference between the mothers of hypothyroid newborns versus normal pregnant women. (Table I) Indeed the immunoglobulins of the former increased significantly the uptake of I^{131}.[12] The underlying mechanism is still unknown but hypothesis can range from the immunoglobulins binding to the thyroid cell membrane and modifying the TSH receptor configuration decreasing its affinity, or a direct action on the iodide pump.

TABLE I

FTRL-5 I^{131} Uptake

		M + S.E.M.
N.S.	(N = 14)	103 ± 4
MCH	(N = 33)	170 ± 10
NPW	(N = 17)	100 ± 3

N.S. = Normal subjects
MCH = Mothers of hypothyroid newborns
NPW = Normal pregnant women

Thymidine Incorporation

The measure of $[^3H]$ thymidine incorporation in the DNA of FRTL-5 cells is a practical and reproducible mean of evaluating the growth promoting effect of TSH.[15] We used this model to evaluate the effect of immunoglobulins from mothers of infant with congenital hypothyroidism on thyroid cell growth.

For the different immunoglobulin preparations to be tested I mg of immunoglobulins was used. Each well also contained 1 μC1/mol of $[^3H]$ thymidine and 1 x 10^{-12} of TSH. After an incubation of 72hrs the incorporation of the thymidine was measured as well as the DNA content in order to express the results as DPM per μg DNA.

When we compare (Table II) the effect of immunoglobulins of normal pregnant women and mothers of infant with congenital hypothyroidism the former seems to have a very significant difference over the other group tested ($p < .001$) whereas the latter were similar to normal women.

From our results it is obvious that a significance difference exists between the effects of immunoglobulin preparations from normal pregnant women (NPW) sera and normal women. There seems to be a substance in the immunoglobulin fraction of proteins from NPW that acts on thyroid cell growth. This substance does not seem to compete with TSH for its receptor site since they do not have any TBII activity. In the serum of mother of hypothyroid infants (MCH) this substance also seems to be absent or they have in their immunoglobulin fraction a factor that cancels the effect of this inhibiting substance. It is quite difficult for the moment to relate those observations to the possible presence of TG1 block activity but we nevertheless are likely to be in presence of some immunoglobulin interfering with the TSH receptor (or another receptor) and influencing the effect of TSH or thyroid growth.

We also looked if the same antibodies could inhibit the binding of TSH on his receptor (TBII activity) using a commercial kit (CRSR Ltd Cardiff U.K.). The resutls were expressed as inhibitions of TSH binding on the purified thyroid membranes. (Fig. 1) The inhibitory activity was measured in the sera of 108, sera of mother of hypothyroid newborns obtained at 3-4 weeks after delivery. Only 7/86 (8%) of the mothers of infants with permanent hypothyroidism showed any TBII activity whereas 4/22 mothers (18%) of

TABLE II

FTRL-5 $[^3H]$ thymidine incorporation index

		% M ± S.E.M.
N.S.	(N = 7)	72 ± 3
MCH	(N = 44)	107 ± 2
NPW	(N = 21)	89 ± 9

(100% = TSH 10^{-12}M)

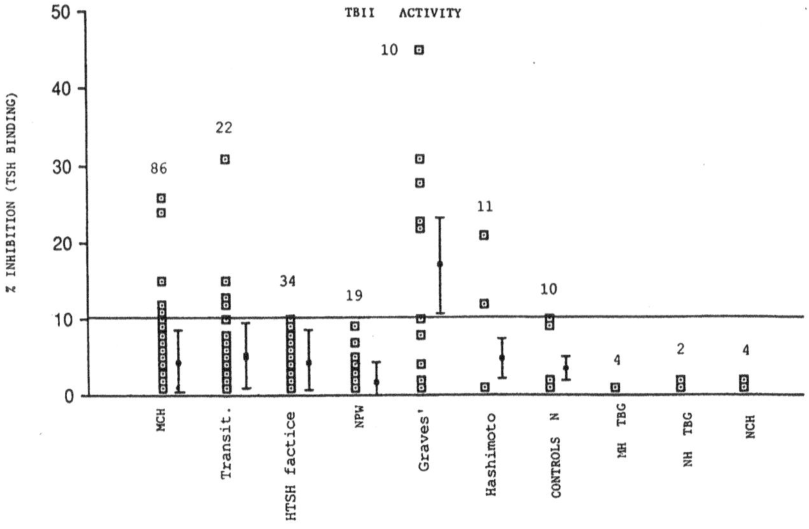

Fig. 1. TBII activity in different thyroid pathology. The final columns represent the mothers of permanent hypothyroid newborns whereas the second represents mothers of newborns with transient hypothyroidism.

newborns with transient hypothyroidism were positive. It should be noted that none of the control (N=30) showed any activity indicating that these antibodies might be implicated in some patients with congenital hypothyroidism more so with the cases of transient hypothyroidism.

In families with multiple cases of hypothyroidism (Table III) one mother had two newborns with transient hypothyroidism whereas the third newborn was normal. It should be noted that the mother had Hashimoto's disease and was CA_2 positive during her second pregnancy. Treatment (L-T4 125 µg I.D.) was started before the second pregnancy and the third infant was euthyroid.

In the second family, the mother was euthyroid with no abnormal thyroid parameters except for a positive CA_2 during her first two pregnancies. Both infants lead permanent hypthyroidism. Her third child was normal but

TABLE III

PREGNANCY

	I					II					III				
	TSH	T4	MC	TBII	CA_2	TSH	T4	MC	TBII	CA_2	TSH	T4	MC	TBII	CA_2
Mother	5	136	1/80,000	-	-	1	226	1/20,000	-	+	0.3	175	1/1,280	-	-
BB	56	136	1/5000	-	-	486	52	1/640	-	-	10	176	1/640	-	-
	6	104				9	105								

PREGNANCY

	I					II					III				
	TSH	T4	MC	TBII	CA_2	TSH	T4	MC	TBII	CA_2	TSH	T4	MC	TBII	CA_2
Mother	8	87	0	0	+	5	86	0	0	+	5	43 (6 months)	0	-	-
											67	201	1/2500		
BB	807	22	0	-	-	830	28	0	-	-	3	188			

the mother became hypothyroid six months after delivery with positive antimicrosomal antibodies. Again these particular cases indicate the possible role of autoimmunity in the etiology of congenital hypothyroidism.

Because of the possible role of bacterial and viral infection (the virus membrane sharing some epitope of the TSH receptor: Yersinia enterolotica)[13] or by revealing some inaccessible antigens (CA_2 in sub-acute thyroiditis),[14] the titers of various viral antibodies were measured. All the mothers studied were negative against Yersenia Enterolotica whereas they had a higher percentage of antibodies against adenovirus versus a control population indicating possibly a role for viral infections (cross reaction with thyroid membrane epitopes?) in the pathogenesis of congenital hypothyroidism. The presence of the CA_2 antibodies in this population is further discussed in another section of this book.

Conclusion

In summary it appears that a majority of the mothers of hypothyroid newborns develop an antibody which act specifically on the fetal thyroid gland either affecting its growth or function. This immune phenomena is either primary or secondary to the presence of an unknown antigen and the antibody appears to be polyclonal in regard to the different method utilized to detect it (TG1-block, CA2 antigen, FRTL-5 model).

Much further work is needed to delineate this phenomenon by either purifying the antibody in order either to make an experimental animal model or to isolate the antigen involved.

REFERENCES

1. R.M. Blizzard, R.W. Chandler and B.H. Landing et al, Maternal autoimmunisation to thyroid as a probable cause of athyreotic cretinism, N. Engl. J. Med. 263:327 (1960).
2. J.H. Dussault, J. Letarte and H. Guyda et al, Lack of influence of thyroid antibodies on thyroid function in the newborn infant and on a mass screening program for congenital hypothyroidism, J. Pediatr. 96:385 (1980).
3. K. Endo, S.M. Amir and S.H. Ingbar, Detection and properties of TSH binding inhibitor immunoglobulins in patients with Graves' disease and Hashimoto's thyroiditis, J. Clin. Endocrinol. Metab. 46:734 (1978).
4. N. Matsuura, Y. Yamada and Y. Nohara et al, Familial neonatal transient hypothyroidism due to maternal TSH-binding inhibitor immunoglobulins, N. Engl. J. Med. 303:738 (1980).
5. H.A. Drexhage, G.F. Bottazzo and L. Bitensky et al, Thyroid growth-blocking antibodies in primary myxoedema, Nature 289:594 (1981).
6. D. Doniach, Hashimoto's thyroiditis and primary myxoedema viewed as separate entities, Eur. J. Clin. Invest. 11:245, (1981).
7. N. Matsuura, Y. Yamada and Y. NOhara et al, Familial neonatal transient hypohtyroidism due to maternal TSH-binding inhibitor immunoglobulins, N. Engl. J. Med. 303:738 (1980).
8. R.D. Van Der Gaag, H.A. Drexhage and J.H. Dussault, Role of maternal immunoglobulins blocking TSH-idnuced thyroid growth in sporadic forms of congenital hypothyroidism, Lancet 1:246 (1985).

9. A.Svejgaard and L.P. Ryder, Disease associations, in: "Histocompatibility Techniques," H.M. Dick and F. Kissemeyer-Neilson, eds, Elsevier/North Holland Biochemical Press, Amsterdam (1979).
10. F.S. Ambesi-Impiombato, L. Parks and H.G.Coons, Culture of hormone-dependent epithelial cells from rat thyroids, Proc. Natl. Acad. Sci. 77:3455 (1980).
11. S.J. Weiss, N.J. Philp and E.F. Grollman, Iodide transport in a continuous line of cultured cells from rat thyroid, Endocrinol. 114:1090 (1984).
12. J.H. Dussault and D.Bernier, ^{125}I Uptake by FRITL5 cells: a screening test ot detect pregnant women at risk of giving birth to hypothyroid infants, Lancet 2:1029 (1985).
13. M. Weiss and S.H. Ingbar, Demonstration of a saturable binding site for thyrotropin in Yersinia enterocolitica, Science 219:1331 (1983).
14. R.D. Van Der Gaag, A.M. Bloot and H.A. Drexhage, The thyroid and autoimmunity, in: "Proceeding of the International Symposium on Thyroid and Autoimmunity," H.A. Drekhage and W.M. Wiersinga, eds, International Congress, Amsterdam (1986).
15. W.H. Dere and B.Rapoport, Control of growth in cultured rat thyroid cells, Moll. Cell. Endocrinol. 44:195 (1986).

TSH-BLOCKING ANTIBODIES AND CONGENITAL HYPOTHYROIDISM

Luca Chiovato, Paolo Vitti, Claudio Marcocci, Gianfranco Fenzi, Lia Giusti, Ferruccio Santini, Pablo Bassi, Mariella Ciampi, Massimo Tonacchera, and Aldo Pinchera

Cattedra di Endocrinologia
University of Pisa
Tirrenia, Pisa, Italy

INTRODUCTION

A role of maternal thyroid antibodies in the prediction and/or the development of congenital hypothyroidism was suggested by previous reports[1-4]. Maternal autoantibodies able to inhibit the TSH-dependent adenylate cyclase[1,2] or cell growth[3] have been described in several newborns with congenital hypothyroidism (CH). Moreover it has been suggested that the presence of immunoglobulin G (IgG) able to increase TSH-stimulated iodine (I) uptake in rat thyroid cells could predict the occurrence of congenital hypothyroidism[4]. To investigate the diagnostic and pathogenetic relevance of humoral autoimmunity in CH, thyroid antibodies were studied in sera of 60 consecutive newborns recalled for examination during our screening program and in sera from the 27 mothers whose children were affected by CH. Additional sera derived from selected cases of hypothyroid newborns born to mothers with Hashimoto's thyroiditis were also studied.

MATERIALS AND METHODS

Sera and immunoglobulin G (IgG)

Sera were collected from 60 consecutive newborns recalled for high spot TSH (> 50 µU/ml) during our screening program for congenital hypothyroidism (total number of screened newborns = 74.230). Among them 25 had permanent and 2 had transient congenital hypothyroidism. Sera were also obtained from the 27 mothers of neonates with CH. Serum samples were also investigated from two selected cases of mothers with Hashimto's thyroiditis (HT) who gave birth to newborns with CH.

The IgG fraction was prepared by the DEAE-Sephadex method and was used at concentrations of 0.5-1.0 mg/ml.

Microsomal (MAb) and thyroglobulin (TgAb) antibodies

MAb and TgAb were determined by passive hemagglutination in serum of all recalled newborns and in that of mothers who gave birth to children with CH.

Assays for antibodies blocking the biological effect of TSH on thyroid cells

IgGs from mothers of CH children and from their newborns were tested for their effects on basal or TSH-stimulated cAMP production, ^{125}I uptake and ^{3}H-thymidine incorporation in a differentiated strain of rat thyroid cells (FRTL-5) maintained in continuous culture[5]. FRTL-5 cells were cultured in 96 well plates under standard conditions and were deprived of TSH for 4-7 days before use.

Simultaneous assay of ^{125}I uptake and ^{3}H-thymidine incorporation. Experiments were initiated by adding to cells 100 µl medium containing 0.5 mM IBMX, 0.2 µCi ^{3}H-thymidine, TSH (1mU/ml) and IgG (0.5 mg/ml). After 48-60 h the medium was removed and washed cells were pulsed with ^{125}I for 40 min at 37 C. Incubation was terminated by aspirating the medium and cells were rapidly washed with ice-cold HBSS. TCA (5%) was added and ^{125}I released in the supernatant was counted[6]. Cells were treated with the dyphenylamine reagent for 12-18 h and aliquots were counted for ^{3}H-thymidine. Results were expressed as % of control cultures incubated without IgG.

cAMP production assay. FRTL-5 cells, precultured in medium deprived of TSH for 4 days, were incubated for 60 min at 37 C with IgG alone (1 mg/ml), TSH alone (40 µU/ml) or TSH + IgG, all dilutions being performed in hypotonic medium[7]. At the end of the incubation cAMP was measured in the extracellular medium as previously described[8]. Results were expressed according to the following formula:

$$1 - \frac{\text{cAMP (TSH + IgG) - cAMP (IgG)}}{\text{cAMP (TSH) - cAMP (control buffer)}} \times 100$$

TSH-binding inhibiting antibody (TBIAb) assay

TBIAb[9] was measured in the IgG fraction using a commercial kit (Mallinckrodt Diagnostica, Dietzenbach, Germany). Briefly, IgG preparations (2 mg/ml) were incubated for 2 h at room temperature with solubilized porcine thyroid membrane in the presence of ^{125}I-TSH. At the end of the incubation ^{125}I-TSH-TSH-receptor complex was precipitated by polyethylene glycol and the pellet was counted.

Cytotoxicity assay

Complement-mediated cytotoxicity was performed on monolayer cultures of human thyroid cells prepared from Graves' goiters and maintained in the same medium containing TSH used for FRTL-5 cells. Briefly: cells were radiolabelled with ^{51}Cr (3-6 µCi/well) at 37 C for 60 min. After washing cells were incubated for 2 h at 37 C with IgGs (1 mg/ml) plus fresh complement. At the end of the incubation the radioactivity present in the supernatant was determined by gamma counting. Results were expressed as specific lysis with respect to normal IgGs.

Table 1

Prevalence of thyroglobulin and microsomal antibodies in sera of newborns recalled for an elevated spot TSH during years 1978-1987.

TSH (μU/ml)	Recalled newborns		Newborns with CH	
	Total	MAb/TgAb (+)	Total	MAb/TgAb (+)
>50<100	28	1 (3.5%)	2[a]	0
>100	32	3 (9.4%)	25	3 (12%)
Total	60	4 (6.6 %)	27	3 (11%)

a= Transient hypothyroidism.

Table 2

Clinical and laboratory findings in the mothers of 3 CH newborns with circulating thyroid antibodies.

Mother	Thyroid status	MAb[a]	TgAb[b]	TBIAb[c]	TBkAb[d]	Thyroid defect (newborn)
BMU	Euthyroid	40^2	10^2	0%	7%	Agenesis
GK	Subclinical hypothyroid	640^2	80^2	5%	40%	Hypoplasia
FG	Euthyroid	320^2	Neg	3%	21%	Agenesis

MAb: microsomal antibody; TgAb: thyroglobulin antibody; TBIAb: TSH-binding inhibiting antibody; TBkAb: TSH-dependent cAMP blocking antibody.

a,b: results expressed as reciprocal of titer.
c: upper limit of normals=25%.
d: upper limit of normals=29%.

RESULTS

Thyroid antibody findings in newborns recalled for CH

Microsomal and/or thyroglobulin antibodies were detected in 4/60 (6.6%) sera of neonates recalled for high spot TSH (> 50 μU/ml) (Table 1). Three out of 4 (75%) recalled newborns with thyroid antibodies had CH, their prevalence over the whole group of CH neonates being 11%. CH in the 3 antibody positive newborns was due to gland agenesis (2 cases) and thyroid hypoplasia (1 case).

Neonatal antibody titers in the 3 children ranged from 1/1600 to 1/409600 for MAb and from neg to 1/1600 for TgAb. The same antibody titers were detected in sera of their mothers who were affected by a previously unrecognized Hashimoto's thyroiditis (Table 2). Two of them were euthyroid and one had a subclinical hypothyroidism which became overt 3 months after delivery. IgG from the latter mother (GK) produced a significant inhibition of TSH-induced cAMP production, ^{125}I uptake and ^{3}H-thymidine incorporation in FRTL-5 cells. IgGs from the remaining two mothers with HT (Table 2) and from the other 24 mothers of CH children did not produce any significant effect on TSH-dependent cAMP production (Fig. 1), ^{125}I uptake or ^{3}H-thymidine incorporation in FRTL-5 cells. It is worth noting that in no case evidence was obtained for a stimulating effect on ^{125}I uptake by maternal IgG. Assays for TBIAb gave negative results with all IgGs from mothers of CH neonates, including those with Hashimoto's thyroiditis (Table 2).

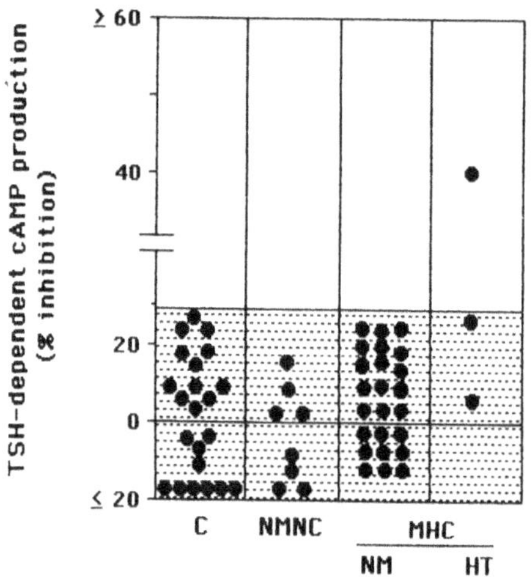

Fig. 1 Results obtained in the TSH-dependent adenylate cyclase inhibition assay with IgGs from: C = normal controls; NMNC = normal mothers of normal children; MHC = mothers of hypothyroid children (NM = normal mothers; HT = mothers with Hashimoto's thyroiditis).

Finally, the IgG of the Hashimoto's mother (GK) which had TSH-blocking activity was the only one that produced a complement-mediated cytotoxic effect on human thyroid cells in primary culture (specific lysis = 11%).

Thyroid antibodies were undetectable in all 3 children born to mothers with Hashimoto's thyroiditis when tested at age 1 year. One of the two children with thyroid agenesis and circulating thyroid antibodies in the neonatal period (born to FG, Table 2) had a nonmonozygotic twin who also suffered from transplacental passage of maternal MAb, but was euthyroid. In this regard it must be pointed out that no functional or cytotoxic action could be demonstrated in the IgG of the Hashimoto's mother of this couple of twins.

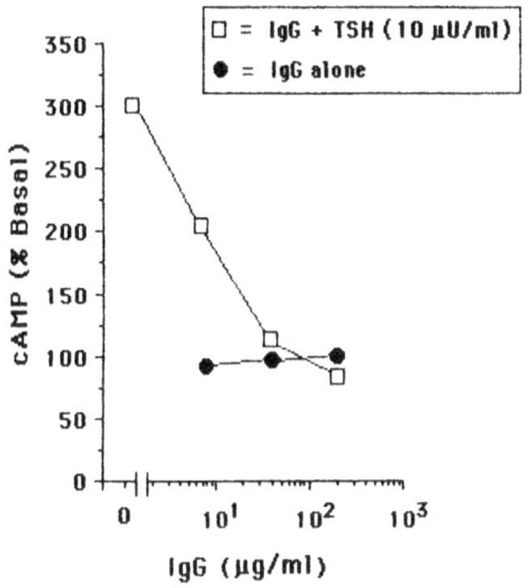

Fig. 2 Effect of the IgG from a mother with atrophic Hashimoto's thyroiditis (case 1) on basal and TSH-stimulated adenylate cyclase in FRTL-5 cells.

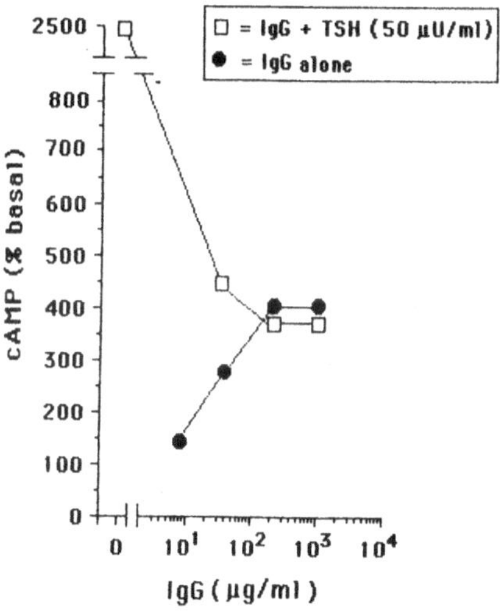

Fig. 3 Effect of the IgG from a mother with atrophic Hashimoto's thyroiditis (case 2) on basal and TSH-stimulated cAMP production in FRTL-5 cells.

Thyroid antibody findings in two selected Hashimoto's mothers of CH newborns

Two additional mother-newborn pairs detected by other screening programs in Italy were studied. Both mothers had atrophic Hashimoto's thyroiditis with hypothyroidism since years before pregnancy and delivered neonates with congenital hypothyroidism that was permanent in one child (case 1) and transient in the other (case 2).

IgG from mother of newborn 1 had no effect when added alone on FRTL-5 cells, but together with TSH produced a complete and dose-dependent inhibition of the hormone-stimulated adenylate cyclase (Fig. 2), ^{125}I uptake and ^{3}H-thymidine incorporation. Child 1 was found to have by echography and scintiscan thyroid tissue normally placed in the neck that was unable to maintain euthyroidism even after 1 year of age. At that time TSH-blocking antibodies transferred from the mother had disappeared from his serum.

IgG from mother of child 2 produced a clear-cut stimulation of cAMP production (Fig. 3), ^{125}I uptake and ^{3}H-thymidine incorporation in FRTL-5 cells, but when added together with TSH it inhibited the action of the hormone in all 3 assays. This TSH-blocking effect of the mother's IgG apparently prevailed in vivo since her child had transient hypothyroidism that remitted after the disappearance of maternal autoantibodies from serum.

DISCUSSION AND CONCLUSIONS

Congenital hypothyroidism due to the transplacental transfer of maternal autoantibodies that block the physiologic action of TSH on thyroid cells and/or inhibit cell growth has been repeatedly described. Instances of both transient[1,2] and permanent[10] congenital hypothyroidism have been reported, but the mechanism by which thyroid impairment occurs remains to be established. Moreover, the precise incidence of autoimmune cases over the whole group of children with congenital hypothyroidism is still unclear. While in several reports a few neonates born to mothers with hypothyroid Hashimoto's thyroiditis have been described[1,2], in the group studied by Van der Gaag et al[3] more than 40% of all newborns with CH were found to have thyroid growth blocking antibodies. Moreover, in the latter study most of the mothers of the affected children were euthyroid and had no clinical sign of autoimmune thyroid disease.

In our study, the prevalence of cases with passive transfer of thyroid autoimmunity, as assessed by measurement of MAb and/or TgAb, was 6.6% (4/60) in the group of newborns recalled for an elevated spot TSH, and 11% (3/27) in those with confirmed CH. In the latter 3 cases clear evidence of a previously unrecognized Hashimoto's thyroiditis was obtained in the respective mother. Antibodies blocking the physiologic action of TSH on thyroid cells were measured with the use of sensitive, well standardized and reproducible methods which employ the FRTL-5 cell strain[5-9]. With these techniques TSH-blocking antibodies could be detected in the IgG of only 1 of the 3 mothers with Hashimoto's thyroiditis who delivered CH neonates. The same IgG preparation was also found to produce complement-mediated cytotoxicity on human thyroid cells. The newborn had permanent CH due to gland hypoplasia that might have been produced by maternal antibodies. None of the other IgGs from mothers who delivered CH children blocked or enhanced the effect of TSH on cAMP production, ^{125}I uptake or ^{3}H-thymidine incorporation in FRTL-5 cells. Thus the frequency of TSH-blocking antibodies in our series of mothers of children with CH was 1/27 (3.7%). These data confirm and extend previous observations reported by our laboratory[11].

TSH-blocking antibodies were also found in the IgGs from two selected Hashimoto's mothers of CH newborns referred from other centers. One of the children had a typical transient hypothyroidism that remitted after the disappearance of maternal antibodies from his serum. In the other case hypothyroidism did not remit after 1 year of age, despite no circulating thyroid antibody could be detected at that time. These data suggest that the transplacental passage of maternal thyroid antibodies could determine both transient and permanent damage of the gland.

Several mechanisms may be implicated for the hypothyroidism induced by TSH-blocking antibodies. While transient congenital hypothyroidism might be explained by temporary inhibition of thyroid function produced by these antibodies, the mechanism responsible for permanent congenital hypothyroidism is more difficult to envisage. It may be postulated that in neonates maternal TSH-blocking antibodies may act by suppressing the development and maturation of the fetal thyroid gland. Other immune mechanisms, as complement-mediated cytotoxicity, may also be involved in congenital hypothyroidism due to passive transfer of thyroid autoimmunity.

In summary, our data indicate that: i) 6.6% of newborns recalled for an elevated spot TSH had circulating MAb and/or TgAb derived from the mother affected by a previously unrecognized Hashimoto's thyroiditis; ii) the prevalence of thyroid antibody-positive cases over the whole CH group was 11%; iii) in the present series of 27 CH children a biological effect of maternal IgG on TSH activity (i.e. inhibition of cAMP production, ^{125}I uptake and ^{3}H-thymidine incorporation) could be demonstrated only in 1 case (3.7%) with permanent hypothyroidism. Two additional cases of CH, one permanent and the other transient, with maternal TSH-blocking antibodies have been documented; iv) IgGs able to increase TSH-stimulated ^{125}I uptake in FRTL-5 cells were not detected in mothers of children with CH.

In conclusion, the present study supports the concept that passive transfer of thyroid autoimmunity may be a cause of both permanent and transient congenital hypothyroidism and that TSH-blocking antibodies may be one of the implicated factors. However, evidence has been obtained that this mechanism is relatively infrequent and accounts only for a minority of all cases of congenital hypothyroidism.

ACKNOWLEDGEMENTS

The Authors are grateful to Prof. G. Giovannelli, Dr. C. Volta (Clinica Pediatrica, University of Parma), Dr. E. Roti (Cattedra di Endocrinologia e Patologia Costituzionale, University of Parma), Prof. G. Bona, Dr. M. Zaffaroni (Istituti Universitari di Pediatria, University of Torino) for providing sera of 2 mothers with Hashimoto's thyroiditis who delivered newborns with congenital hypothyroidism.

REFERENCES

1. N. Matsuura, Y. Yamada, Y. Nohara, J. Konishi, K. Kasagi, K. Endo, H. Kojima, and K. Wataya, Familial neonatal transient hypothyroidism due to maternal TSH-binding inhibitor immunoglobulins. N. Engl. J. Med. 303: 738 (1980).
2. M. Iseki, M. Shimizu, T. Oikawa, H. Hojo, K. Arikawa, Y. Ichikawa, N. Momotani, and K. Ito, Sequential serum measurements of

thyrotropin binding inhibitor immunoglobulins G in transient familial neonatal hypothyroidism. J. Clin. Endocrinol. Metab. 57: 384 (1983).

3. R.D. Van Der Gaag, H.A. Drexhage, and J.H. Dussault, Role of maternal immunoglobulins blocking TSH-induced thyroid growth in sporadic form of congenital hypothyroidism, Lancet 1: 246 (1985).

4. J.H. Dussault, and D. Bernier, ^{125}I uptake by FRTL-5 cells: screening test to detect pregnant women at risk of giving birth to hypothyroid infants, Lancet 2:1019 (1985).

5. F.S. Ambesi-Impiombato, L.A.M. Parks, and G.H. Coon, Culture of hormone dependent epithelial cells from rat thyroids, Proc. Natl. Acad. Sci. (U.S.A.) 77: 3455, (1980).

6. C. Marcocci, W.A. Valente, A. Pinchera, S.M. Aloj, L.D. Kohn, and E.F. Grollman, Graves' IgG stimulation of iodide uptake in FRTL-5 rat thyroid cells: A clinical assay complementing FRTL-5 assay measuring adenylate cyclase and growth stimulating antibodies in autoimmune thyroid diseases, J. Endocrinol. Invest. 6: 463 (1983).

7. L. Chiovato, P. Vitti, A. Lombardi, G. Lopez, F. Santini, E. Macchia, G.F. Fenzi, C. Mammoli, S. Battiato, and A. Pinchera, Detection and characterization of autoantibodies blocking the TSH-dependent cAMP production using FRTL-5 cells. J. Endocrinol. Invest. 10: 383 (1987).

8. P. Vitti, W.A. Valente, F.S. Ambesi Impiombato, G.F. Fenzi, A. Pinchera, and L.D. Kohn, Graves' IgG stimulation of continuously cultured rat thyroid cells: a sensitive and potentially useful clinical assay. J. Endocrinol. Invest. 5: 179 (1982).

9. E. Macchia, G.F. Fenzi, F. Monzani, F. Lippi, P. Vitti, L. Grasso, L. Bartalena, L. Baschieri, A. Pinchera, Comparison between thyroid stimulating and TSH binding inhibiting immunoglobulins of Graves' disease, Clin. Endocrinol. (Oxf.) 15: 175 (1981).

10. R.D. Van Der Gaag, H. Frisch, M. Weisel, G. Wick, and H.A. Drexhage, Congenital hypothyroidism in a Turkish family: the role of immunoglobulins blocking the trophic effects of TSH and maternal-foetal relationship. Acta Endocrinol. 11: 44 (1986).

11. C. Marcocci, L. Giusti, L. Chiovato, M. Ciampi, G.F. Fenzi, and A. Pinchera, Screening for risk of delivery of a hypothyroid baby, Lancet 2: 403 (1986).

DISCUSSION

<u>DELANGE (Brussels)</u> : I presume that the cause of congenital hypothyroidism was thyroid agenesis, or an ectopic thyroid, or hypoplasia. The diagnosis was based on scintigraphy, which in turns depends upon thyroid uptake. Since the uptake can be modified by some of these antibodies, my question is : at what age was the scan performed ? At the time of diagnosis or at the time of re-evaluation later ? Shouldn't we also perform thyroid echography systematically in such patients ?

<u>PINCHERA (Pisa)</u> : Thyroid scans were performed after the age of one year, after the antibodies had disappeared. I should also mention that the assays for thyroid blocking antibodies were performed serially and there was a fall in antibody levels similar to that observed for microsomal antibodies. So far, we have not seen a case in which microsomal or blocking antibodies were still present after one year of age. Therefore, thyroid uptake could not have been impaired by the presence of antibodies. Echography was used in 2 cases and confirmed the scintigraphy data.

<u>FISHER (Torrance)</u> : Dr. Pinchera, is it your view that there is an increased prevalence of antimicrosomal and antithyroglobulin antibodies in your infants with permanent congenital hypothyroidism ? And this taking into account the prevalence of antibodies in the normal surrounding Italian population ?

<u>PINCHERA (Pisa)</u> : All we are saying is that these infants with CH, born to mothers with autoimmune thyroiditis, had antithyroid antibodies of various kinds (microsomal, TSH-blocking, etc...) transferred to them from the mother. This does not imply a causal relationship, however. In addition, we have seen a number of infants with microsomal antibodies but without CH. In fact, such infants represent the vast majority but they were not included in the present study. The second point is that 5 infants with CH had microsomal but not blocking antibodies. Only one of the latter exhibited evidence of cytotoxicity which I do not believe to be associated with the blocking antibodies. I think that there is a "family" of antibodies, including a "subfamily" of antibodies to the TSH receptor : some block the binding and the action of TSH, some block the binding but not the action and the other way around.

In our hands, when we find antibodies with TSH blocking activity, they block all effects of TSH : adenylate cyclase, iodide uptake, thymidine incorporation. Growth blocking antibodies are also "thyroid function" blocking antibodies. Similarly, growth promoting antibodies also usually have "thyroid function" promoting activities.

AUTOIMMUNITY. PANEL DISCUSSION

Moderator : L. Braverman

FISHER (Torrance) : Dr. Matsuura, what is the percentage of newborns with transient CH whose mothers had TSH-blocking antibodies ?

MATSUURA (Sapporo) : 3 % of the congenital hypothyroid babies'mothers had blocking antibodies.

FISHER (Torrance) : That's approximately in agreement with Dr. Pinchera. Thank you Dr. Matsuura. Is it the same in Quebec Dr. Dussault ?

DUSSAULT (Quebec) : TBII positive mothers would be 7 %, and microsomal antibodies about 2 %. Blocking antibodies, however, were found in 40 % of CH babies.

BRAVERMAN (Worcester) : So your numbers Dr. Drexhage are in total contradiction with those of Drs. Matsuura and Pinchera.

DREXHAGE (Amsterdam) : I would like to make a general comment. Using the data from several studies, I think it is fair to state that in infants with CH, microsomal antibodies are found in 3-15 % which corresponds to the figures found in most normal populations of adults. Thyroglobulin antibodies are present in 5-10 % and TBII in 0-10 %. These are the TSH-receptor antibodies described by Matsuura and Pinchera, which block the action of TSH at the TSH receptor. They use FRTL-5 cells which, in my opinion, is a lousy system and this allows them to find 3-7 % of CH babies with what they describe as "blocking antibodies". We have used a totally different system, the cytochemical bioassay, which allows us to detect 70 % of babies and their mothers with positive antibodies. Our assay is much more sensitive. At birth, there is a good correlation between positive babies and positive mothers. After 3-5 years, the correlation is lost. The measurement is performed with about 0.5 ml in serum; IgG are purified using a protein-A sepharose column.

FISHER (Torrance) : I can not understand how one order of magnitude differences in positivity could be explained solely on the basis of differences in sensitivity.

DREXHAGE (Amsterdam) : With the FRTL-5 or other relatively insensitive assyas, TBII are found in about 30 % of patients with primary myxoedema. With the cytochemical bioassay, the positivity is 80-90 %. Therefore, there is a clear gain in sensitivity with the latter method.

SACK (Tel Aviv) : I am not only confused but I am alarmed because of this high frequency of blocking antibodies. Up to now, we have thought that there is a very small risk of CH for the newborn infants of women who have Hashimoto thyroiditis. What we hear today seems different. As far as I understand autoimmune diseases, they are not "B"-cell but "T"-cell mediated. How do you explain that T-lymphocytes from the mother will be transmitted to the babies ? Passive transfer of auto-antibodies is not equivalent to transfer of auto-immune disease. I, therefore, suggest that perhaps these high frequencies of positive blocking antibodies are just a side-effect with no relation to CH.

FISHER (Torrance) : I agree with Dr. Sack. Those mothers don't have other evidence or auto-immune thyroid disease by the standard criteria that we use. So what kind of unusual thyroid disease are we looking at ?

BRAVERMAN (Worcester) : Dr. Drexhage, I shall add one more point on this line. What you are saying is that all Hashimoto's women, either pregnant or non-pregnant, probably do have growth inhibitory immunoglobulins and yet all Hashimoto's women do not have hypothyroid babies.

DREXHAGE (Amsterdam) : Correct. Most of the ladies with primary hypothyroidism in whom we find blocking antibodies give birth to perfectly normal healthy babies. It's an exception (maybe 20 %) that such a baby has a transient form or a permanent form of congenital hypothyroidism. So the blocking antibodies which act in vivo together with the cytotoxic antibodies may not be the cause of the disease or the sole factor to precipitate the disease. It might be a marker or perhaps an additional phenomenon. I don't want to state that this particular antibody is the cause of the disease.

PINCHERA (Pisa) : If we try to put things together, those who use what Dr. Drexhage refers to as a "lousy cell system" do find a fairly good agreement between 10 % positivity in CH and evidence of thyroid auto-immune disease refering to to passive transfer of antibodies. Second, only part of these subjects have a blocking antibody : about 3 % in the case of the Japanese and about 3 to 5 % in the Italians. Therefore, I agree with Dr. Sack. Humoral immunity cannot explain congenital hypothyroidism. On the other hand, it is important to think about the mechanisms involved in auto-immune atrophic thyroiditis. Atrophic thyroiditis is related to destruction and/or hypofunction of the gland by various types of antibody-mediated mechanisms. In patients who have regenerating effects mediated by TSH, there will be a delayed hypothyroidism preceded by a period of euthyroid goitrous Hashimoto's disease. Those who have blocking antibodies have destruction by other types of antibodies but have no regenerating effects mediated by TSH because of the blocking antibodies. Presumably that is the situation in atrophic thyroiditis.

In children, blocking antibodies have usually been found in patients with transient hypothyroidism, and in these cases there is a consistency between the presence of antibodies and the transient type of hypothyroidism. You also have subjects with permanent congenital hypothyroidism who also have this type of antibody. In these infants, the mothers usually have some kind of autoimmune thyroid disorder. I personally have doubts that blocking antibodies represent the cause of permanent CH in these patients. The fact that these patients have other types of antibodies could perhaps indicate that they have cytotoxic antibodies which coact with blocking antibodies in producing the effect. So, humoral antibodies would not explain everything but would be an indicator. Finally, Dr. Drexhage, I cannot understand how, by using your sensitive technique, you find 60 % positivity in CH while the positivity in adults with atrophic thyroiditis is similar to the 20 % found with other techniques.

DREXHAGE (Amsterdam) : I do not agree. With our assay, when we test patients with atrophic thyroiditis, we find 80 % positivity. And I do not think it is a question of sensitivity. I think we talk about two qualitatively different things. Perhaps the FRTL-5 cell system is not suitable to detect these antibodies.

FISHER (Torrance) : Then, it could be an antibody against an IGF-1 receptor and not aginst the TSH-receptor ?

DREXHAGE (Amsterdam) : Certainly. Yes. It would make some sense because it is presumed that early thyroid growth is not TSH-dependent.

CZERNICHOW (Paris) : Don't you think that it is actually time to go to more functional studies ?

DREXHAGE (Amsterdam) : We have just started to do such experiments.

BROWN (Worcester) : We have injected a potent thyroid growth blocking antibody into methimazole-treated goitrous female rats and we were unable to demonstrate any growth blocking effect. These are preliminary data, however.

FISHER (Torrance) : Another point of interest is the adenovirus antibody. It has not been entirely discarded as a possible cause of CH, even though there is no epidemiological data to indicate that we are dealing with a viral-mediated disease. The data is compatible with the view that Dr. Drexhage is looking at a different receptor-mediated effect and hence we expect never to see any correlation if that is the case.

ADDISON (Manchester) : I'd like to ask Dr. Drexhage a little bit about the persistance of the second colloid antigen and growth antibodies in the children after birth.

It seems to me that the interpretation of the persistance of these antibodies is that these children have been exposed and immunized in utero. What is the earliest time in utero that an antibody can be produced to an exogenous stimulation. This is important if we imply that a virus may be the cause and the antibody only an epi-phenomenon. This could provide a potential explanation for the persistance of the antibody. And if so, does the timing of the appearance of an immunoglobulin in the baby from its own sources ties in with the development of the thyroid gland ?

DREXHAGE (Amsterdam) : B-cells develop, I think during the 4-5th month of gestation. These initial B-lymphocytes produce IgM. The switch from IgM to IgG mostly occurs after birth when the neonate is confronted with a variety of environmental factors. But it can happen before that in case of intra-uterine infection.

DUSSAULT (Quebec) : A final comment. There is a variation in the frequency of CH from year to year. After 10 years, you come up with a mean prevalence of 1/4000 but we had years with a prevalence of 1/2500 and other years where it was 1/7000. Also, there is a seasonal variation, at least in Quebec. I could close the laboratory probably from May till September without missing many newborns with CH.

PART 3. DIAGNOSIS

CONGENITAL DEFECTS ASSOCIATED WITH CONGENITAL HYPOTHYROIDISM

John Lazarus, Ieuan Hughes*, Peter Harper and Stephen Bamforth

Departments of Medicine and Paediatrics*
University of Wales College of Medicine
Heath Park, Cardiff, Wales

INTRODUCTION

Congenital hypothyroidism (CH) has an incidence of about 1 in 4000 live births.[1] The aetiology of the condition is poorly understood although it is recognised that athyreosis occurs in approximately 30% of infants and various degrees of ectopia of the thyroid gland in the remaining cases.[2] In 1986 we reported that 20% of a group of CH infants identified in one screening programme had congenital malformation in addition to CH.[3] In view of this finding and a confirmatory report from one centre in the USA[4] we have attempted to establish the possible relationship between CH and congenital defects worldwide.

Methods

A questionnaire enquiring about the association of congenital abnormalities seen in infants with CH was sent to 15 centres worldwide who were known to be actively involved in screening for neonatal hypothyroidism. Ten centres were in Europe, 1 in USA, 2 in Canada, 1 in South America and 1 in Japan. The following items of information were requested: number of infants born with CH (excluding those with transient hyperthyrotropinaemia) during a defined time period; incidence of CH during this period; number of infants with congenital abnormalities; details of the type of abnormality i.e. cardiac, chromosomal, integument, spina bifida or other. Finally the incidence of congenital abnormalities in the screened area was requested.

Results

Twelve out of the 15 questionnaires were returned, 9 from Europe, one each from USA and Canada and one from Japan. While this survey was in progress, Fernhoff et al[4] published data on congenital defects in 100 CH infants in Georgia, USA and their data have been used in the analysis.

The number of infants screened per centre ranged from 50,400 to 1,941,146 (mean 483,819) and the incidence of CH ranged from 1/2135 to 1/6783 (mean 1/4090). Of the 1538 infants with CH congenital abnormalities were noted in 123 (8.0%). In individual centres the percentage of associated abnormalities ranged from 0 to 19.2. There was no evidence of any relationship between the total number of infants screened and the percentage of associated abnormalities in those with CH (Table 1).

Categorisation of the congenital abnormalities showed that 35 (28.5%) were cardiac, 15 (12.2%) were chromosomal, 11 (8.9%) were abnormalities of the integument, 3 infants (2.4%) had spina bifida and 59 infants (48%) had a variety of defects such as metabolic, locomotor disturbances and others. The incidence of various forms of abnormality in the different areas questioned is shown in Table 2. A number of centres were not able to provide any information. A representative review of the literature shows a variable incidence of malformations (Table 3).

Discussion

In 1985 we reported two cases of malformation of the hair in infants with CH.[11] This finding led us to survey all cases of CH diagnosed in the Welsh Hypothyroid Screening Programme since its inception. From 1981 to 1985 the incidence of associated congenital malformation was 20%. None of these infants had thymic hypoplasia, a condition that has previously been described in association with hypothyroidism in infancy.[12,13]

The incidence of congenital malformations in newborn live infants is not precisely known and probably varies with geography. Major abnormalities are said to occur in approximately 1.5 - 3% of live newborns. The data from the present survey suggests that the incidence of congenital malformation in congenitally hypothyroid children is significantly higher than normal. The reason for the wide variation in reported incidence of abnormalities in CH (from 0 to 23%) may reflect genetic, racial and environmental factors as well as variation in ascertainment rates in different countries.

Even though the number of infants screened (over 6m) is large the small number of abnormalities of any particular type is not enough to draw meaningful conclusions. However, cardiac malformations which are reported to occur in 0.8% of live births were found in 12.2% of CH babies in this study.

This study may act as a pointer to further studies in depth to confirm that there is an increased frequency of abnormalities in CH and to investigate the mechanisms. There are no data on the pathogenesis of the CH in this study. It would be tempting, for example, to suggest that certain types of embryological defect in the development of the thyroid gland (e.g. agenesis) may be associated more with other embryological maldevelopments. Another factor which may be of relevance is the iodine status of the mother and infant. Striking variations in thyroidal and urinary iodine have been documented, for example, between Belgium and Canada in

Table 1. Incidence of Congenital Anomalies in CH Infants

	Yrs of survey	No. of Infants screened	Incidence of CH	No. of Infants with CH	No. of Infants with Abnormalities (%)
Italy (Pisa)	1982-85	50,400	1:2800	18	2(11.1)
Holland	1978-81	70,459	1:2135	33	3(9.1)
Belgium	1974-86	125,255	1:3055	41	0(0.0)
Canada (Ontario)	1979-85	168,000	1:4000	42	3(7.1)
Sweden	1980-82	190,400	1:2800	68	5(7.4)
Switzerland (Central, West & South)	1977-82	203,268	1:3909	52	10(19.2)
France (Toulouse)	1977-87	289,224	1:4017	72	1(1.4)
Japan (Osaka)	1975-85	352,000[1]	1:4000	88	3(3.4)
Finland	1980-87	400,000[1]	1:3497	114	1(0.9)
Scotland	1979-86	455,000	1:3700	123	6(4.9)
USA - Georgia[2]	1979-86	678,300	1:6783	100	23(23)
USA - New England	1976-86	1,366,200	1:4600	297	30(10.1)
England, Wales & Northern Ireland	1982-84	1,941,146	1:3962	490	36(7.4)
Total		6,289,652	1:4090	1538	123(8.0)

[1]Estimated figures
[2]Data adapted from Fernhoff et al 1987[4]

Table 2. Incidence of Congenital Abnormalities in Screened Areas - Questionnaire Replies

Site	Incidence
New England	3%[1]
Ontario	0%[2]
Ontario	0.045%[3]
Ontario	0.083%[4]
Ontario	0.062%[5]
Wales	3.9%[1]
Pisa	0.66%[1]
Switzerland	unknown
Belgium	2%[1]
Toulouse	4.6%[1]
Sweden	1%[1]

[1]Major abnormality
[2]Cardiac
[3]Downs
[4]Spina Bifida
[5]Anencephaly

Table 3. Incidence of Congenital Abnormalities in Live Newborns

Abnormality	Incidence (%)	Reference
Major	1.5-2.5	Robertson & Forfar 1978[5]
Major	2.6	Holmes 1976[6]
All	7	McIntosh1954[7]
Major	1.7	Stevenson et al 1950[8]
Major	2.0	McKeown & Record 1960[9]
Minor	2.0-10.0	Robertson & Forfar 1978[5]
Cardiac	0.2-0.6	Robertson & Forfar 1978[5]
Cardiac	0.8	US Dept Health 1985[10]
Chromosomal	0.12	Robertson & Forfar 1978[5]
Spina Bifida	0.01-0.34	Robertson & Forfar 1978[5]

Acknowledgements

We thank all the responders to the questionnaire for their help.

preterm infants.[14] So far there are no data in human relating iodine status to the incidence of congenital abnormality.

References

1. R. Illig, Primary TSH screening, in: Congenital Hypothyroidism, eds. JH Dussault & P Walker, Marcel Dekker Inc., Butterworths, New York, Basel, London, p179-187 (1983).
2. T.P. Foley, Jr., Sporadic congenital hypothyroidism, in: Congenital Hypothyroidism, eds. JH Dussault & P Walker, Marcel Dekker Inc., Butterworths, New York, Basel, London, p231-259 (1983).
3. S. Bamforth, I. Hughes, J. Lazarus and R. John, Congenital anomalies associated with hypothyroidism, Arch Dis Child 61: 608-609 (1986).
4. P.M. Fernhoff, A.L. Brown, and L.J. Elsas. Congenital hypothyroidism: Increased risk of neonatal morbidity results in delayed treatment. Lancet i: 490-491 (1987).
5. J.G. Robertson, and J.O. Forfar, Prenatal paediatrics, in: Textbook of Paediatrics, eds. JO Farfar & GC Arneil, 2nd edition, Churchill Livingstone, Edinburgh, p76-103 (1978).
6. L.B. Holmes, The malformed newborn - practical perspectives, Boston Developmental Disabilities Council, Boston (1976).
7. R. McIntosh, K.K. Merritt, M.R. Richards, M.H. Samuels, and M.T. Bellows, The incidence of congenital malformations: A study of 5,964 pregnancies, Pediatrics 14: 505-522 (1954).
8. S.S. Stevenson, J. Worcester, R.G. Rice, 677 congenitally malformed infants and associated gestational abnormalities. Pediatrics 6: 37-50 (1950).
9. T. McKeown, R.G. Record, Malformations in a population observed for five years after birth, Ciba Foundation Symposium on Congenital Malformations, ed. Sir G. Wostenholme, J & A Churchill, London (1960).
10. Congenital Malformations Surveillance. CDC, PHS, USA Dept Health and Human Services 23-25 (1985).
11. S. Bamforth, J. Lazarus, I. Hughes, D. Bradley, R. John, and P. Harper, Congenital abnormalities in infants identified through the Welsh Regional Hypothyroid Screening Programme, in: Iodine Deficiency Disorders and Congenital Hypothyroidism, ed. G Medeiros-Neto, RMB Maciel, A Halpem, Ache Publ., Sao Paulo, p256-260 (1986). Report of Proc. of Satellite Meeting (held in Brazil 1985).
12. R. Hong, R. Gatti, J.C. Rathburn, R.A. Good, Thymic hypoplasia and thyroid dysfunction, N Engl J Med 470 (1970).
13. W.E. Decoteau, F. Murphy, J.W. Gerrard, Athyrotic cretinism associated with thymic hypoplasia and amyloidosis of lymphoid tissue. Can Med Assoc J 111: 51-53 (1974).
14. F. Delange, A. Dalhem, P. Bourdoux, R. Lagasse, Increased risk of primary hypothryoidism in preterm infants. Pediatrics 105: 462-469 (1984).

DISCUSSION

GLINOER (Brussels) : If hypothyroidism was, in some way, the cause for the associated abnormalities which you reviewed, then you would expect to see some relationship with the associated abnormalities found in cases with peripheral resistance to thyroid hormones. Is that the case ?

REFETOFF (Chicago) : The incidence of related congenital abnormalities in thyroid hormone resistance is probably higher than in normal newborns. However, strictly related to thyroid hormone deficiency, I can only think of deafness. All the other abnormalities are variable and probably reflect a more major defect at the chromosomal level.

SACK (Tel Aviv) : Is there a possible relationship between the cardiac abnormalities and hypothyroidism in some of those babies due to iodine overload from angiography ?

LAZARUS (Cardiff) : I have no data on iodine overload in these infants.

SACK (Tel Aviv) : One should look at those infants with low IQ even when they have been treated very early. We have seen 2 children with a syndrome that we call the "Cohen syndrome" : they have a short filtrum, a receding mandibule, some have myopia, and they have very long fingers and feet. The differential diagnosis is with the Marfan syndrome. They are also mentally retarded. Hence, hypothyroidism plus another syndrome that includes mental retardation could explain some of the infants with very low IQ in spite of early treatment.

PANTELAKIS (Athens) : We have a single screening center for the whole country and we have now screened the Greek population for 9 years. We have screened about 1,200,000 babies and the incidence of CH is 1/3500 neonates. I have personally followed 95 % of these infants with CH including now about 300 infants. My experience is that associated congenital abnormalities are not more frequent in these infants compared to the rest of the population.

ADDISSON (Manchester) : I'd like to ask a related question. In our series of patients with CH in the

Northwest of England, we noticed a very markedly increased death rate, not necessarily associated with congenital anomalies. We had 6 deaths among our first 100 patients and two additional deaths missed on screening for administrative reasons. The cause of death was highly variable. It occurred in infants with permanent CH, at several months of age. The infants were of course all treated except for the 2 missed cases.

RESISTANCE TO THYROID HORMONES AND SCREENING

FOR HIGH THYROXINE AT BIRTH

Samuel Refetoff*, Myriam Charbonneau+, David H. Sarne*, Junta Takamatsu* and Jean H. Dussault+

*Departments of Medicine and Pediatrics, The University of Chicago, Chicago, Illinois and +Laval University School of Medicine and the Quebec Network of Genetic Medicine, Quebec, Canada

Resistance to thyroid hormone encompasses a heterogeneous group of conditions characterized by reduced responses of target tissues to a supply of thyroid hormone which under normal circumstances would be excessive. The defects have always been partial, because if complete, they would probably be incompatible with life.

Twenty years ago, a syndrome combining deaf-mutism, stippled epiphyses, goiter and abnormally high level of serum protein-bound iodine (PBI), was described in two of five sibs born of consanguineous parents.[1] Another sib, with elevated PBI first measured at 8 weeks of age, and who appeared to have stigmata of congenital hypothyroidism, proved to be also affected.[1,2] Subsequent studies provided support to the postulated pathogenesis of the defect, namely, target organ refractoriness to thyroid hormone.[2] Indeed, the compounds circulating in the blood of the affected sibs, which were responsible for the high PBI level, proved to be the authentic L-isomers of thyroxine (T4) and triiodothyronine (T3). These hormones were synthesized and secreted in excessive amounts and penetrated the peripheral tissues where they were metabolized at an accelerated rate through the normal pathways. These high levels of endogenous thyroid hormone, as well as the administration of supraphysiological doses of L-T4 and L-T3, failed to produce the expected metabolic effects. As a matter of fact, in the presence of toxic levels of thyroid hormone, some tissues such as the bones, manifested signs of hypothyroidism while others, such as the heart, exhibited a mild degree of thyrotoxicosis.[2] However, by and large, the patients appeared to be euthyroid, as if the partial resistance to thyroid hormone was compensated by the excessive supply of the hormone.

Because the level of thyroid gland activity and the resultant amount of thyroid hormone produced and secreted are driven by pituitary thyrotropin (TSH), the supply of hormone to all body tissues is dependent upon the degree of resistance to thyroid hormone feedback regulation at the level of the thyrotroph. This explains, in part, the heterogeneity in the clinical manifestations and level of responses to thyroid hormone among various tissues in the same individual. The classification of thyroid hormone resistance syndromes (Table 1) is biased by the availability of sensitive tests which measure the responsiveness of the pituitary thyrotrophs to thyroid hormone and the scarcity of specific tests which provide such information at the level of most peripheral tissues.

Table 1. Syndromes of Tissue Resistance to the Action of Thyroid Hormone

1. Generalized resistance to thyroid hormone (GRTH)
2. Pituitary resistance to thyroid hormone (PRTH)
3. Peripheral tissue resistance to thyroid hormone (PTRTH)

In GRTH, the untreated patient should be euthyroid at the expense of high concentrations of thyroid hormone, the level of which is set by the thyrotrophs, sharing the defect. In reality, the level of resistance is not uniform in all peripheral tissues so that, for example, a subject with a normal metabolic rate may have growth retardation as well as tachycardia. In PRTH, the resistance to thyroid hormone is confined to the pituitary gland. As in GRTH, the feedback regulation is set up at a higher thyroid hormone level, which in this case produces symptoms and signs of thyrotoxicosis. The feature which sets PRTH apart from other forms of thyrotoxicosis is the preserved TSH response to thyrotropin releasing hormone (TRH). The existence of PTRTH, sparing the pituitary, can be seriously questioned not only because there have been no other reports since the publication of a single case in 1981,[3] but also because of the concept suggesting all tissues to be involved except for one. If the condition truly existed, its detection would be extremely difficult considering that affected individuals would have normal TSH and thyroid hormone levels. Thus, the diagnosis would have to be based on measuring the effects of thyroid hormone on peripheral tissues, which is not simple due to the scarcity of specific tests.

Because the clinical presentation of GRTH is variable, its identification requires a high degree of suspicion. Patients are most likely to present with a high serum T4 level in the absence of clear signs of thyrotoxicosis. A normal basal serum TSH level measured by one of the newer sensitive and more reliable methods[4] is a logical first step prior to the initiation of more involved diagnostic procedures. A suppressed basal TSH is compatible with one of the more common causes of hyperthyroidism such as Graves' disease, thyrotoxicosis facticia, or toxic nodular goiter. If TSH is not suppressed, the presence of inherited or acquired abnormalities of thyroid hormone-binding serum proteins should be considered. Measurement of serum T3 levels and testing family members at this point can be quite helpful. Total T3 would be high in TBG excess[5] and normal in familial dysalbuminemic hyperthyroxinemia.[6] These two conditions as well as the presence of thyronine autoantibodies should be further confirmed by specific tests. Reduced conversion of T4 to T3 may occasionally give rise to elevation of total and free T4 but not T3 levels. This can occur in a variety of nonthyroidal illnesses or during the administration of some drugs.[7] A familial form of hyperthyroxinemia presumably due to a defective 5'-monodeiodination has been also described.[8,9]

Having excluded thyroid hormone transport abnormalities, the combination of elevated serum levels of thyroid hormone and a non-suppressed TSH, narrows the differential diagnosis to a handful of relatively uncommon conditions termed by Gershengorn and Weintraub "inappropriate secretion of TSH".[10,11] These include autonomous hypersecretion of TSH by a pituitary adenoma, PRTH and GRTH. The distinction between TSH-producing adenomas and PRTH can be difficult since both give rise to thyrotoxicosis. The TSH response to the stimulation by TRH and suppression by T3, as well as measurement of the free α/total TSH ratio are most helpful.

Data from 45 patients with such tumors and 30 patients with PRTH (Table 2) show that TSH in PRTH is more likely to increase in response to TSH and to decrease in response to supraphysiological amounts of T3. More consistently, the αTSH/TSH ratio is above one in tumors but not in PRTH.

Table 2. Type and Frequency of Responses of TSH and Relative Elevation of αTSH in Patients with TSH-Induced Thyrotoxicosis

	TSH PRODUCING TUMOR		PITUITARY RESISTANCE TO THYROID HORMONE	
	Number	%	Number	%
Total Number of Patients	45	100	30	100
TRH Administration				
Stimulation	12/30[a]	40	15/17	88
No change	19/30	60	2/17	12
T3 Administration				
Suppression	5/14[b]	36	14/15	93
No change	9/14	64	1/15	7
αTSH/TSH Ratio[c]				
Above 1	15/16	94	2/13	15
Below 1	1/16	6	11/13	85

Data given as number and percent of each type of response over number of patients tested.
[a]Two patients tested during methimazole treatment, one of whom did not respond prior to treatment
[b]Only partial or transient suppression
[c] αTSH may be elevated in post-menopausal women

The most vexing problem in the diagnosis of GRTH is the lack of specific and reliable tests to assess the responses of peripheral tissues to thyroid hormone. Measurement of sex hormone-binding globulin (SHBG) in serum has been proposed as an index of hepatic response to the biological effect of the hormone.[12] While a positive correlation exists between SHBG levels and the free T4 index, only two-thirds of thyrotoxic patients with FT_4I levels comparable to those of patients with GRTH could be identified on the basis of their basal SHBG values, considerably depreciating the diagnostic value of this test.[13] However, when the incremental rise of SHBG (Δ SHBG) was measured in response to graded doses of T3, each given for three days, the following picture emerged (Fig. 1). All 21 subjects with hypo-, hyper-, and euthyroidism responded to both 100 and 200 μg of T3. In contrast, only 1 of 10 patients with GRTH had a comparable response and a paradoxical decline was observed in more than half.

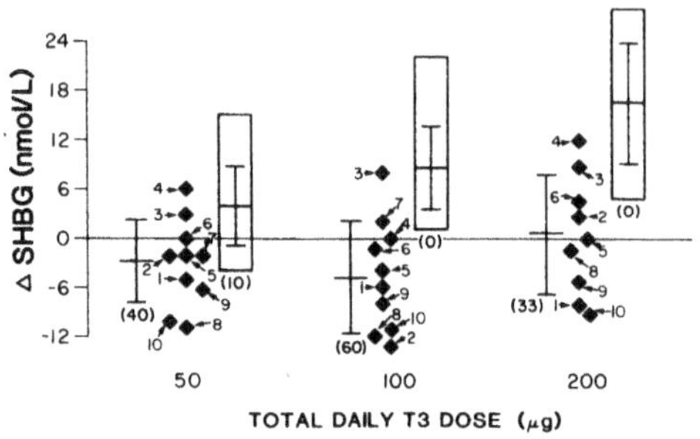

Fig. 1. Comparison of the serum SHBG responses to graded doses of T3 in patients with GRTH and nonresistant subjects. The numbered diamonds indicate individual patients with GRTH. The open bars represent the ranges of values for nonresistant subjects (combining hypothyroid, normal, and thyrotoxic subjects). The vertical lines represent the mean (±SD) values for the resistant and nonresistant subjects at each dose level. The numbers in parenthesis indicate the percentage of subjects in each group with a value significantly (>2 nmol/L) below that before T3 administration.

These results reinforce the diagnostic value of measurements carried out under the influence of exogenous hormone. In our experience, among 24 parameters measuring thyroid hormone action on peripheral tissues before and after the administration for three days of three doses of T3 (50, 100 and 200 μg/day), the more useful are the changes of the basal metabolic rate, serum SHBG, cholesterol, triglyceride, creatine phosphokinase, ferritin, urinary carnitine and magnesium and urea nitrogen and body weight under constant calorie and nitrogen intake. Interestingly, as with SHBG, paradoxical responses of ferritin and creatine phosphokinase to thyroid hormone were commonly observed in patients with GRTH.

The possibility that the syndrome represents a genetic defect at the level of the hormone effector was postulated even before evidence for the existence of a nuclear T3 receptor became available. Thus, it is not surprising that the hypothesis was tested soon after the development of a method to measure such a putative receptor. Using circulating mononuclear cells from an affected member of the original family with the syndrome, we were able to demonstrate a 10-fold decrease in T3-binding affinity by nuclear proteins.[14] Our subsequent studies combined to those of other laboratories showed that, in 32 affected subjects from 21 families studied, abnormalities in the affinity or capacity of the nuclear T3-receptor were present in mononuclear cells of one-third and in fibroblasts, in one-half of cases (Table 3).

Table 3. Studies on Nuclear T3 Binding in Patients with GRTH

TISSUE	AFFINITY (Ka)	CAPACITY (Bmax)	NUMBER OF CASES	NUMBER OF FAMILIES
Circulating mononuclear cells	↓	↑	4 } 32%	3
	N	↓	3 }	3
	N	N	15	8
	TOTAL		22	13[a]
Cultured skin fibroblasts	↓[b]	↓[b]	4 } 50%	2
	↓	↑	1 }	1
	N + 2nd binding site	N	2	2
	N	N	7	6
	TOTAL		14	11
Total Subjects and Families in Whom Nuclear T3 Binding was Studied			32[c]	21[c]

[a]One family had affected subjects with and without binding abnormalities
[b]Data generated from kinetic studies of T3 nuclear uptake
[c]Both mononuclear cells and fibroblasts were studied in 4 subjects from 3 families.
N = normal ↑ = increased ↓ = decreased

Attempts to establish alternative methods for the tissue diagnosis of resistance to thyroid hormone were met with difficulties. Measurements included, the respiratory control of mitochondria from muscle,[2] measurement of thyroid hormone effect on low density lipoprotein metabolism,[15] protein phosphorylation,[16] glycosaminoglycan,[17] and fibronectin[18] synthesis. Of these, the suppressive effect of T3 on fibronectin synthesis by cultured skin fibroblasts appears to be most promising. Indeed, this effect was blunted in fibroblasts obtained in 6 of 7 patients with GRTH as compared to those from normal subjects (Fig. 2). The one patient that showed a normal response had the most mild clinical manifestations of the syndrome.

The prevalence of the syndrome remains unknown. However, since its initial description in 1967, the number of patients uncovered has steadily increased to over 200. Of the 142 published cases, 126 belonged to 38 families with two or more members affected. Six cases appeared to be sporadic and in 10 others no information concerning inheritance was available because of adoption or failure to study family members. Seven were the product of consanguineous unions and two were affected twins. These data firmly establish the inherited nature of the disorder.

In an attempt to determine the incidence of resistance to thyroid hormone, 282,870 newborns from the Province of Quebec, Canada were screened for high T4 by analysis of blood collected on filter paper. Results are shown in Table 4. 1:7,645 had T4>20 μg/dl and 1:11,315, a T4>22 μg/dl. Follow-up serum samples were obtained at 18-35 days post partum from 20 of the 37 infants with T4 >20 μg/dl and their mothers.

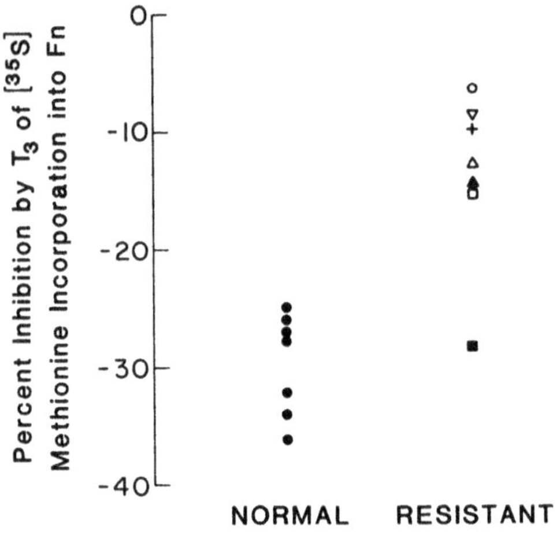

Fig. 2. The inhibitory effect of T3 on fibronectin (Fn) synthesis in fibroblasts from patients with GRTH compared to those from normal subjects. Cultures were exposed for 3 days to 10^{-7} M T3 (free T3, 2 x 10^{-9} M), and Fn synthesis was measured. The results are expressed as percent inhibition compared to that in the same fibroblasts cultured without added T3.

Table 4. Prevalence of High T4 at Birth According To Selection of Cut-Off Criteria

T4 CUT-OFF VALUE (μg/dl)	TOTAL CASES (Breakdown by Sex)	PREVALENCE	FOLLOW-UP CASES (%)
$\geq$20	37 (24F, 13M)	1:7,645	20 (54%)
$\geq$22	25 (16F, 9M)	1:11,315	12 (48%)
20 $\leq$T4<22	12 (8F, 4M)	1:23,573	8 (66%)

In four infants, a cause for the high T4 concentration at birth could be established. Three baby boys had familial TBG excess and one baby girl autoimmune neonatal thyrotoxicosis. Neonatal acquired TBG excess appeared to be the cause of hyperthyroxinemia in the remaining 16 infants since no inherited cause of TBG excess could be found by testing the parents and some of their sibs. The condition appears to be transient since in three of the infants followed over a period of 11 months, total T4 and TBG concentrations returned to normal. The possibility that the failure to uncover resistance to thyroid hormone is due to either a low prevalence of the disorder or to normal T4 levels at birth in patients with GRTH were considered. However, over the period of one year we have identified 2 newborns with GRTH in the state of Illinois where the birth rate is approximately 150,000/year. A retrospective analysis showed that both infants had high T4 values with normal TSH at birth. These observations suggest that our experience in the Province of Quebec reflects a difference in the prevalence of the disorder among ethnic groups. In fact, not a single patient of the 200 cases with GRTH so far identified were of French extraction.

ACKNOWLEDGEMENT - Supported in part by U.S. Public Health Service grant DK 15070.

References

1. S. Refetoff, L. T. DeWind, and L. J. DeGroot, Familial syndrome combining deaf-mutism, stippled epiphyses, goiter and abnormally high PBI: Possible target organ refractoriness to thyroid hormone, J. Clin. Endocrinol. Metab., 27:279 (1967).
2. S. Refetoff, L. J. DeGroot, B. Benard, and L. T. DeWind, Studies of a sibship with apparent hereditary resistance to the intracellular action of thyroid hormone, Metabolism, 21:723 (1972).
3. M. M. Kaplan, S. L. Swartz, and P. R. Larsen, Partial peripheral resistance to thyroid hormone, Am. J. Med., 70:1115 (1981).
4. G. G. Klee and I. D. Hay, Assessment of sensitive thyrotropin assays for an expanded role in thyroid function testing: Proposed criteria for analytic performance and clinical utility, J. Clin. Endocrinol. Metab., 64:461 (1987).
5. W. A. Burr, D. B. Ramsden, and R. Hoffenberg, Hereditary abnormalities of thyroxine-binding globulin concentration, Quart. J. Med., 49:295 (1980).
6. R. Rajatanavin and L. E. Braverman, Euthyroid hyperthyroxinemia, J. Clin. Invest., 6:493 (1983).
7. L. Wartofsky and K. D. Burman, Alterations in thyroid function in patients with systemic illness: The "euthyroid sick syndrome". Endocrine Rev., 3:164 (1982).
8. H. R. Maxon, K. D. Burman, B. N. Premachandra, I.-W. Chen, A. Burger, P. Levy, and L. P. Georges, Familial elevation of total and free thyroxine in healthy, euthyroid subjects without detectable binding protein abnormalities, Acta Endocrinol., 100:224 (1982).
9. N. Kleinhaus, J. Faber, L. Kahana, J. Schneer, and M. Scheinfeld, Euthyroid hyperthyroxinemia due to a generalized 5'-deiodinase defect, J. Clin. Endocrinol. Metab., 66:684 (1988).
10. M. C. Gershengorn and B. D. Weintraub, Thyrotropin-induced hyperthyroidism caused by selective pituitary resistance to thyroid hormone: A new syndrome of "inappropriate secretion of TSH", J. Clin. Invest., 56:633 (1975).
11. B. D. Weintraub, M. C. Gershengorn, I. A. Kourides, and H. Fein, Inappropriate secretion of thyroid stimulating hormone, Ann. Intern. Med., 95:339 (1981).
12. D. S. Cooper, P. W. Ladenson, B. C. Nisula, J. F. Dunn, E. M. Chapman, and E. C. Ridgway, Familial thyroid hormone resistance, Metabolism, 31:504 (1982).
13. D. H. Sarne, S. Refetoff, R. L. Rosenfield, and J. P. Farriaux, Sex hormone-binding globulin in the diagnosis of peripheral tissue resistance to thyroid hormone: The value of changes after short term triiodothyronine administration, J. Clin. Endocrinol. Metab., 66:740 (1988).
14. J. Bernal, S. Refetoff, and L. J. DeGroot, Abnormalities of triiodothyronine binding to lymphocyte and fibroblast nuclei from a patient with peripheral tissue resistance to thyroid hormone action, J. Clin. Endocrinol. Metab., 47:1266 (1978).
15. A. Chait, R. Kanter, W. Green, and M. Kenny, Defective thyroid hormone action in fibroblasts cultured from subjects with the syndrome of resistance to thyroid hormone, J. Clin. Endocrinol. Metab., 54:767 (1982).
16. H. Nakamura, S. Refetoff, L. J. DeGroot, and A. L. Horwitz, T3 effects on protein phosphorylation in cultured skin fibroblasts from normal subjects and patients with peripheral tissue resistance to thyroid hormone, The Second Asia and Oceania Thyroid Association Meeting, Tokyo, Japan, August 19-22, 1982.

17. Y. Murata, S. Refetoff, A. L. Horwitz, and T. J. Smith, Hormonal regulation of glycosaminoglycan accumulation in fibroblasts from patients with resistance to thyroid hormone, *J. Clin. Endocrinol. Metab.*, 57:1233 (1983).
18. P. Ceccarelli, S. Refetoff, and Y. Murata, Resistance to thyroid hormone diagnosed by the reduced response of fibroblasts to the triiodothyronine induced suppression of fibronectin synthesis, *J. Clin. Endocrinol. Metab.*, 65:242 (1987).

IODINE NUTRITION AND CONGENITAL HYPOTHYROIDISM

François Delange

University of Brussels, Belgium. Depts. of Pediatrics and Radioisotopes, Hôpital Saint-Pierre, 322, rue Haute
B - 1000 - Brussels, Belgium

INTRODUCTION

In Western countries with an adequate dietary supply of iodine the incidence of congenital hypothyroidism is 1/3000 to 1/4000 newborn infants (1). This frequency justifies systematic screening, which is extremely efficient and has an elevated benefit-cost ratio (2-3). In such countries CH is due principally to congenital anomalies in the development of the thyroid (4); it is unpredictable and unpreventable.

In contrast, in conditions of severe iodine deficiency and endemic goiter the incidence of congenital hypothyroidism can be as high as 1/10 newborns (5). And yet screening programs followed by individual therapy of the detected cases are virtually non-existant. This is due to the socio economic limitations usually characteristic of developing countries with severe iodine deficiency. In endemic goiter areas congenital hypothyroidism is due to iodine deficiency. Consequently, it is potentially predictable and preventable. However, little has been achieved on a world scale to eradicate iodine deficiency and its consequences (6).

Finally, whathever its origin, prolonged thyroid failure occurring in humans during fetal and early postnatal life can result in irreversible brain damage and mental retardation (7).

The purpose of this paper is to review available data on the relationship between iodine nutrition, essentially iodine deficiency, and neonatal hypothyroidism in humans, seen essentially in the light of the problems raised in public health.

SEVERE IODINE DEFICIENCY AND CONGENITAL HYPOTHYROIDISM IN ENDEMIC GOITER

The part of the world where the most dramatic picture of neonatal hypothyroidism due to iodine deficiency in the environment has been reported is the Ubangi endemic goiter area in Central Africa (8, 9). It is very isolated and poor. The economy is self subsistent. The diet is monotonous and severely deficient in iodine. Goiter affects 75 % of the population and the prevalence of visible goiter is 51 %.

The iodine intake is only 15 ug/day, one tenth of the minimum recommended dietary allowance. The ffects of iodine deficiency are aggravated by a state of chronic SCN overload resulting from consumption of poorly detoxified cassava, rich in goitrogenic compounds.

In these extreme conditions, clinically euthyroid adults have low serum concentrations of T4, elevated TSH but also elevated T3, indicating moderate thyroid failure partly compensated by elevated T3. Some adults, however, exhibit a biochemical picture of extremely severe thyroid failure with almost undetectable serum levels of T4 and T3 and extremely elevated TSH (10). They are clinically grossly hypothyroid with apathy, myxedema causing puffy features, and all the signs and symptoms of thyroid failure.

Additional clinical features suggest that extreme hypothyroidism was already present during early life : the patients show marked retardation of growth and sexual maturation and a marked delay in bone maturation. They are also severely mentally retarded. On the basis of these observations in adult cretins, the hypothesis was proposed that myxedematous cretinism in Zaire was the long term consequence of thyroid failure occurring during the perinatal life (11).

This was confirmed by the results of pilot studies on neonatal thyroid screening in the Ubangi area (12, 13). The serum concentrations of TSH and T4 on cord blood in unselected newborns showed marked variability between individuals and many values were distinctly abnormal. As many as 11.2 % of the newborns had cord serum TSH above 100 uU/ml and cord serum T4 below 3 ug/dl, indicating severe congenital hypothyroidism. In contrast, such a picture was observed in only .025 % of a newborn population in Brussels.

A similar situation was observed for blood samples collected on filter paper on day 5 and at the time of control examinations (14) : 9 % of 5 day-old infants had serum TSH above 100 uU/ml and T4 below 3 ug/dl. This indicates that the observations made on cord blood did not represent an artefact due for example to the stress of delivery.

The correction of maternal iodine deficiency during pregnancy by a single injection of iodized oil prevented the occurrence of congenital hypothyroidism in the newborns (15, 16). This confirms the role played by iodine deficiency in the incidence of congenital hypothyroidism in the Ubangi area.

The Himalayan belt in northern India is another part of the world where severe congenital hypothyroidism resulting from iodine deficiency has been evidenced. In the Himalayas, myxedematous cretinism also exists although to a lesser extent than in Zaire (17). Kochupillai et al. (18) reported that in such areas, 4 % of the newborns had cord serum TSH above 100 uU/ml and cord serum T4 below 3 ug/dl.

One of the major aspects of neonatal hypothyroidism in endemic goiter areas is that, in addition to clearcut cases of severe cretinism, iodine deficiency results in impairment of intellectual capacity in clinically euthyroid individuals who do not exhibit the other signs of endemic cretinism (Review in 5, 7). Figure 1 shows that individuals born in iodine deficient areas to iodine deficient mothers have a frequency distribution of IQ shifted towards low values as compared to controls with the same socio economic characteristics but with no iodine deficiency. This indicates that overt cretinism is only the top of the iceberg of the long term consequences of iodine deficiency during the perinatal period on brain development.

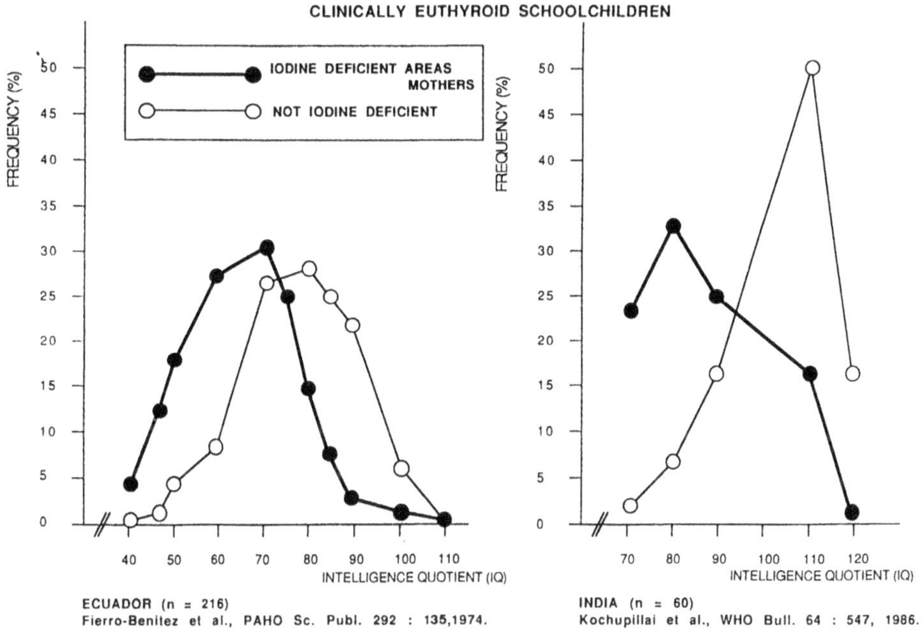

Figure 1. Comparison of the frequency distributions of Intelligence Quotients (IQ) in clinically euthyroid, iodine deficient schoolchildren, and in non iodine deficient controls in Ecuador S.A. (19) and India (20).

IODINE DEFICIENCY AND NEONATAL HYPOTHYROIDISM IN EUROPE

Impairment of neonatal thyroid function due to iodine deficiency is not limited to exotic areas in developing countries but is a public health issue in some parts of Europe today.

For example, comparative studies were conducted in three areas in the hills surrounding the volcano Etna in Sicily (21) : the city of Catania, at sea level with no goiter and a normal iodine intake (control area), a moderate endemic goiter area and a severe one, with decreasing iodine intake. In spite of the severity of the endemias and the degree of iodine deficiency, thyroid function in adults, evaluated on the basis of serum TSH and T4, was barely modified. In contrast, serum TSH and T4 in cord blood were shifted towards high and low values respectively in the endemic areas, with an increase in the recall rate under suspicion of congenital hypothyroidism from .16 % in the control area up to 9.78 % in the hyperendemic area. These results suggest that newborns are more sensitive than adults to the effect of iodine deficiency.

A recent publication by the European Thyroid Association (22) summarized the data presently available on iodine intake in Europe. It indicated that in many European regions or countries, iodine intake was far below the value of 100 ug per day classically recommended (23) and was accompanied by a significant problem of goiter, especially in some parts of Germany, Spain and Italy.

Table 1. Comparison of indexes of iodine nutrition in adults and newborns and of the iodine content of breastmilk in various parts of Europe. Compiled from Scriba et al. (22), Delange et al. (24), Heidemann et al. (25), Mallol et al. (26), Etling and Gehin-Fouque (27), Delange et al. (28) and Siracusano et al. (29).

COUNTRY or REGION	ADULTS		INFANTS AT SCREENING FOR CH	
	Urinary excretion of I (ug/day)	I content of breastmilk (ug/dl). Mean ± SEM(n)	City	Median urinary I conc. (ug/dl)
The Netherlands	70 - 120		Rotterdam	16.2
Finland	250		Helsinki	11.2
Sweden	200	9.3 (60)	Stockholm	11.0
Sicily (non endemic)	113		Catania	7.1
Switzerland	93		Zurich	6.2
Spain (Madrid)	89	7.7 ± 0.9 (69)		
France	33 - 75	8.2 ± 0.5 (68)	Lille	5.8
Belgium (Brussels)	51	9.5 ± 0.6 (91)	Brussels	4.8
Italy (Rome)	37		Rome	4.7
Fed. Rep. Germany North	35		Berlin	2.8
South	20	2.5 (41)	Freiburg	1.2
Sicily (endemic)	22	2.7 ± 0.3 (47)		
Germ. Dem. Rep. South	16	1.2 ± 0.1 (55)	Iena	0.8

These data were recently updated. Table 1 compares the data presently available in Europe on an evaluation of iodine intake in adults, the iodine content of breastmilk and the urinary concentrations of iodine in healthy newborn infants on day 5, used as an index of their state of iodine nutrition. There are marked regional variations in the iodine supply in adults in Europe, and several regions, especially southern Germany, are clearly iodine deficient. There are marked differences in the iodine content of breastmilk and in the iodine supply of newborn infants in Europe (24), which are remarkably parallel to the changes in the iodine intake of the general population. The results in infants indicate that, in spite of the massive use of industrialized formula milk for newborns, environmental factors still play an important role in iodine nutrition in newborn infants in Europe today.

It has been shown that the changes in iodine intake in various parts of Europe have no influence on the incidence of permanent sporadic congenital hypothyroidism, which is between 1/3000 to 1/4000 throughout Europe (24). In contrast, the changes are accompanied by marked differences in the frequency of serum TSH above 50 uU/ml on day 5, that is in the recall rate of newborn populations at screening.

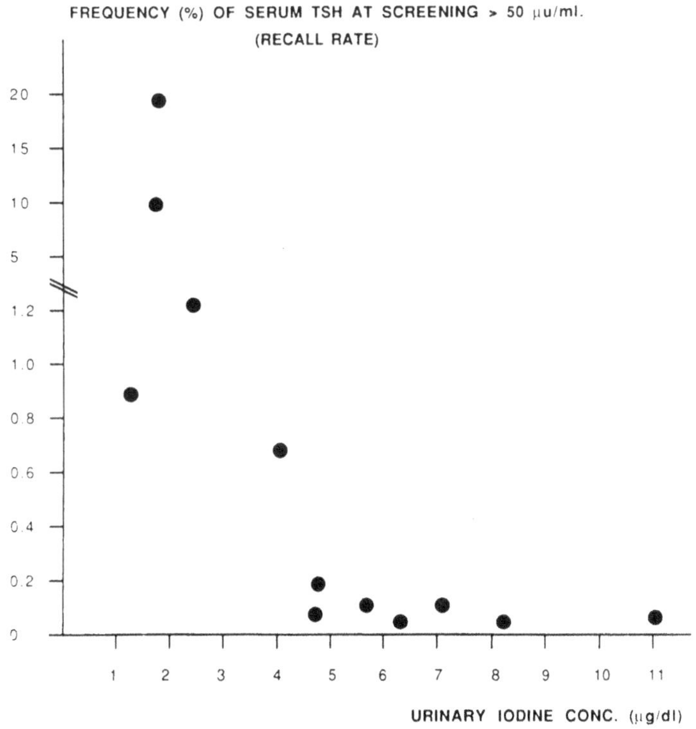

Figure 2. Relationship between the recall rate at screening for congenital hypothyroidism and an index of iodine nutrition in newborn populations.

Figure 2 shows that there is a clearcut inverse relationship between the recall rate at screening and the iodine supply of newborn populations. Each point results from the analysis of 50 to 200 urine samples in unselected healthy newborns and in screening data from 50000 to 200000 unselected newborns, except in areas of severe endemic goiter where the samples were much smaller. The recall rate was below .2 %, usually below .1 %, as long as the urinary iodine remained above a critical threshold of about 4 to 5 ug/dl. When this threshold was reached, there was a sudden and dramatic increase in the recall rate with a 100 fold increase in the extreme conditions found in Central Africa. In non endemic areas the control examinations were normal in all recalled infants except in the very few with confirmed sporadic hypothyroidism, and, as stated earlier, the incidence of permanent congenital hypothyroidism was similar in all areas. This indicates that the frequency of very transiently elevated TSH at screening, i.e. the "false positives", is inversely related to the iodine intake of newborn populations and represents very transient alteration of thyroid function. In contrast, in spite of the changes in the iodine supply, thyroid function in adults is unmodified, except in areas with overt endemic goiter.

Consequently, these data clearly confirm that newborn infants are more sensitive than adults to the antithyroid effects of iodine deficiency.

Table 2 shows that relatively iodine deficient newborn infants are also particularly sensitive to the antithyroid effects of iodine excess.

The table shows that one single cutaneous application of Povidone-Iodine in mothers at the time of delivery for epidural anesthesia or cesarian section results in a marked increase in the iodine content of breastmilk on day 5. This indicates that iodine, possibly iodide, is liberated from the disinfectant and is concentrated by the mammary gland. Breastfed infants born to these iodine overloaded mothers have a sharp increase in their urinary iodine, indicating that they are also iodine overloaded. This overload results in a thirty-fold increase in the recall rate at screening. Again, hyperthyrotropinemia was very transient because the control examinations were normal in all recalled infants.

These data indicate a minor and transient Wolff Chaikoff (32) effect in borderline iodine deficient infants suddenly submitted to iodine overload.

As indicated by several groups (33-35), iodine overload is a frequent and often unrecognized cause of transient alteration of thyroid function in newborn infants in Europe, especially when the baseline iodine intake is low.

PUBLIC HEALTH DIMENSIONS OF IODINE DEFICIENCY DISORDERS

It is now established that at least eight hundred million people are exposed to iodine deficiency in Asia, Africa and Latin America and are at risk of developing some of the consequences of iodine deficiency (36). Among these, 3.15 million are affected by cretinism, the extreme form of irreversible mental retardation resulting from iodine deficiency. In addition, an even larger but unknown number of individuals are irreversibly mentally handicapped because of iodine deficiency. Figures from Europe should increase the number of individuals at risk by another 20 to 30 million.

The disorders resulting from iodine deficiency are theoretically entirely preventable. In practice, they are not because of the great gap between knowledge and application which is a characteristic of developing countries with chronic nutritional disorders.

Table 2. Influence of maternal iodine overload on the state of iodine nutrition and recall rate at screening for congenital hypothyroidism in breastfed infants. * Mean ± SEM (n) ** Median. Compiled from Chanoine et al. (30, 31).

GROUPS	MOTHERS	BREASTFED INFANTS DAY 5	
	I content of breastmilk (μg/dl)	Urinary I conc. (μg/dl)	Frequency of serum TSH > 50 μU/ml (Recall rate,%)
No iodine overload			
Normal controls	9.5 ± 0.6 (91)*	4.8 (196)**	0.1 (2358)
Maternal iodine overload			
Skin disinfection with PVP-I for			
Epidural anesthesia	16.0 ± 1.4 (39)	27.5 ± 4.8 (27)	3.2 (980)
Cesarian section	126.7 ± 42.1 (10)	183.5 ± 57.8 (11)	2.7 (321)

Because of the immensity of the problem on a world scale and in order to bridge this gap, an international body was created a couple of years ago (6). It is the International Council for Control of Iodine Deficiency Disorders (ICC IDD). It has an Executive Committee, a board of about 30 members and regional coordinators for the different WHO regions of the world. The primary objective of ICC IDD is to promote effective public health programs in order to eradicate iodine deficiency disorders and consequently to improve quality of life, educability of children and productivity of labour.

CONCLUSIONS

Severe iodine deficiency as seen in some developing countries can induce neonatal hypothyroidism. The degree of thyroid failure is similar to the one observed in sporadic congenital hypothyroidism but its frequency is up to 300 times higher. The long term consequences are endemic cretinism and endemic mental retardation.

Moderate iodine deficiency as seen in some European countries can induce transient alterations of thyroid function and regulation in newborns, inducing an unnecessarily elevated recall rate at screening.

Neonates are more sensitive than adults to the antithyroid effects of iodine deficiency. Iodine deficient newborns are hypersensitive to the antithyroid effects of iodine excess.

The long term consequences of a state of moderate iodine deficiency during the neonatal period on mental development are not evaluated.

Appropriate action is urgently needed to correct both severe and moderate iodine deficiency on a world scale. However, endemic iodine excess and its possible consequences should be avoided.

REFERENCES

1. G.N. Burrow and J.H. Dussault, Neonatal Thyroid Screening, Raven Press, New York, pp. 1-322 (1980).
2. C. Laberge, Organization and cost benefits of mass screening programs, in : Neonatal Thyroid Screening, G.N. Burrow and J.H. Dussault eds., Raven Press new York, pp. 189-197 (1980).
3. F. Caucal, T. Lebrun, J.C. Sailly, J.P. Farriaux and J.L. Dhondt, Une évaluation économique du rapport bénéfices/coûts du dépistage néonatal de la phénylcétonurie et de l'hypothyroïdie. Centre de Recherches Economiques, Sociologiques et de Gestion (CRESGE); Association Régionale pour le Dépistage et la Prévention des maladies métaboliques et des Handicaps de l'enfant Publ., Lille, pp. 1-69 (1987).
4. P. Rochiccioli, Thyroid dysgenesis, in : Pediatric Thyroidology, F. Delange, D.A. Fisher and P. Malvaux eds., Karger Publ., Basel pp. 154-173 (1985).
5. F. Delange, Adaptation to iodine deficiency during growth : etiopathogenesis of endemic goiter and cretinism, in : Pediatric Thyroidology, F. Delange, D. Fisher and P. Malvaux, eds, Karger Publ., Basel, pp. 295-326 (1985).
6. B.S. Hetzel, J.T. Dunn and J.B. Stanbury, The prevention and control of Iodine Deficiency Disorders. Elsevier Publ., Amsterdam, pp. 1-354, (1987).

7. F. Delange, Relation of thyroid hormones to human brain development, in : Fetal brain disorders. Recent approaches to the problem of mental deficiency, B.S. Hetzel and R.M. Smith eds, Elsevier North Holland Publ., pp. 285-296 (1981).
8. A.M. Ermans, N.M. Nbulamoko, F. Delange and R. Ahluwalia, Role of cassava in the etiology of endemic goitre and cretinism. International Development Research Centre, Ottawa, pp. 1-182 (1980).
9. F. Delange, F.B. Iteke and A.M. Ermans, Nutritional factors involved in the goitrogenic action of cassava. International Development Research Centre Publ., Ottawa, pp. 1-100 (1982).
10. R. Lagasse, P. Bourdoux, P. Courtois, P. Hennart, G. Putzeys, C. Thilly, M. Mafuta, Y. Yunga, A.M. Ermans and F. Delange, Influence of the dietary balance of iodine/thiocyanate and protein on thyroid function in adults and young infants, in : Nutritional factors involved in the goitrogenic action of cassava, F. Delange, F.B. Iteke and A.M. Ermans eds, International Development Research Centre Publ., Ottawa, pp. 34-39 (1982).
11. F. Delange, A.M. Ermans, H.L. Vis and J.B. Stanbury, Endemic cretinism in Idjwi Island (Kivu Lake, Republic of the Congo) J. Clin. Endocrinol. Metab. 34 : 1059-1066 (1972).
12. C.H. Thilly, F. Delange, R. Lagasse, P. Bourdoux, L. Ramioul, H. Berquist and A.M. Ermans, Fetal hypothyroidism and maternal thyroid status in severe endemic goiter, J. Clin. Endocrinol. Metab. 47 : 354-360 (1978).
13. F. Delange, C. Thilly, P. Bourdoux, P. Hennart and A.M. Ermans, Influence of dietary goitrogens during pregnancy in humans on thyroid function of the newborn, in : Nutritional factors involved in the goitrogenic action of cassava, F. Delange, F.B. Iteke and A.M. Ermans eds, International Development Research Centre Publ., Ottawa, pp. 40-50 (1982).
14. P. Courtois, F. Delange, P. Bourdoux and A.M. Ermans, Significance of neonatal thyroid screening tests in severe endemic goiter (Abstract n° 81), Ann. Endocrinol. 43 : 51A (1982).
15. C.H. Thilly, F. Delange, J. Golstein-Golaire and A.M. Ermans, Endemic goiter prophylaxis by iodized oil : a reassessment, J. Clin. Endocrinol. Metab. 36 : 1196-1204 (1973).
16. F. Delange, Anomalies in physical and intellectual development associated with severe endemic goiter, in : Towards the eradication of endemic goiter, cretinism and iodine deficiency, J.T. Dunn, E.A. Pretell, C.H. Daza and F.E. Viteri eds., PAHO Scientific Publication 502, Pan American Health Organization, Washington DC, pp. 49-67 (1986).
17. F. Delange, P. Valeix, P. Bourdoux, R. Lagasse, J.P. Ermans, C. Thilly and A.M. Ermans, Comparison of the epidemiological and clinical aspects of endemic cretinism in Central Africa and in the Himalayas, in : Fetal brain disorders. Recent approaches to the problem of mental deficiency., B.S. Hetzel and R.M. Smith eds, Elsevier North Holland Publ., Amsterdam, pp. 243-263 (1981).
18. N. Kochupillai, N. Gupta, M.M. Godbole and M.M.S. Ahuja, Qauntitative endocrine abnormalities in cord blood of newborns from severely iodine deficient regions with prevalence of endemic cretinism, in : Iodine nutrition, thyroxine and brain development, N. Kochupillai, M.C. Karmakar and V. Ramalingaswami eds., Tata McGraw-Hill publ., New Dehli, pp. 200-204 (1986).
19. R. Fierro-Benitez, I Ramirez, E. Estrella and J.B. Stanbury, The role of iodine deficiency in intellectual development in an area of endemic goiter, in : Endemic goiter and cretinism : continuing threats to world health, J. Dunn and G.A. Medeiros-Neto eds., Pan American Health Organization Scientific Publication n° 292, pp. 135-142 (1974).

20. N. Kochupillai, C.S. Pandav, M.M. Godbole, M. Metha and M.M.S. Ahuja, Iodine deficiency and neonatal hypothyroidism, Bull. WHO 64 : 547-551 (1986).
21. L. Sava, F. Delange, A. Belfiore, F. Purrella and R. Vigneri, Transient impairment of thyroid function in newborn from an area of endemic goiter, J. Clin. Endocrinol. Metab. 59 : 90-95 (1984).
22. Report of the Subcommittee for the study of endemic goiter and iodine deficiency of the European Thyroid Association, Goitre and iodine deficiency in Europe, Lancet i : 1289-1293 (1985).
23. F. Delange, Physiopathology of iodine nutrition, in : Trace elements in nutrition of children, R.K. Chandra ed., Nestlé Nutrition Vevey, Raven Press, New York Publ., pp. 291-299 (1985).
24. F. Delange, P. Heidemann, P. Bourdoux, A. Larsson, R. Vigneri, M. Klett, C. Beckers and P. Stubbe, Regional variations of iodine nutrition and thyroid function during the neonatal period in Europe, Biol. Neonate 49 : 322-330 (1986).
25. P.H. Heidemann, P. Stubbe, H. Thal, R. Hellström, N. Skanke and A. Larsson, Influence of iodine prophylaxis on thyroid function and iodine excretion in newborn infants and their mothers : comparison between Sweden and Germany, Acta Endocrinol. Vol. 3 Suppl. 274, pp. 47-48 (1986).
26. J. Mallol, G. Morreale de Escobar and F. Escobar del Rey, Thyroid hormone and iodine intakes of bottle-fed versus breastfed babies, Endocrinologia 35 : 12-20 (1988).
27. N. Etling and F. Gehin-Fouque, Iodinated compounds and thyroxine binding to albumin in human breast milk, Pediatr. Res. 18 : 901-903 (1984).
28. F. Delange, P. Bourdoux, J.P. Chanoine and A.M. Ermans, Physiopathology of iodine nutrition during pregnancy, lactation and early postnatal life, in : Vitamins and minerals in pregnancy and lactation, H. Berger ed., Raven Press Publ., New York, pp. 205-214 (1988).
29. M.F. Siracusano, F. Vermiglio, S. Battiato, F. Ardita, F. De Luca and F. Trimarchi, Concentrazione di iodo nel latte materno in due aree a differente apporto iodice in Sicilia, Quarte giornate italiane della tiroide, Messina, Taormina 11-13/12/86, Abstract p. 55.
30. J.P. Chanoine, P. Bourdoux, A. Pardou and F. Delange, Iodinated skin disinfectants in mothers at delivery and impairment of thyroid function in their breastfed infants, in : Frontiers of thyroidology, G.A. Medeiros-Neto and E. Gaitan eds., Plenum Press New York Publ., pp. 1055-1060 (1986).
31. J.P. Chanoine, M. Boulvain, P. Bourdoux, A. Pardou, M.V. Van Thi, A.M. Ermans and F. Delange, Increased recall rate at screening for congenital hypothyroidism in breastfed infants born to iodine overloaded mothers, Arch. Dis. Child. : In Press.
32. J. Wolff, Iodide goiter and the pharmacologic effects of excess iodide, Amer. J. Med. 47 : 101-124 (1969).
33. J.P. Chabrolle and A. Rossier, Goitre and hypothyroidism in the newborn after cutaneous absorption of iodine, Arch. Dis. Child. 53 : 495-498 (1978).
34. D. L'Allemand, A. Grüters, P. Heidemann and P. Schürnbrand, Iodine-induced alterations of thyroid function in newborn infants after prenatal and perinatal exposure to Povidone iodine, J. Pediatr. 102 : 935-938 (1983).
35. D. L'Allemand, A. Grüters, P. Beyer and B. Weber, Iodine in contrast agents ans skin disinfectants is the major cause of hypothyroidism in premature infants during intensive care, Hormone Res. 28 : 42-49 (1987).
36. B.S. Hetzel, An overview of the prevention and control of Iodine Deficiency Disorders, in : The prevention and control of iodine deficiency disorders, B.S. Hetzel, J.T. Dunn and J.B. Stanbury eds., Elsevier Publ., Amsterdam, pp. 7-31 (1987).

DISCUSSION

MORREALE (Madrid) : My question concerns the areas with severe iodine deficiency where babies are born with low T4 and high TSH and have hypothyroidism. If they are treated early with T4, do you have information on the CNS defects corrected or irreversably present ?

DELANGE (Brussels) : There is no data to answer your question. One might suspect, by similarity with sporadic congenital hypothyroidism, that the long term prognosis is the same. Also, in some cases, neonatal hypothyroidism is only transient, spontaneously reverting to normal. One last point is that when you correct iodine deficiency in the mothers, the babies are born normal and develop normally.

ERMANS (Brussels) : I was interested by the correlation between the recall rate and the magnitude of urinary iodine excretion. In fact, the recall rate was only slightly modified for very large changes in urinary excretion. Furthermore, the recall rate dramatically rises for a concentration which is below 4 μg/dl. It suggests that for a critical degree of iodine deficiency there is a complete inadaptation of the system. Would you care to comment about this puzzling point because adaptation to iodine deficiency usually is considered a progressive phenomenon without a threshold.

DELANGE (Brussels) : This threshold of about 4 to 5 μg/dl in urine corresponds to an iodine intake of about 50 μg per day in adults. Adaptation to iodine deficiency means increased thyroidal clearance of iodide with the attempt to maintain a normal absolute iodine uptake. In adults, the iodine content of the thyroid is the same in Brussels and in the United States. In newborns, it is not the same. We have data (not yet published) on the iodine content of the thyroid gland of newborn infants who died for non-thyroid reasons. Comparing data from Toronto and Brussels (there are also other data from Cardiff and from East Germany), there are marked differences. The iodine content of newborn thyroid glands in Brussels is 3 times lower than in Canada. I think therefore that adults have their whole lifetime to adapt to iodine deficiency while newborns do not and they rapidly reach a critical point with increased TSH.

WILLGERODT (Leipzig) : In our area with moderate iodine deficiency, we have had an incidence in newborns of 4 % of visible goiters during recent years. In 45 % of these newborns, we find high TSH values and reduced T4 levels. Iodine excretion in healthy adults in our area averages 32 µg/g creatinine and in newborns 1.7 µg/dl. For the last 3 years we have had iodine prophylaxis in GDR and during the past 2 years, we have not seen any visible goiter in newborn infants and iodine excretion has increased to 3.2 µg/dl. These data emphazize that the newborn thyroid has an increased sensitivity to the lack of iodine.

CHEN (Taipei) : Could you comment about possible iodine overload in countries where the salt is iodized ?

DELANGE (Brussels) : This question raises many problems related to the administration of iodide by means of iodized salt : how much do you add; in what form; what is the stability; how is it to be distributed, etc... ? There is a risk of a sudden increase in iodine intake in these populations, and it has been documented that during the first few months or years after a sudden increase in the iodine intake in a general population you can induce "epidemics" of hyperthyroidism. This hyperthyroidism is transient and disappears within 2 years. To answer your question, then, there is some risk but is relatively minor. This risk must be balanced with the risks of not correcting hypothyroidism.

MORREALE (Madrid) : What is the minimal iodine intake to recommend for the preparation of formula milk ?

DELANGE (Brussels) : The classical recommendation for newborn infants is 7 µg iodine/kg body weight/day, which represents a concentration in formula milk of 5 µg/dl. Some have less, very few have more and most formulae have irregular iodine contents. My personal opinion is that 5 µg/dl is too low, based on the fact that iodine balance studies made in newborn infants (term and preterm) show that in order to have a positive iodine balance the requirement is in the range of 30 µg/kg/day. Below that, a preterm infant does not have a positive iodine balance because of relatively high urine iodine loss. For full-term infants, I think we should increase the 5 µg/dl formula milk to at least 7 to 10 µg/dl to ensure that a positive iodine balance is reached and maintained within safe limits.

ORGIAZZI (Lyon) : In the Zaire study, I noticed that the urinary excretion of thiocyanate in hypothyroid children was higher than in the euthyroid children, despite living in the same environment. Do you suggest that in addition to iodine deficiency, there could be a defect in the clearance of some goitrogen in children with more profound hypothyroidism ?

DELANGE (Brussels) : No. I suggest that the cretins have a higher dietary intake of cassava just because they sit there and eat whatever comes in the reach of their hands ! It would therefore be a consequence of hypothyroidism, rather than a possible cause. On the other hand, it is not impossible that thiocyanate has a teratogenic effect by itself. However, we have no hard data on this.

EKINS (London) : Do you have any data on TSH values in pregnant women in these regions as compared to normal women?

DELANGE (Brussels) : Yes. Claude Thilly has reported on that in the JCEM : TSH levels in pregnant women were elevated. In other words, these women were "hypothyroid" during pregnancy.

EKINS (London) : We have some limited data from Yugoslavia using an "ultrasensitive" TSH assay indicating that during pregnancy the blood TSH concentration fell rather remarkably, as compared with non-pregnant women in the same region. This suggests some other source of thyroid stimulation.

DELANGE (Brussels) : In Zaire, we did not observe a difference in serum TSH in pregnant and non pregnant women living in the same environmental conditions.

SCINTIGRAPHY IN CONGENITAL HYPOTHYROIDISM

A.M. Ermans, J. Verelst, J.P. Chanoine *, and F. Delange

Departments of Radioisotopes and Pediatrics
University St. Pierre Hospital
Brussels

* Aspirant at Fonds National de la Recherche Scientifique, Belgium

Once the diagnosis of congenital hypothyroidism (CHT) is established by the screening tests, it is mandatory that the etiology, the severity and the prognosis of the disease should be confirmed as soon as possible. It is necessary to distinguish between patients with transient hypothyroidism, dyshormogenesis, ectopic thyroid or complete athyreosis. The Second International Conference on Neonatal Thyroid Screening (1) reported that 90% of primary congenital hypothyroidism is caused by thyroid dysgenesis, about two-thirds being thyroid ectopia and one third agenesis . These conditions are easily identified on the basis of the morphological abnormalities shown in thyroid scintigraphy. Thyroid scintigraphy is at present used extensively throughout all the screening centres in the world (2-7).

Three questions are discussed in the present paper:

1) Which radiopharmaceuticals should be used for scintigraphy, in the detection of congenital hypothyroidism?
2) What is the optimal methodology?
3) Is the irradiation risk induced by thyroid scintigraphy acceptable in neonates suspected of CHT?

Choice of tracer

^{123}I NaI and $^{99m}Tc\ O^{4-}$ are the two main radiopharmaceuticals used for thyroid scintigraphy (2-6); both have to be injected intravenously. The usual doses and the main properties of the two tracers are summarized in table 1. ^{123}I NaI being organified in the gland, its thyroid uptake is higher and permits optimal detection of any active tissue. $^{99m}Tc\ O^{4-}$ is concentrated in the thyroid gland but also in the salivary glands which may constitute a source of misinterpretation, principally in the presence of a thyroid remnant of very small size (7). Interference from the salivary glands is however easily reduced by bottle-feeding the baby immediately before the test. Because of its higher thyroid uptake and the lower surrounding background , ^{123}I NaI is at present considered as the optimal tracer to use in the

TABLE 1.

RADIOPHARMACEUTICALS USED FOR THYROID SCINTIGRAPHY IN THE DETECTION OF CONGENITAL HYPOTHYROIDISM.		
TRACER	123-I NAI	99M-TC-PERTECHNETATE
HALF-LIFE	13.3 HOURS	6.6 HOURS
24 hrs. THYROID UPTAKE (ADULTS)	30%	<4%
ORGANS VISUALIZED IN THE NECK	THYROID	THYROID AND SALIVARY GLANDS
THYROID METABOLISM	CONCENTRATED AND ORGANIFIED	ONLY CONCENTRATED
USUAL DOSES (IV)	25μc	200 μc
ENERGY	160 kev	140 kev
AVAILABILITY	CYCLOTRON PRODUCT (T 1/2 : 13.3 hrs)	99 Mo GENERATOR (T 1/2 : 66 hrs)
POSSIBLE CONTAMINANTS	124-I AND 125-I	0
COST/DOSE	EXPENSIVE	NEGLIGIBLE

diagnosis of CHT (6,7). In addition, recent information has suggested the interest of using ^{123}I NaI for the collection of quantitative data (6). In the case of an organification block, sequential measurements of thyroid uptake performed after IV administration of perchlorate do indeed give evidence of a prompt release of the radioiodide taken up by the gland.

^{123}I NaI has to be obtained directly from a cyclotron a few hours before the scintigraphy is undertaken. It is therefore not permanentely available in most laboratories. This is the main reason for which a ^{99m}Tc pertechnetate continues to be used on a very large scale (2); it is indeed directly produced from the elution of its precursor (99Mo-Molybdene) which has a half-life of 66 hours. ^{99m}Tc pertechnetate is, moreover, the basic radiopharmaceutical used in any nuclear medicine laboratory and is permanently available. This is important in view of the ability to perform the test at any time and to allow the doctors to give adequate information immediately to the family of the neonate suspected of CHT.

Finally, the low radiation level induced by ^{123}I NaI is only achieved in the case of a complete radionuclide purity. In certain production methods, some impurities constituted by ^{125}I and ^{124}I may increase appreciably the irradiation risk because of the long half life of these compounds (6).

In conclusion, ^{123}I NaI and ^{99m}Tc O^{4-} are both efficient radiopharmaceuticals to perform scintigraphy in the diagnosis of CHT. ^{123}I NaI is often prefered because it is a more physiological tracer; in practice the choice will be made on the basis of the own experience of the laboratory, its possibilities of connections with a cyclotron and of the modalities of its contacts with the screening units.

Methodology

Thyroid scintigraphy is performed 15-30 minutes after administration of the tracer by means of a rectilinear scanner or by a scintillation camera. In both devices, adequate immobilisation of the neonate during the test is achieved without sedatives by the use of a bed designed to form a vacuum by means of aspiration.

- Rectilinear scanner

This popular device can be easily applied and provides satisfactory resolution. The test takes 15-20 minutes. Its main advantage is the possibility of delineating very accurately the anatomical position of the thyroid tissue in relation to various reference points: the tongue, the trachea, the mouth or the nose. A limitation to the use of this method is the impossibility of processing the images.

This method is very reliable in the detection of any ectopic thyroid remnant either with a lingual or with an hyoid localisation. Diagnosis can easily be confirmed by continuing with a lateral view. The absence of any concentration of the tracer indicates that complete athyreosis is highly probable. Because of the limited resolution of the device and background interference, the technique may, in some cases, be unable to differentiate between complete athyreosis or the presence of very small thyroid remnants. In practice, this limitation does not appear important from the diagnostic point of view because both situations may be considered as having an identically severe prognosis. In addition, the possibility of showing a very small thyroid remnant is possibly related to the time at which scintigraphy is performed; it has indeed been shown that in some cases of ectopia a marked involution can take place during the first weeks of life (8).

- Scintillation camera

The subject is examined with a large field of view gamma camera fitted with a general-purpose parallel-hole collimator and connected on line to a computer system of 32-K memory. Analogic images are obtained from the display of the gamma camera and, at the same time, data are collected by computer with a static acquisition in a 128 x 128 matrix by inserting zoom in the input to reduce the field of view. A series of observations is carried out for the purpose of studying the anterior and lateral areas of the neck. The data collection is terminated when 200 K counts are reached; digital images are displayed on a video screen. Images are also registered after the radioactive markers have been positioned in order to have reference points on the localisation of the thyroid. This method allows good statistics to be obtained in a few minutes, notably reducing the problem of positioning and prolonged immobilisation of the young patient.

Resolution is not better than the rectilinear scanner but the possibility of data processing constitutes an obvious advantage. The field seen by the camera is however often larger than the one scanned by the rectilinear scanner so that the exact localisation of the thyroid remnant is sometimes less precise.

TABLE 2

99m-Tc AND 123-I NaI INDUCED RADIATION.		
TISSUE	99m-TcO4 (*)	123-I Na(**)
THYROID	0.68 rads	4 rads
TOTAL BODY	0.16 rads	0.23 rads
(*) 200 μc (**) 25 μc		

Moreover, it must be taken in mind that the absence of iodine uptake into the thyroid gland and in the salivary and gastric glands, may be due to a congenital lack of iodine transport. This congenital defect is however very rare (9,10)

As described recently (11), ultrasonography is performed systematically in our centre before scintigraphy on the neonates suspected of CHT. If the morphology of the gland is found normal, scintigraphy is not performed; indeed we consider that, in the absence of dysgenesis, scintigraphy does not contribute consistently to the etiological diagnosis.

Radiation risk

Thyroid scintigraphy is achieved at the cost of a certain radiation risk derived from the radiopharmaceutical used (12-16). Irradiation of the thyroid gland and of all other organs in the body ("total body") induced by the usual doses of ^{99} Tc O_4^- (200μc) and of ^{123}I NaI (25 μc) given to the neonate suspected of CHT is shown in table 2. The radiation dose estimates for the newborn are strongly dependent on the assumed thyroid uptake. After the second week, the newborn thyroid uptake is considered to be similar to the adult.

The thyroid dose calculated by Hedrick and Milavickas (16) who assume 50% uptake and a thyroid mass of 1g is 0.48 rad/μc for ^{123}I but is only 0.16 rads/μc if one assumes 27% uptake and a thyroid mass of 1.5g (13).

The risk of development of a malignancy induced by radiation is known from observations of atomic-bomb survivors in Hiroshima and Nagasaki and of the Marshall Islands inhabitants irradiated by fallout from test explosions of atomic weapons. The probability of radiation induced by the radiopharmaceuticals used for thyroid scintigraphy in the doses used for newborns is estimated to 3×10^{-6} for ^{99} Tc O_4 and 2×10^{-5} for ^{123}I NaI. Available information (17) failed to show that the risk factor in newborn differs from that in adults. Beekhuis and Piers (15) calculated that the extra malignancies induced by scintigraphy in The Netherlands where 60 children are suspected of CHT every year is about one malignancy every thousand years. This figure will be about 7 times lower using ^{99m} Tc O_4^-.

It is therefore stated that thyroid scintigraphy performed in newborn with ^{123}I NaI and ^{99m} Tc O_4^- for the evaluation of CHT does not noticeably increase the spontaneous tumour frequency and is acceptable if clinically indicated (12-16).

The above mentioned estimates were performed assuming thyroid glands with a normal size and a normal metabolism. Iodine kinetics in the ectopic glands is probably markedly modified because of the reduction of size and because the increased rate of uptake and of release of the radiopharmaceutical induced by the augmented TSH stimulation. This kinetics is presently poorly documented for obvious ethical reasons (6). This information wil be probably very soon available through quantitative studies performed with ^{123}I NaI.

Conclusion

In conclusion, thyroid scintigraphy provides in most of newborns suspected of CHT the crucial information concerning the etiology and the prognosis of the disease and permits to take, without any delay, a decision regarding the need for a long term treatment with thyroid hormones. This immediate information not only improves the prognosis of the disease but also avoids unnecessary treatment and allows the doctors to give immediately a clear statement to the parents. Rectilinear scanner and scintillation camera are both adequate devices to perform these tests. ^{123}I NaI and ^{99m} Tc O_4^- are the radiopharmaceuticals of choice and both are able to give adequate morphological information providing the laboratories of nuclear medicine have acquired sufficient experience and that the personnel is trained in the management of newborns.
^{123}I NaI is often prefered to ^{99m} Tc O_4^- but it is less easily available and is inducing a slighly higher irradiation of the thyroid gland. In any case irradiation induced by both radiopharmaceuticals does not, at the doses used for the diagnosis of CHT, increase appreciably the induction of thyroid malignancy and this negligeable risk seems largely compensated by the striking advantage of the method.

BIBLIOGRAPHY

1) P. Rochiccioli,
Thyroid dysgenesis,
Pediat. adolesc. Endocr., 14:154 (1985).

2) A. Cassio, E. Turba, A. Balsamo, A. Ballestrazzi, R. Pozzato and M. Marinelli,
In vivo isotope study of the thyroid with $^{99m}TcO_4$ in neonatal congenital hypothyroidism,
Eur. J. Nucl. Med., 9:351-353 (1984).

3) J. Guillet, B. Basse-Cathalinat, G. Soubiran, G. Guillet and P. Blanquet,
Etudes isotopiques in vivo de la fonction thyroïdienne en pédiatrie : ^{99m}Tc, ^{123}I ou ^{131}I?
Ann. Pediatr., 28:621-626 (1981).

4) D.A. Price, R.M. Ehrlich and P.G. Walfish,
Congenital hypothyroidism. Clinical and laboratory characteristics in infants detected by neonatal screening,
Arch. Dis. Child., 56:845-851 (1981).

5) E.J. Schoen, L.V. Dos Remedios and M. Backstrom,
Heterogeneity of congenital primary scintigraphy : the importance of thyroid scintigraphy,
J. Perinat. Med., 15:137-142 (1987).

6) T.E. Hilditch and H.J. Jackson,
Quantitative ^{123} iodide scintigraphy and radiation dosimetry in infants with congenital hypothyroidism,
Eur. J. Nucl. Med. 11:132-135 (1985).

7) J.C. Savoie and A.F. Leger,
Decreased salivary gland accumulation of pertechnetate in neonatal hypothyroidism,
J. Nucl. Med., 22:832 (1981).

8) F. Delange, personal communication (1988).

9) J.B. Stanbury and E.M. Chapman,
Congenital hypothyroidism with goitre. Absence of an iodide-concentrating mechanism,
Lancet, 1:1162-1165 (1960).

10) J. Wolff, R.H. Thompson and J. Robbins,
Congenital goitrous cretinism due to absence of iodide-concentrating ability,
J. Clin. Endocrinol., 24:699-707 (1964).

11) V. Toppet, J.P. Chanoine, M. Spehl and F. Delange,
Discrepancy between the results of thyroid echography and thyroid scintigraphy in congenital hypothyroidism,
Ped. Res., 23:107 (1988).

12) NCRP Report n° 73,
Protection in nuclear medicine and ultrasound diagnostic procedures in children,
Bethesda, MD: NCRP, 1983:15.

13) H.N. Wellman and R.T. Anger,
Radioiodine dosimetry and the use of radioiodines other than I-131 in thyroid diagnosis.
Semin. Nucl. Med., 1:356-378 (1971).

14) J.G. Kereiakes, H.N. Wellman, G. Simmons et al.,
Radiopharmaceutical dosimetry in pediatrics,
Semin. Nucl. Med., 2:316-327 (1972).

15) H. Beekhuis and D.A. Piers,
Radiation risk of thyroid scintigraphy in newborns,
Eur. J. Nucl. Med., 8:348-350 (1983).

16) W.R. Hendricks and L.R. Milavickas,
Reevaluation of the newborn thyroid dose from radioiodine,
J. Nucl. Med. 28:1208-1209 (1987).

17) G.W. Dolphin,
Radiation carcinogenesis, in: "Recent results in cancer research, thyroid cancer",
W. Duncan, ed., Springer Verlag, Berlin, p. 23, (1980).

CERVICAL ULTRASOUND IN CONGENITAL HYPOTHYROIDISM

M. Hassan*, C. Garel*, J. Leger**, and
P. Czenichov**

*Dep. de Radio., **Unite d'Endocrino Pediatrique
Hop R. Debre, 48, Bd Serurier 75019 Paris, France

As far as congenital hypothyroidism (C.H) is concerned, the usefulness of cervical ultrasound (C.U.S) has not been established yet, although some investigators had called attention to its potential benefits as far back as 1984. (1, 2)

Theoretically, the aims of C.U.S in CH may be threefold :

1 - To help to determine the precise anatomical diagnosis : athyrosis, ectopic thyroid, goitrous, hypoplasia or transient hypothyroidism.

2 - To allow a safe follow up of patients under treatment.

3 - To offer an approach of the gross pathology in C.H.

Our purpose in this paper is to emphasize some of the important issues related to the C.U.S studies and to present preliminary results.

TECHNICAL FACTORS

High resolution real-time devices and some technical skills are essential to perform accurate examinations. 7.5 and 5 mHZ probes are adequate respectively for the neck and the tongue. We have used linear array transducers (Hitachi - Japan), the 5 mHZ unit being convex and suiting well the shape of submental area. An equivalent fluid path should be interposed between the transducer and the patient's skin to eliminate near-field artefacts and to allow a clear delineation of the details of the superficial planes. We have been using Sonar-aid (Geistlich-Switzerland) for a few months with satisfactory results. Sedation is not necessary, but the babies must be firmly handled and there would be some advantage to examine them just after feeding. The use of a pacifier is questionable as vigourous sucking may disturb the procedure. The thyroid is best imaged in transverse cuts in the neonate whereas the tongue is well depicted in both transverse and sagittal slices, the last one being more suitable to precisely localize the foramen coecum area. The examination should be video-taped.

Careful attention should be given to these technical factors as they are of paramount importance in the issue of the procedure.

NEEDS OF STANDARDS

Adult thyroid ultrasound appearance is well-known, as this procedure is accepted as the most sensitive non invasive method to define the internal anatomy of the gland. However, experience in children is far less extensive and studies in the newborn are very scarce (3). Normal data concerning the thyroid, the tongue and the upper cervical area are not available. We believe this must be a compulsory prerequisite if the aims are to establish a gold-standard method in assessment in C.H. Such a work is currently in progress in our department. At the moment no definitive statement can be made since our series are still too small. However some data begin to come into view concerning the anatomy of the tongue and the thyroid.

Very few papers are devoted to ultrasound anatomy of the tongue. We have found that Shawker's work published in 1984 was the most interesting (4). This study only deals with adults examined in a resting state. It must be stressed that the general shape of the tongue varies with its movements. Particulary, its posterior part contracts, changing from a flat to a more convex postero-superior base. Nevertheless, the ultrascund image seems to be constant from one patient to another. The landmark of the mid-longitudinal plane is the hyoid bone, easily detected thanks to its acoustic shadow. The muscles, geno hyoidieus and mylohyoidieus, just over the subcutaneous plane, lie anteriorly and appear as a hypoechogenic stripe of a few millimeters of thickness. We will focus on the posterior aspect of the tongue because this is the site of foramen coecum where the ectopic thyroid is supposed to be found.

Starting from the hyoid bone, forward and slightly upward there is a tiny hyperechoic triangle contiguous to the posterior aspect of the muscular plane. This structure represents the tendinous insertion on the hyoid bone of the different muscles in the area. In the transverse plane it appears as an echoic "pseudo nodule" in the midline between the two digastric muscles. The tongue itself, just above, shows a clover-like area of hyperechogenicity which extends more or less forward joining the posterior ridge of the genoglossus muscle. Consequently this part of the tongue appears heterogeneous in ultrasound studies. Finally the surface of the tongue is perfectly seen from its posterior aspect to the tip.

The echo anatomy of the thyroid is more familiar. The echo pattern seems not to differ whatever the age is from the well-known adult appearance. Some changes do occur concerning the outline of the gland. During infancy and early childhood the borders appear rather concave anteriorly and laterally. Progressively and with puberty there is an antero-lateral bulging of each lobe especially the right one. Linear measurements were obtained as follows : width 7 to 16 mm, thickness 6 to 12 mm ; height 20 to 45 mm.

PERSONAL RESULTS IN PATIENTS WITH CH

We have been studying two groups of patients for the last year.

1 - First group :

It includes sixteen children between the ages of 5 and 12 years and one baby of 8 months old, who have been previously classified as ectopic thyroid through scintiscanning performed at the moment of diagnosis.

The baby has been treated for 5 months ; the others have been given LT4 for many years and with the exception of one case since the age of 1 month. All children have been off treatment for one month at the time of this examination (see chapter : "Thyroglobulin and congenital hypothyroidism").

All these cases showed similar striking appearance in the lower cervical area. Echogenic tissue with a triangular shape was found on each lateral aspect of the trachea, medial to the carotid artery, in every child. In 16 patients this tissue was strongly hyperechogenic as compared to the muscles of the neck-wall. In one patient it was iso echogenic. These echoes appeared quite different from normal thyroid echo pattern, even if at first glance they could mimmick thyroid lobes. A careful evaluation with adequate tools reveals a linear and stratified pattern of the echoes instead of the homogeneous characteristic high level echoes. The greatest width of each "lobe" was included between 3 mm (8 months old) and 10 mm (8Y old). 15 out of the 17 cases were included between 5 and 8 mm. The thickness ranged from 3 to 7 mm. As the boundaries of these structures were rather ill-defined, longitudinal measurement did not appear reliable enough. We could notice moreover, a particular position of the carotid arteries, anterior to the jugular veins in 76 % of the cases.

The nature of these structures is poorly understood. We presume that they correspond to fat and/or fibrosis on account of the hyperechogenecity. This tissue occupies the normal thyroid area, together with the carotid arteries.

In this group, we could never find any structure in the tongue nor along the thyroglossal duct that could suggest an ectopic gland.

2 - Second group

Five one-month old babies were studied because of positive testing for C.H. For each child scintiscanning was performed before the C.U.S. Two were classified as athyrosis : one, in fact, turned out to have a normal thyroid gland (hypothyroidism due to maternal iodate administration) ; for the second patient as well as for the three others of this small group, the findings were similar to those of the first group : that is to say fatty fibrosis appearance in the lower cervical area and no evidence of ectopic gland in the three cases classified as ectopic.

DISCUSSION

Several remarks can be made from our series.

- We personally have never been able to detect any ectopic thyroïd tissue either at the base of the tongue, or at any point along the thyro-glossal duct in patients classified as ectopic, whatever they were newborn recently diagnosed, or children treated for a few years.

- We believe that this inability could hardly be related to technical factors, since the procedure actually delineated the anatomy with exquisite details. Had an ectopic gland sharp outlines and a certain size (more than 2/3 mms) it would have been detectable through ultrasound studies, unless ectopic tissue showed an echopattern quite different from the normal gland and similar to the somewhat heterogeneous one at the base of the tongue.

The morphologic data concerning these conditions are rare. Lingual thyroid refers to an entire thyroid gland that has failed to descend to its normal cervical location ; in most cases it lies within the tongue, just under the mucosa, at the normal location of the foramen coecum. It may be located a little lower just above or in front of the hyoid bone. Ectopic thyroid refers to a developing gland which had begun to descend but which had failed to reach its normal location. Such arrrested descent leaves the gland at, or just below, the hyoid bone, or in the prelaryngeal area. Very rarely, hyperdescent leads the gland to a substernal site. Ectopic thyroid is considered to be much less common than lingual thyroid. The morphological appearance of the two entities have been essentially described in patients whose symptoms were related to the mass and not to intense marked hormonal deficiency, specially in the early childhood. It would not be suprising consequently that the patients who are currently detected by modern screening, and further classified as ectopic through isotope studies do exhibit, in fact, a different anatomical pattern. An hypothesis can be made that the tissue detected in that way may consist of tiny clusters of thyroid cells without any macroscopical organization which could permit an approach by imaging methods. To the best of our knowledge we are aware of only 2 studies which support the ability of ultrasound to demonstrate ectopic thyroid in patients with C.H detected through hormonal screening. In 1985, Miller (5) has reported 2 neonates with C.H. detected by blood screening. Prior to C.U.S, both babies were evaluated by scintiscanning (I 123 or Tc 99 m pertechnetate) which demonstrated lingual thyroid in each case. After this identification U.S examination was performed and showed what the author called a "discrete mass with a smooth internal texture consisting of low-level echoes... at the base of the tongue". He noted, moreover, that "the appearance of this thyroid tissue was similar to that in other persons with normally situated glands".He presented two ultrasound pictures that we cannot consider as demonstrative. The Lille group recently (6) published a more extensive work based on 60 newborns referred to them after positive neonatal screening. In this series ultrasound procedure prooved to demonstrate ectopic tissue in the tongue in up to 79 % of the group classified as ectopic. No false positive has been noted. Those images are not convincing because we reckon such figures can be obtained in every normal child. Their results will be discussed during the workshop.

On late february, Muir and the Toronto group (7) reported their disappointing experience about C.U.S as a mean to classify C.H. As we did, they could not identify any lingual gland in a group of 13 newborn prooved to be ectopic through radionuclide scan. These findings are in agreement with our own results. However the authors considered that as a failure in ultrasound imaging and hoped that advances in technology will allow for improved reliability in future. We personnaly think that current state of the art of U.S. is high enough to widely demonstrate that, at least in most cases, the tissue detected by radio scanning on the upper neck of the so-called ectopic group, does not indeed correspond to any organized anatomical structure macroscopically detectable.

With respect to athyrois, hypo and hyperplasia their results are more difficult to discuss owing to the lack of pictures or descriptive details in the article. We can hardly guess that some problems did occur in the assessment of thyroid volume.

2 - The concept of thyroid hypoplasia

It seems to us very difficult, if not impossible, to accurately assess a thyroid hypoplasia through ultrasound study. To be able to discuss such an entity would imply an extensive control study of normal neonates. Furthermore, accurate measurements are often difficult at this age.

Dammaco et al (8) have reported 5 hypothyroid infants whose glands were normally located but were "hypoplastic with reduced echogenicity compared with those of the controls". These patients were found to have a progressive decrease of their serum thyroglobulin. The areas of transverse section were measured but no values were given in that paper. The published picture can hardly be assessed. We find it difficult at the moment to ascertain actual hypoplasia.

3 - The "empty" thyroid area

This is a very interesting finding in ultrasound study in children with C.H. With some experience the appearance of what can be called an "empty thyroid area" is easy to distinguish from normal glands even in the newborn, provided that the examination is done with adequate equipment. This appearance was similar in newborns and children under treatment for many years.

The exact nature of the high level echoes observed remains questionable although one can assume that they probably represent fibro-fatty tissue and tiny vessels. One interesting point to consider is the shape of this tissue which could suggest that some thyroid gland had been present in the cervical area at one moment during prenatal life.

Indeed further studies are necessary to clarify these findings constantly observed in the cervical area, in both athyrosis and ectopic groups.

4 - Normal and enlarged thyroid

Finding a normal thyroid in a newborn referred for suspected C.H. through screening programs should alert on the possibility of a false diagnosis or of transient hypothyroidism. In some selected centers, C.U.S may become a part of the routine newborn examination. Finally, enlarged thyroid is rarely observed in the newborn. These cases are exquisitely documented by ultrasound. They must often correspond to maternal iodate intake and more rarely to hormonal biosynthesis failure.

To conclude, our experience with C.U.S in C.H. may be summarized as follows :

1 - We have not been able to demonstrate any morphological equivalent of the functional thyroid cells shown by scintiscanning in the group classified as "ectopic" despite high anatomic resolution.

2 - In the lower neck, the U.S appearance is quite similar in patients classified as athyrosis or as ectopic. There is a fibro-fatty tissue with a configuration that could perhaps suggest a secondary atrophy rather than an agenesis or failure of migration.

3 - For such purposes, top level procedures should be supplied and a thorough criticism is necessary in the interpretation of the U.S findings.

REFERENCES

1 - POYHONEN L., LENKO H.L.,
Ultrasonography in congenital hypothyreosis.
Acta Paediatr. Scand., 1984, 73, 523-526

2 - PASTO M., CARPENTER GG.,
Ultrasound evaluation of newborn thyroid structure : visualisation without radiation.
Petriatr. Res., 1984, 18, 173 A (abstract 463)

3 - BACHRACH LK., DANEMAN D., DANEMAN A., MARTIN DJ,
Use of ultrasound in childhood thyroid disorders.
The Journal of Pediatrics, 1983, 103, 547-552

4 - SHAWKER TH., SONIES BC., STONE M.,
Soft tissue anatomy of the tongue and floor of the mouth : an ultra sound demonstration.
Brain and language, 1984, 21, 335-350

5 - MILLER JH.,
Lingual thyroid gland : sonographic appearance
Radiology, 1985, 156, 83-84

6 - YTHIER H., FARRIAUX J.P, LEMAITRE L., MARCHANDISE X.,
(26 references)
Apport de l'échographie au diagnostic étiologique des hypothyroïdies congénitales.
Arch. Fr. Pediatr, 1987, 44, 691-5

7 - MUIR A., DANEMAN D., DANEMAN A., EHRLICH R.,
Thyroid scanning, ultrasound and serum thyroglobulin in determing the origin of congenital hypothyroidism.
AJDC,1988, 142, 214-216

8 - DAMMACO F., DAMMACO A., CAVALLO T., SANSONNA S.,
Serum thyroglobulin and thyroid ultrasound studies in infants with congenital hypothyroidism.
The Journal of Pediatrics, 1985, 106, 451-453

NEONATAL ECHOGRAPHIC FINDINGS
IN CONGENITAL HYPOTHYROIDISM

Marianne Spehl(1), Véronique Toppet(1)
Jean-Pierre Chanoine(2-3)* and François Delange(2,3)

Departments of Radiology(1), Pediatrics(2) and
Radioisotopes(3) of the Hopital St.Pierre
Free University of Brussels (Belgium)

INTRODUCTION

Echography (US) offers an alternative to isotope study to differentiate among children presenting with biological neonatal congenital hypothyroidism (CH)those with anatomical anomalies from those with function impair.

It is a non invasive technique that can be repeated without harm to the neonates and that is recognized as innocuous by the parents who all experimented echography during pregnancy. It allows a precise study of thyroid gland anatomy and gives an accurate quantification of thyroid volume. But, thyroid echography in neonates is a delicate and challenging procedure that requires high resolution US machines and skilled echographers. We report here our personal experience based on the examination of neonates recalled to our hospital under suspicion of CH.

ULTRASOUND TECHNIQUE AND PATIENTS

Thyroid echography was performed in all neonates without sedation, usually after feeding, preferably in a quiet surrounding (quiet room, lights down, calm sonographer). Patients lay supine, appropriate hyperextension of the neck was obtained using a coin shaped plastic foam cushion. The use of a pacifier was helpful in discriminating thyroid tissue from adjacent structures as the thyroid moved up and down during swallowing.

*Aspirant of the "Fonds National de la Recherche Scientifique" (F.N.R.S., Belgium)

Ultrasonography has to be performed with the highest possible frequency probes (7.5 MHZ to 10.0 MHZ) and appropriate short focus. The contact surface of the probe has to be as small as possible, as the neck of a newborn baby is rather small even in hyper-extension. In our experience, the best results were obtained in newborn with a "peroperative small parts" type of probe (7.5 MHZ linear electronic probe with a 2.5 cm long contact surface) especially as far as sagittal scans were concerned. A longer probe of a more "classical" type (7.5 MHZ linear array 5.0 cm long) was satisfactory on transverse scans but required a slight tilt of the head to be placed sagittally on the babies neck. Depending of the neck anatomy and the babies fatness, a "reston" type coupling material was used to avoid losing skin contact during examination. All examinations were performed on a Toshiba Sonolayer V-SSA-100A.

US examination always started with transverse scans: it was better tolerated by the child and thyroid tissue was better discriminated from adjacent structures than on sagittal scans. Width and thickness of the thyroid lobes were measured on the largest possible slice. Parasagittal scans were performed from the outer part of the neck towards the trachea along the jugular vein and carotid artery axis. The length of each thyroid lobe was measured on the longest possible slice.

The volume of the thyroid was calculated by using the formula of an ovoid: 0.52 x W x L x T for each lobe and then summing up both lobes.

14 newborns in whom the first blood screening tests gave cause to suspect CH (serum TSH above 50mU/l or serum TSH between 20 and 50mU/l together with a serum T4 below 65nM/l) and 50 normal newborns used as controls (after having obtained the informed consent of their parents) were examined. Echographies were performed on the day of recall, always prior to scintigraphy, by either of two experienced echographists, together with obstetrical and neonatal history, control thyroid function tests, physical examination and bone age according to the method of Senecal. Normal thyroid tissue was looked for at the anatomical site of the thyroid gland and whenever not found, thyroid scintigraphy was performed to detect ectopic tissue or confirm agenesis.

US APPEARANCE OF ANATOMICALLY NORMAL THYROID GLAND IN NEONATES

RIGHT LEFT

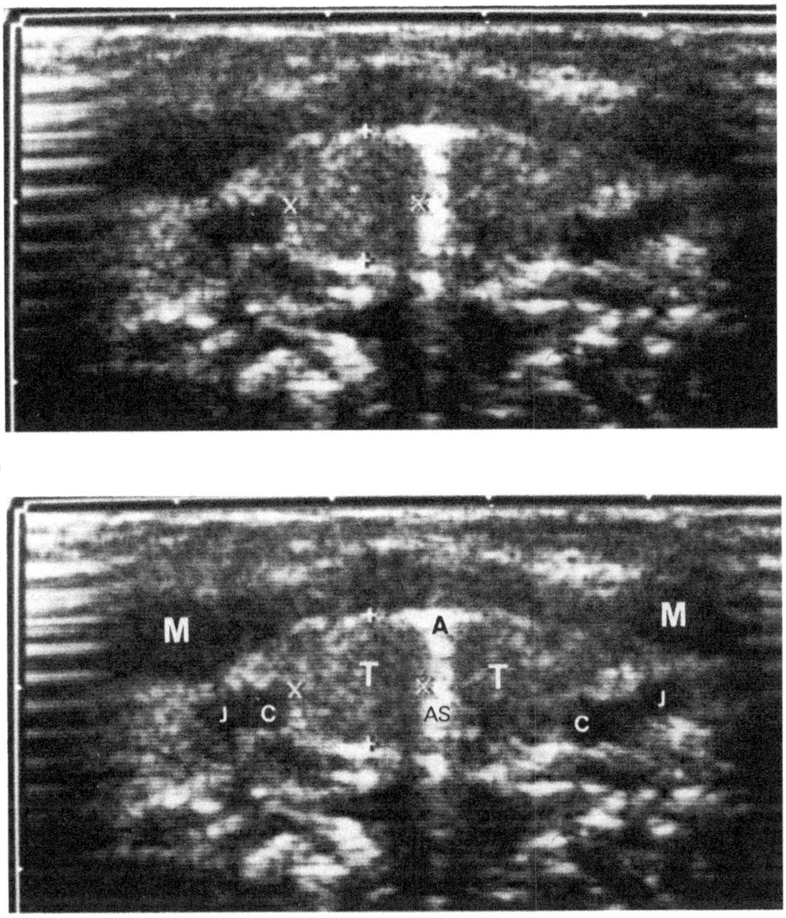

FIG.1. Transverse scan of the neck (7.5MHZ 5cm long linear array transducer) at the thyroid level.

T : normal thyroid lobe
M : muscles
C : carotid artery
J : jugular vein
A : air within the trachea
AS: acoustic shadow behind the trachea
(x....x) width of thyroid lobe
(+....+) thickness of the thyroid lobe

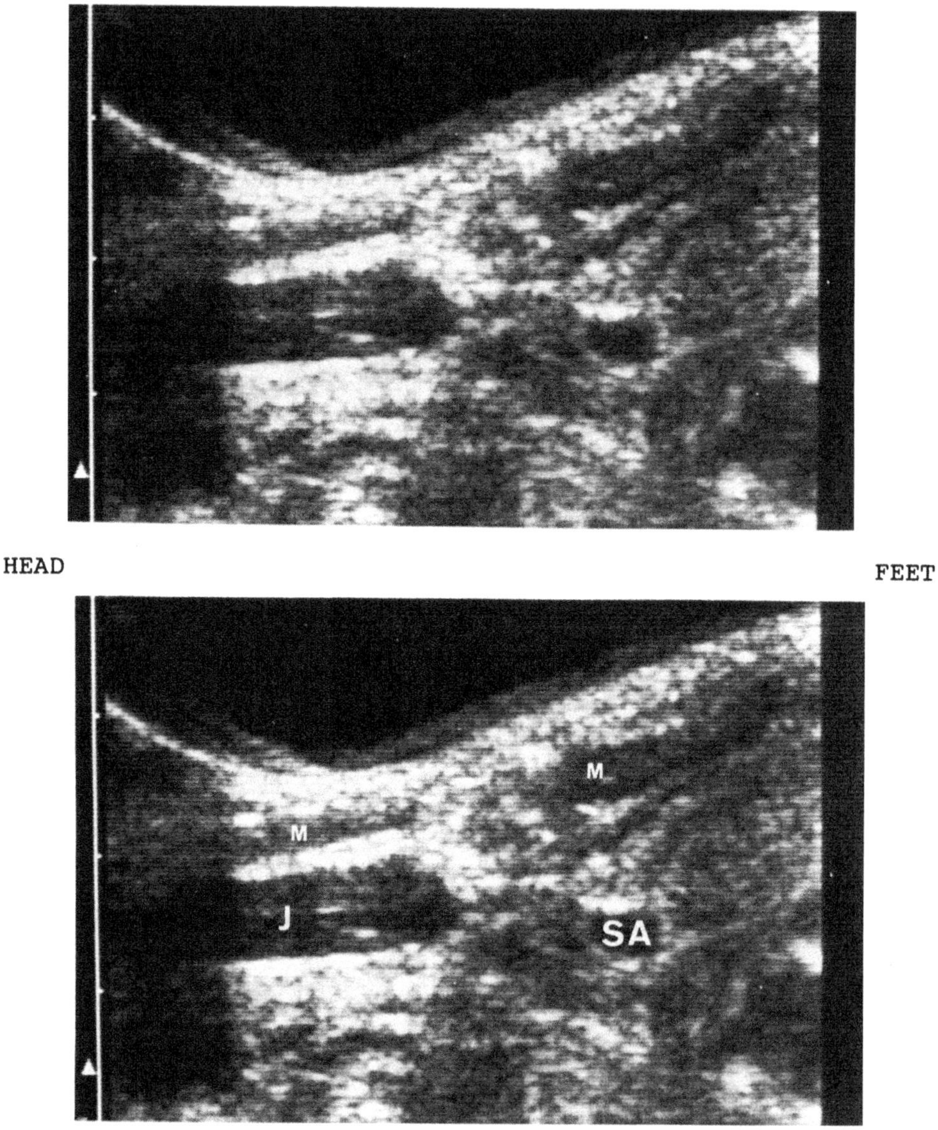

FIG.2. Parasagittal scan at the outer part of the neck of a normal neonate along the jugular axis

J : jugular vein
SA: subclavian artery
M : muscles

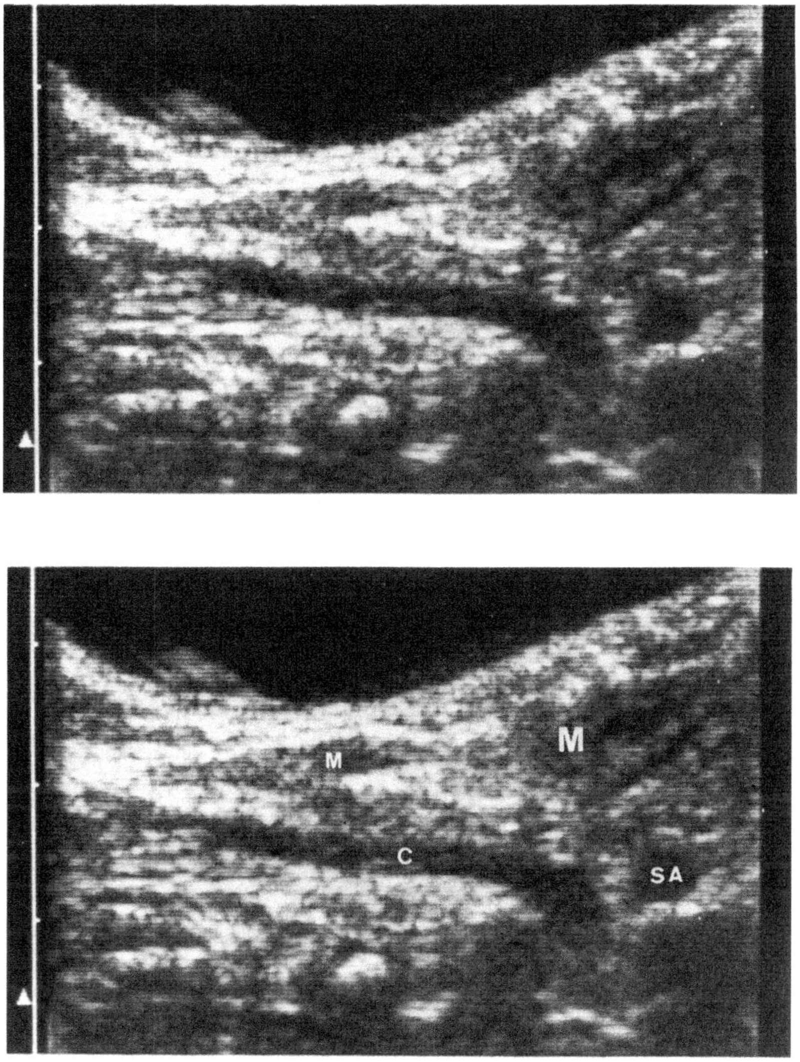

FIG.3. Parasagittal scan along the carotid axis

C : carotid artery
SA: subclavian artery
M : muscles

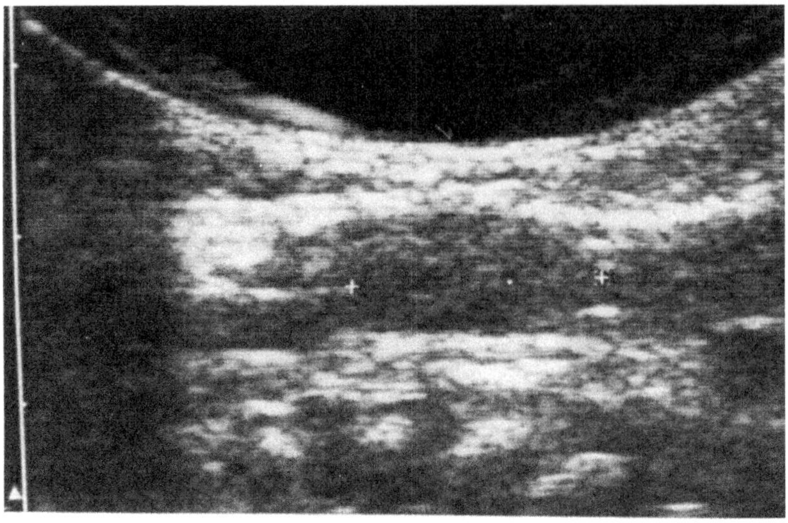

HEAD FEET

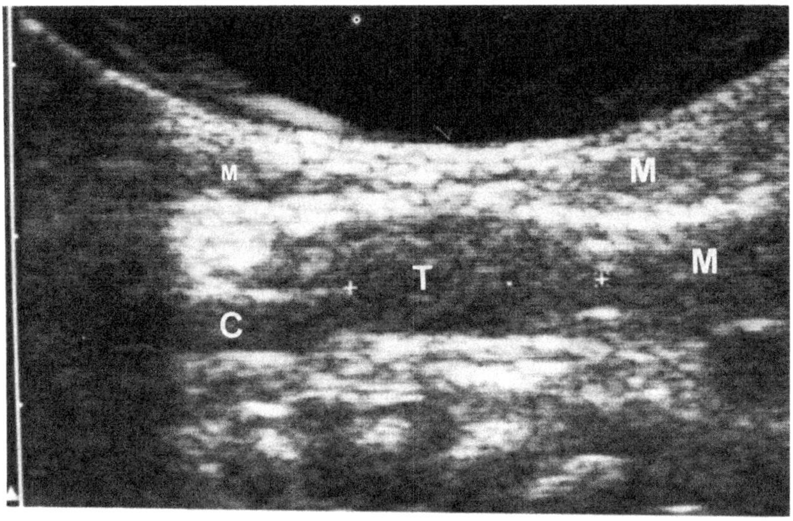

FIG.4. Parasagittal scan of a thyroid lobe (5cm long 7.5 MHZ probe)

T : thyroid lobe
M : muscles
(+....+) length of the thyroid lobe

The length of the probe necessitated a slight tilt of the head which sometimes caused sterno cleido mastoidien muscle to slide a little over the thyroid.

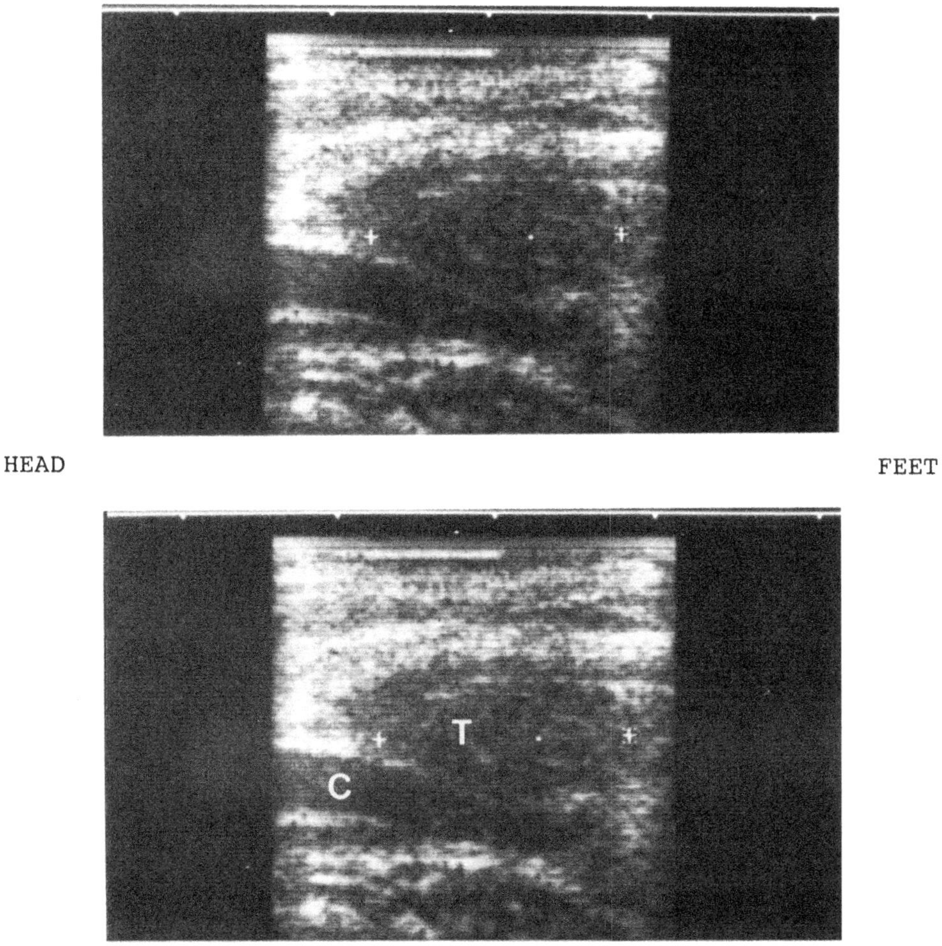

FIG.5. Parasagittal scan of a thyroid lobe with a 2.5cm long 7.5MHZ probe

T : thyroid
C : carotid artery
(+....+) length of the thyroid lobe

Thyroid imaging is easier because of the short contact surface of the probe.

In normal control neonates, the mean volume of the thyroid gland calculated by echography was 831 +/- 383 (SD) mm3.

Among the 14 newborns recalled because of abnormal first blood screening tests, 8 were "false positive" screening tests and echography showed normal thyroid tissue in normal position. The volume of those thyroid (1080 +/-353(SD) mm3) was not significantly different from that of the controls.

US APPEARANCE OF THE THYROID AREA IN THYROID AGENESIS OR ECTOPIA

In all our cases of agenesis or ectopia, echography of the thyroid area showed the same image: small hyperechogenic masses located posteriorly at the anatomical site of the thyroid. Their hyperechogenic characteristic was easy to recognise on transverse scans next to the vascular markers (carotid artery and jugular vein).

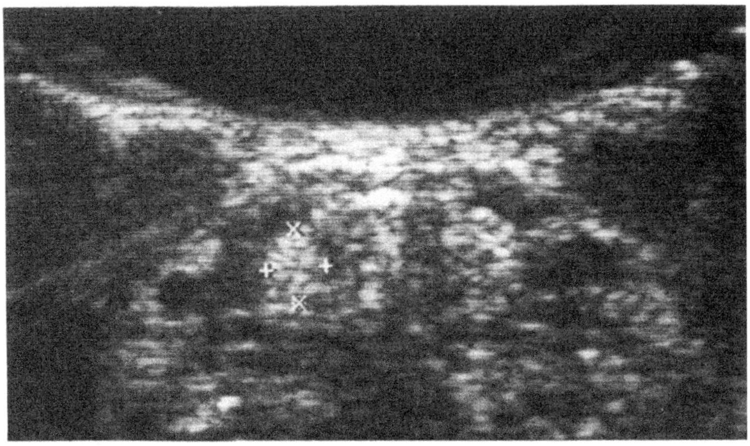

RIGHT LEFT

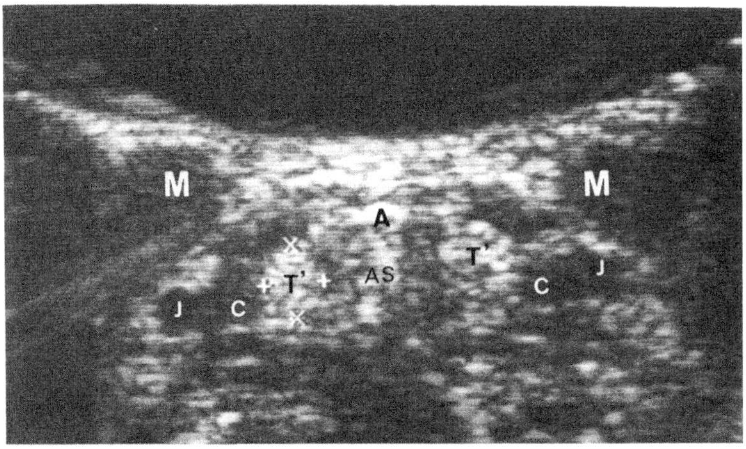

FIG.6. Transverse scan of the neck (7.5MHZ 5cm long transducer) at the thyroid level of a 15 day old CH baby.

T': hyperechogenic tissue
M : muscles
C : carotid artery
J : jugular vein
A : air within the trachea
AS: acoustic shadow behind the air
(+....+) width of hyperechogenic mass
(x....x) thickness of hyperechogenic mass

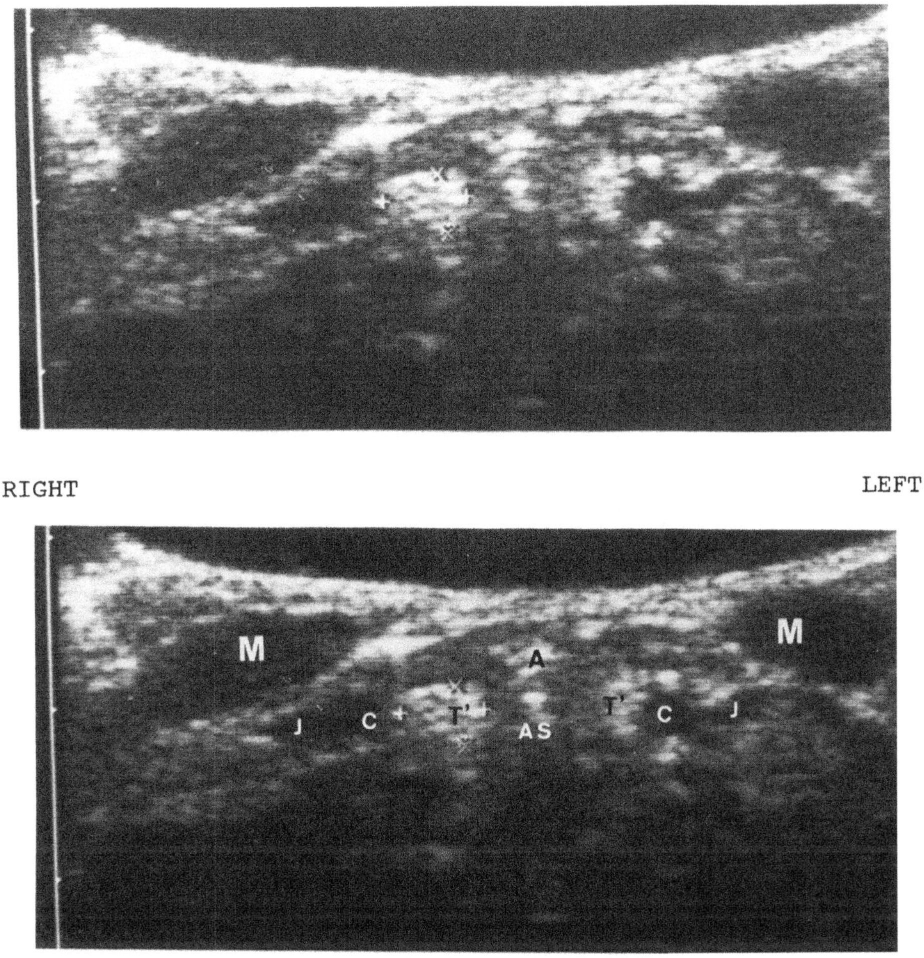

FIG.7. Transverse scan of another 15 day old CH baby showing the same image: small hyperechogenic masses posteriorly located in the thyroid area.

T': hyperechogenic tissue
M : muscles
C : carotid artery
J : jugular vein
A : air within the trachea
AS: acoustic shadow behind the air
(+....+) width of the hyperechogenic mass
(x....x) thickness of the hyperechogenic mass

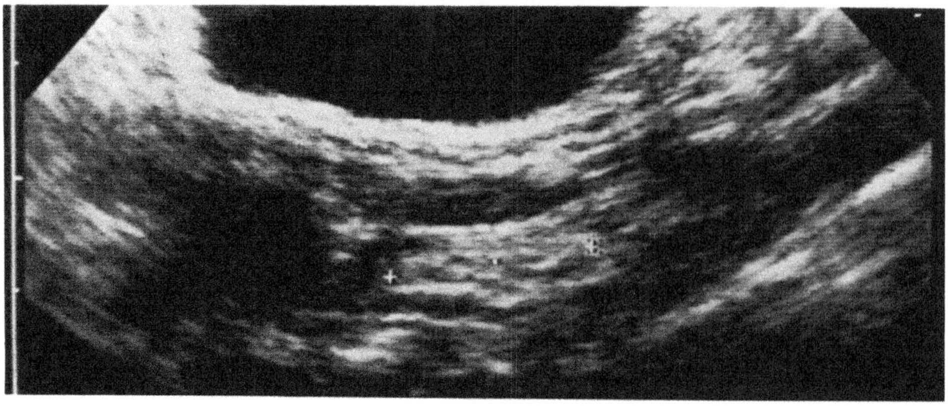

HEAD FEET

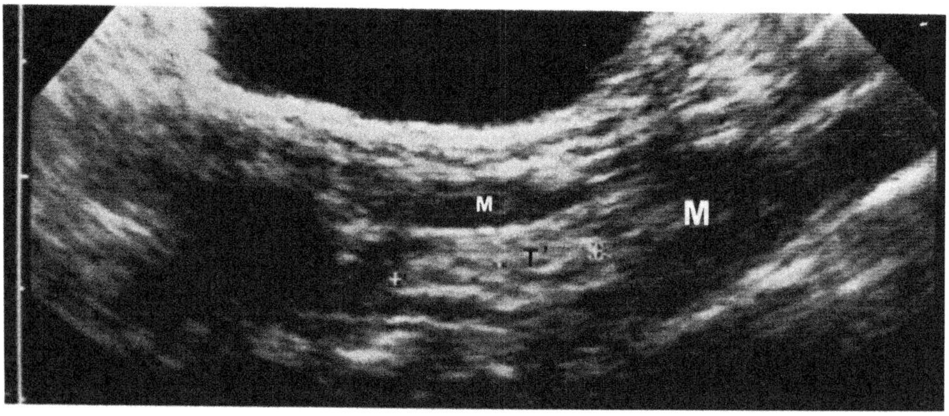

FIG.8. Parasagittal scan through hyperechogenic mass in the thyroid area of a 15 day old CH baby. Mass is very difficult to demonstrate probably because of its closeness to trachea and air shadowing. The tilt of the head causes the sterno cleido mastoidien muscle to slide a little over the mass.

T': hyperechogenic tissue
M : muscles
(+....+) length of the mass of hyperechogenic tissue

The volume of these masses of hyperechogenic tissue (322 +/- 180(SD)mm3) was significantly smaller than the volume of the thyroid in normal newborns.

The nature of this hyperechogenic tissue is unknown. It does not grow with age: we have had the opportunity to study by echography older infants and young adults whose CH had been detected years before and found the same hyperechogenic masses, of the same size, at the same place.

We think that it could represent remnants of the ultimobranchial bodies that have long been described by anatomists in the thyroid area of deceased athyreotic subjects as small cystic structures, surrounded by conjonctive tissue and located deeply along the trachea.

REFERENCES

Bachrach L.K., Daneman D., Daneman A. and Martin D.J., 1983, Use of ultrasound in childhood thyroid disorders, J. Pediatr., 103:547.

Chanoine J.P., Toppet V., Van Vliet G., Lagasse R., Bourdoux P., Spehl M., and Delange F., 1987, Can thyroid echography replace thyroid scintigraphy in congenital hypothyroidism, Ped. Res., 23: 107 (Abstract).

Dammaco F., Dammaco Z., Cavallo T., Sansonna S., Bafundi N., Torelli C., Frezza E., Vitale F. and Griseta D, 1985, Serum thyroglobulin and thyroid ultrasound studies in infants with congenital hypothyroidism, J. Pediatr., 106:451.

Harach H, 1986, Solid cell nests of the human thyroid in early stages of post natal life, Acta Anat., 127:262.

MacCallum W.G. and Fabian M., 1907, On the anatomy of myxoedematous idiot, John Hopkins Hosp. Bull., 198:341.

Muir A., Daneman D., Daneman A. and Ehrlich R., 1988, Thyroid scanning, ultrasound and serum thyroglobulin in determinating the origin of congenital hypothyroidism, AJDC, 142:214.

Pöyhönen L. and Lenko H.L., 1984, Ultrasonography in congenital hypothyreosis, Acta Paediatr. Scand., 73:523.

Senecal J., Grosse M.C., Vincent A., Simon J. and Lefrèche J.N., 1977, Maturation osseuse du foetus et du nouveau-né, Arch. Fr. Pédiat., 34:424.

Van Vliet G., Glinoer D., Verelst J., Spehl M., Gompel C. and Delange F., 1987, Cold thyroid nodule in childhood: is surgery always necessary ?, Eur. J. Pediatr., 146:378.

Ythier H., Farriaux J.P., Lemaitre L. and Marchandise X., 1987, Apport de l'échographie au diagnostic étiologique des hypothyroidies congénitales, Arch. Fr. Pédiatr., 44:691.

THYROGLOBULIN AND CONGENITAL HYPOTHYROIDISM

Paul Czernichow and Juliane Leger
Hopital des Enfants-Malades
149 rue de Sevres 75015 Paris, France

Thyroglobulin (Tg) is a glycoprotein which is released in minute amounts into the circulation by the thyroid gland. Under physiological conditions it can be measured in the serum of most individuals by modern RIA. The serum Tg concentration is elevated in various thyroid disorders (1). Since Tg is secreted by the thyroid gland only, its presence in the circulation should therefore indicate the presence of thyroid tissue. This report concerns the evaluation of Tg measurement in infants with congenital hypothyroidism (CH). Our results indicate that Tg RIA provides useful information about the presence or absence of functional thyroid tissue. It also allows easy evaluation of the function of remnant thyroid tissue in patients with CH who have been treated for several years.

I. THYROGLOBULIN IN NORMAL INFANTS AND CHILDREN

Thyroglobulin seems to be present very early in the development of the thyroid gland. The protein has been detected in human thyroid cells by several techniques as early as 29 days of gestation (2-3). However there are to our knowledge no data on the serum value of this protein during the ontogenesis.Several authors have reported serum Tg values in cord blood (4-9) ; the values are summarized Table I. Although there is a wide range of variation, the reported values are always significantly higher than the normal values for children, adults or pregnant woman. The reasons for these higher values are not known. The Tg concentrations in full term newborns and in preterm babies are identical and the mode of delivery does not influence cord blood Tg values (9).

After birth there is a further elevation of the serum Tg level. As reported by Pezzino et al (6) there is a prolonged rise after delivery. The increased levels were significant at 6h and were maintained for 15 hours. However the exact cause of this prolonged elevation of Tg is not known. An increase release of Tg by the thyroid gland is the most likely explanation although other mechanisms such as a reduced clearance of the protein from infant's plasma after delivery cannot be excluded. It is well known that the newborn thyroid is being stimulated and is actively secreting. After birth there is a burst of TSH release followed by a persistent elevation of thyroid hormone. The rise in Tg reflects this secretion, and demonstrates that the thyroid gland is able to actively secrete Tg when stimulated by TSH. However some data raise questions as to the validity of this explanation. Roti et al (7) studied the effect of TRH on the fetal pituitary-thyroid axis, by injecting mothers with TRH at various times (0-820 min) before delivery and measuring the TSH, thyroid hormones and Tg levels in the cord blood. Cord blood TSH and T3 were significantly increased by the TRH injections but there was no change in Tg concentration. This observation that, despite elevated TSH and T3 secretions, there was no rise in Tg indicates a dissociation between the secretions of Tg and thyroid hormone.

After this first period the serum level of Tg gradually declines (5, 10) and approaches the levels prevalent during the first decade of life by the end of the first year (fig 1).

II. SERUM THYROGLOBULIN IN INFANTS WITH CONGENITAL HYPOTHYROIDISM BEFORE TREATMENT

The absence of thyroid gland or ectopic thyroid tissue can be detected by radioisotope scanning, but many groups have avoided using this technique on very young infants. Since the presence of Tg in serum indicates the existence of secreting thyroid tissue, measurement of this protein could help to determine the type of thyroid defect in infants with CH.

Patients and Methods

A group of 100 children with CH were studied. The infants were screened in several centers in France using filter paper TSH measurements. Patients were clinically evaluated and hypothyroidism diagnosed on the basis of TSH, T4 and T3 measurements. Thyroid scanning with radioiodine ^{123}I was performed on 93 patients. On 5 cases technetium 99 was used. Two patients underwent no radioisotopic explorations. On the basis of this investigation infants were classified as patients with undetectable thyroid tissue (group I, n=15), ectopic gland (group II, n=59) and patients with goiter or eutopic gland (group III, n= 26). Serum Tg was measured using the

Table I . Serum thyroglobulin levels in cord blood

Authors	Number of samples	Value (ng/ml ± SEM)
Osotimehin et al (1978)	191	57 ± 1.7
Ket et al (1981)	158	73 (median) (range 7-269)
Pezzino et al (1981)	16	24.4 ± 3.7 (mothers 12.1±3.1)
Roti et al (1981)	124	138 ± 4.2
Black et al (1982)	56	37 ± 3
Ericson et al (1987)	103	110 (range 23-320)

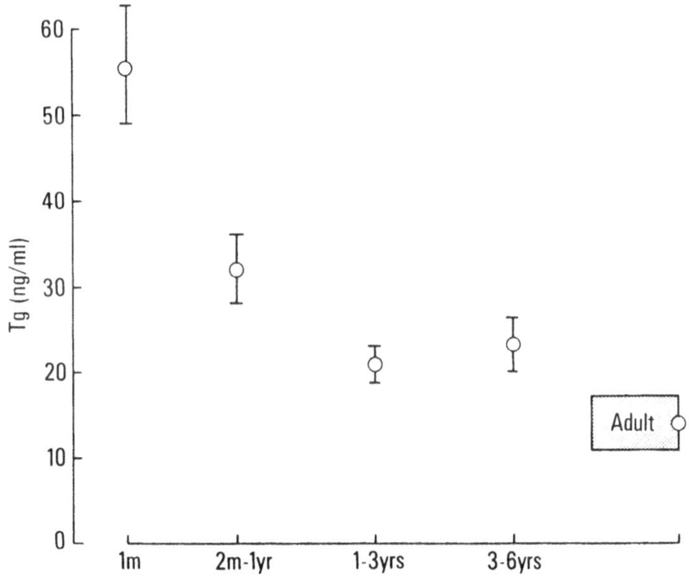

Figure 1 . Decline in serum thyroglobulin (Tg) levels in normal children and adults method of van Herle(10). The lower limit of sensitivity was 2.5 ng/ml. The presence of circulating Tg antibodies was ruled out by the tanned red cell agglutination technique and by RIA using Protein A. Blood was collected before treatment and serum was kept at -20°C before assay. The results of this investigation have been reported previously (12-13).

Results and discussion

Values for serum Tg are given in table II and fig 2.

In infants with undetectable thyroid tissue, Tg was below the limit of sensitivity of the assay in 12 cases and was 6, 6.4 and 9 ng/ml in 3 cases (all 3 infants had had thyroid scanning showing no uptake of radioiodine ^{123}I).

Tg serum levels varied widely in group II and III. In infants with ectopic glands, 1/3 of all cases had normal Tg values, 1/3 had elevated Tg serum concentrations and 1/3 of all cases had value below the normal range. Tg was undetectable in 2 cases. Tg was also undetectable in 3 infants with eutopic glands. In the entire group no correlation was found between Tg and TSH, T4 or T3.

As reported by others, Tg serum values vary widely in CH (4, 8). Elevated Tg concentration which may reached 20 times the normal values are found. This large variation in Tg values demonstrates that plasma Tg levels measurements are of no value in the diagnosis of CH.

However, the results obtained from this large serie of patients in whom the thyroid defect was precisely determined by scintiscan do demonstrate that Tg measurement can be used to distinguish between cases in which thyroid tissue is present or absent. Values above 10 ng are a good indication of the presence of secreting thyroid tissue. Undetectable levels, as described by others (5-14), were associated, in 3 cases, with a gland in normal position and goiter. This indicates that some children with errors of thyroid hormone metabolism have abnormal Tg synthesis or transport. Tg was also undetectable in 2 infants with ectopic glands. In these cases the remnant tissue may have been too small to be visualized by radioisotope scanning.

We have compared the diagnostic value of T4 or Tg measurement in the group of ectopic or athyreotic patients (table III). Patients were classified, according to the radionuclide scanning results, as infants with absent (20%) or ectopic gland (80 %). Cut-off points of 2.5 ng/l for Tg and 1.5 ug/100ml for T4 were used to distinguish between the 2 types of defect. It is clear that T4 measurement grossly exagerate the percentage of patients with no thyroid tissue. The classifications based on Tg measurement and radionuclide scanning were very similar.

Table II . TSH, T4, T3 in a large group of patients with congenital hypothyroidism of different types (mean ± SEM)

Patients	TSH uu/ml	T4 ug/ml	T3 ng/ml	Tg ng/ml
I. Athyreotic	555±87*	1.04±.2◊	39.4±13.8§	2.5±.7
n =	(15)	(15)	(9)	(15)
II. Ectopic	228±26	4.25±.5	134±11.5	111.4±18
n =	(56)	(52)	(35)	(59)
III. Eutopic	188±29	3.0±.7	128.4±23.1	196±61
n =	(23)	(19)	(16)	(26)

*p <.001 vs II and III ; ◊ p < .001 vs II, p<0.02 vs III
§ p< 0.001 vs II, p<0.02 vs III

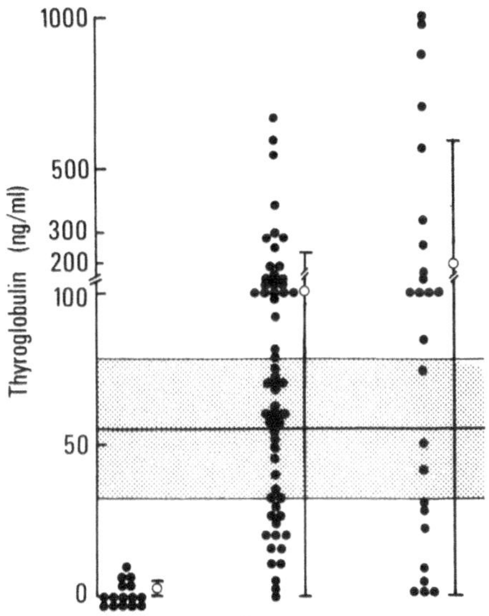

Figure 2 . Thyroglobulin levels in patients with congenital hypothyroidism. Patients were classified according to radioisotope scanning in 3 groups : athyreotic (left), ectopic (middle) or eutopic (right)

Table III . Proportion of patients with ectopic or absent thyroid gland in a group of 63 infants with congenital hypothyroidism (goiter and eutopic glands excluded) according to different criteria used

Type of defect	Radionuclide scanning	T4	Tg
Athyreotic	20 %	32 %	15 %
Ectopic	80 %	68 %	85 %

Criteria used for this classification T4 < 1.5 ug/100 ml and TG < 2.5 ng/ml athyreotic ; T4 > 1.5 ug/100 ml and Tg > 2.5 ng/ml ectopic.

Thus, Tg measurement should be added to the other parameters to provide a better description of the type of thyroid defect. Radioisotope scanning and thyroglobulin (Tg) measurement are both of value in the diagnosis of congenital hypothyroidism (CH). Scanning provides a rapid diagnosis of CH and an indication of the anatomical abnormality involved. But it requires good technical expertise and operators especially trained in the examination of newborn infants. Measurement of Tg becomes the most suitable technique when scintigraphy is not used, or is inadequately adapted to these very young patients. A Tg level above 10 ng/ml in an infant with elevated TSH indicates the presence of thyroid tissue. Undetectable Tg and the absence of goitre provides a diagnosis of athyreosis with a very low margin of error.

Tg measurement and scanning evaluate two different aspects of thyroid function and, as such, both techniques are complementary.

III. CONTROL OF THYROGLOBULIN SECRETION IN PATIENTS WITH ECTOPIC THYROID GLAND

As discussed above, serum Tg may be elevated in patients with ectopic or eutopic thyroid gland. The exact mechanisms responsible for this high serum Tg is not clear. Elevated serum TSH and stimulation of the thyroid gland is probably the primary factor. However other mechanisms should also be considered including a decreased clearance or an increased rate of secretion of Tg during this period of life.

We have studied the changes in serum Tg in a group of infants with ectopic thyroid gland during treatment and after a short interuption of L-T4 administration in an attempt to provide a better understanding of the TSH dependance of Tg secretion.

Table IV . Serum TSH, FT4 and Tg concentrations during several intervals of treatment in 17 children with congenital hypothyroidism due to ectopic thyroid gland (mean ± SEM)

	Before treatment (<1 month)	Interval of therapy (Month) 0.5-2	2-10	10-34	> 34
TSH uu/ml	245 ± 65 (15)	48 ± 1 (9)	5.7 ± 0.9 (11)	4.5 ± 0.9 (8)	4.5± 1.3 (10)
T4 ug/dl	4.7 ± 0.8 (14)	12.4±0.7 (9)	12.7± 0.8 (10)	13± 0.9 (7)	
Tg ng/ml	98.3± 32.8 (17)	71.8± 20.7 (12)	18.5± 3.6* (15)	12.9± 4.9 (9)	1.4± 1§ (14)
Control (Tg only)	55.3± 6.7◊ (12)		33.5± 4.4 (8)	21.4± 2.2 (16)	23.5± 3 (15)

◊ age range 20 to 35 days
§ $p < 0.001$ * $p < 0.02$ vs control

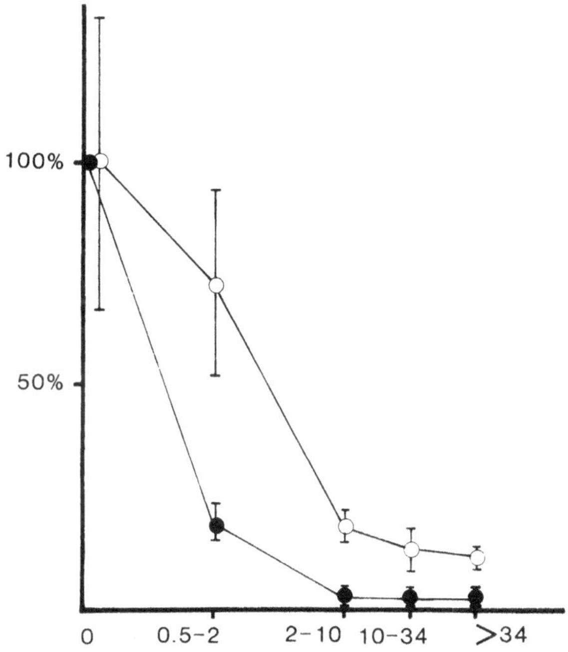

Figure 3. Variation of serum TSH (0) and Tg (0) in children with congenital hypothyroidism during thyroxine therapy. Values on the vertical axis are expressed in percent of initial value at the time of diagnosis. The intervals of treatment in months are shown on the horizontalaxis

Patients and protocole

A group of 17 children with ectopic thyroid glands were studied for 4 years. All had measurable serum Tg levels when evaluated during the first weeks of life (table IV). L-T4 therapy was initiated before 1 month of age and patients were seen at regular intervals for evaluation as indicated in table IV and fig 3.

A separate group of 16 children with ectopic gland was studied to evaluate Tg secretion after long term treatment. They have been treated and followed in our clinic for more than 5 years (7.6 $\pm$ 0.5 yrs). Children were evaluated under adequate therapy (Period I) and after 4 weeks of L-T4 withdrawal (Period II). Informed consent was obtained from the parents, and serum Tg, TSH, FT4 and FT3 were measured during adequate therapy and at the end of period II.

Results and discussion

Mean values for serum Tg and TSH levels in treated patients for the several treatment intervals are shown in table 4 and are represented as percent of the initial values in fig 3. Serum T4 increased with treatment and mean values remained in the high normal range. It can be seen that TSH decreased rapidly and became normal after 2 months of therapy. A parallel decrease in serum Tg was observed although the rate of decline was slower than the fall in serum TSH. After 2 months of therapy the mean serum Tg level was significantly lower than the mean value for normal control children. Serum Tg remained detectable in the serum of treated children at all treatment intervals.

In a separate group of children with ectopic gland, L-T4 withdrawal causes a significant decrease in FT4 and FT3. As shown figure 4, TSH (uu/ml) increased from 5.6 $\pm$1.5 to 90.4 $\pm$9.6 (mean $\pm$ SEM). The basal serum Tg level was 8.3 $\pm$ 1.8 ng/ml, a value lower than normal control, and rose to 25 $\pm$ 3 ng/ml at the end of period II. There was no correlation between Tg and TSH, FT4 or FT3 during this period.

The results of the longitudinal study seem to indicate that the decline of serum Tg values in patients with ectopic gland is at least partly due to the decrease in TSH secretion. However, it should be kept in mind that in normal infants (5) and children (14) there is a gradual decline in serum Tg and one might speculate that our observation in hypothyroid children reflects this normal process of progressive decline in serum Tg. Nevertheless, the decline of serum Tg paralleled a progressive decrease in serum TSH concentrations during LT4 therapy. The converse was true when LT4 therapy was interupted. The high concentration of serum TSH is therefore the primary factor associated with the elevated serum Tg. However, we did not observe a significant correlation between the serum TSH and Tg levels in this group of children and such a correlation might be expected if TSH was the principal cause of the elevated serum Tg. Other mechanisms should also be considered, including the metabolic clearance rate of Tg and its rate of secretion by the thyroid gland. A decreased clearance of Tg may explain a high serum Tg level in hypothyroidism, but to our knowledge there are no data available on this point and we conducted no measurements of Tg clearance.

Thus, an increased rate of secretion by an ectopic tissue remnant remains a possible cause. It has been shown in patients with nontoxic goiter (16), in some cases of thyroiditis (17) and in some patients with benign and malignant thyroid tumors (18) that increased serum Tg levels may be observed in the absence of an elevated serum TSH. In this situation an increased rate of Tg secretion by an abnormal tissue may be involved. A similar mechanism may operate in children with ectopic tissue. Furthermore, it has been shown that the serum Tg of premature neonates is elevated compared to that of term infants (15). Hence, the ectopic glands in our patients may have remained in a fetal state. Finally, our findings may reflect abnormal secretion by dystrophic tissue or an abnormal developmental process in the ectopic glands. Although histological descriptions of these glands are scarce, available data show that tissue of ectopic thyroid is abnormal and dysplasic (19-21).

The persistence of measurable serum Tg in treated patients suggests that thyroid secretion occurs to some extent in patients with treated congenital hypothyroidism. It has been shown that the serum Tg level falls to undetectable levels when thyroid tissue is totally suppressed. This can be demonstrated in patients with thyrotoxicosis factitia (21), in whom complete suppression of thyroid function is associated with undetectable serum TSH levels, low radioiodine uptake, and undetectable serum Tg concentrations.

In our group of patients, the L.T4 dosage was adjusted to maintain serum TSH levels within a normal range, and the serum TSH level in these children was never undetectable. The ectopic tissue was therefore continuously stimulated, presumably accounting for the persistent measurable Tg serum levels. The concept of functioning remnant thyroid tissue is further emphasized by the secondary rise in serum Tg observed when the thyroxine replacement dose was reduced. This increase in serum Tg indicates that thyroid tissue was present in these children and, when stimulated, could secrete more Tg and possibly T4. Our results indicate that the measurement of serum Tg can be a useful tool for evaluating residual thyroid function in patients with congenital hypothyroidism after treatment.

<u>In conclusion</u>

Tg measurement has allowed to describe 2 opposite situation. Very low or undetectable serum levels are found when tissue is absent or in infants with goiter and inborn

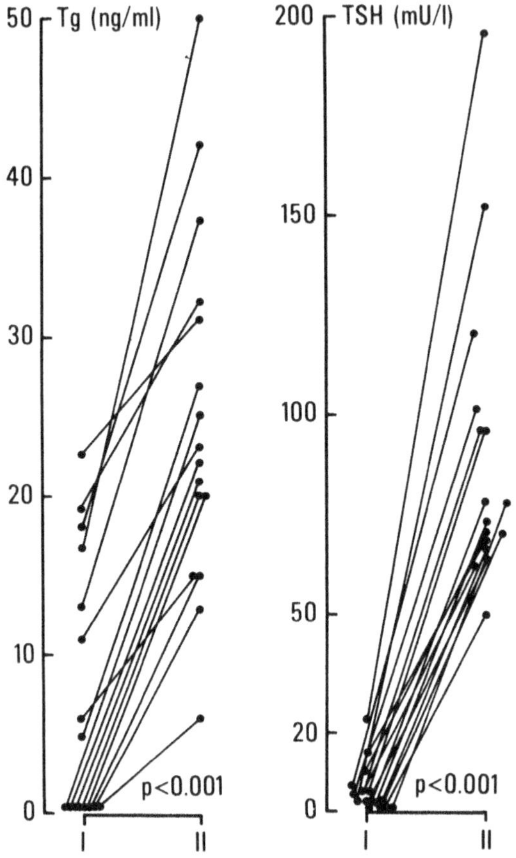

FIGURE 4 . Variation in serum Tg and TSH during adequate therapy (period I) and LT4 withdrawal (period II) in a group of 16 children with congenital hypothyroidism due to ectopic gland.

errors of thyroid hormone metabolism. High or even very high values are found in children with ectopic or eutopic gland. The reason(s) for these elevated values are not clear. TSH stimulation of the gland seems to be the primary factors since L.T4 treatment decreases Tg serum levels. Tg measurement and radioisotope scanning are complementary investigation which allowed a better description of this disease.

Acknowledgments

We thank Dr Schlumberger and Fragu for their continuous support in Tg evaluations and T. Sanchis for her expert technical help.

REFERENCES

1. A. J. van Herle, G. Vassart, J. E. Dumont, Control of thyroglobulin synthesis and secretion, N Engl J Med 301 : 307 (1979)
2. P. Olin, R. Ekholm, S. Almquist, Biosynthesis of thyroglobulin related to the ultrastructure of the human fetal thyroid gland, Endocrinology 87: 1000 (1970)
3. D. Gitlin and A. Biasucci, Ontogenesis of immunoreactive thyroglobulin in the human conceptos. J Clin Endocrinol Metab 28 : 849 (1969)
4. B. Osotimehin, E. G. Black, R. Hoffenberg, Thyroglobulin concentration in neonatal blood : a possible test for neonatal hypothyroidism, Br Med J 2: 1467 (1978)
5. J. L. Ket, J. J. M. de Vijlder, H. Bikker, M. H. Gons, W. H. H. Tegelaers, Serum thyroglobulin levels: the physiological decrease in infancy and the absence in athyroidism, J Clin Endocrinol Metab 53: 1301 (1981)
6. V. Pezzino, S. Filetti, A. Belfiore, S. Proto, G. Donzelli, R. Vigneri, Serum thyroglobulin levels in the newborn, J Clin Endocrinol Metab 52 : 364 (1981)
7. E. Roti, A. Gnudi, E. L. Braverman, G. Robuschi, R. Emanuele, P. Bandini, L. Benassi, A. Pagliani, and C. H. Emerson, Human cord blood concentrations of thyrotropin, thyroglobulin, and iodothyronines after maternal administration of thyrotropin-releasing hormone. J Clin Endocrinol Metab 53 : 813 (1981)
8. E.G. Blak, S. J. Bodden, J. A. Hulse, R. Hoffenberg, Serum thyroglobulin in normal and hypothyroid neonates, Clin Endocrinol 16 : 267 (1982)
9. U. B. Ericsson, S. A. Ivarsson, and P. H. Persson, Thyroglobulin in cord blood. The influence of the mode of delivery and the smoking habits of the mother, Eur J Pediatr 146 : 44 (1987)
10. J. Léger, A. Tar, M. Schlumberger, and P. Czernichow, Control of thyroglobulin secretion in patients with ectopic thyroid gland. Ped Res 23 : 266 (1988)
11. A. J. van Herle, R. P. Uller, N. L. Matthews, J. Brown Radioimmunoassay for measurement of thyroglobulin in human serum, J Clin Invest 52 : 1320 (1973)

12. P. Czernichow, M. Schlumberger, R. Pomarède, P. Fragu, Plasma thyroglobulin measurements help determine the type of thyroid defect in congenital hypothyroidism, J Clin Endocrinol Metab 56 : 242 (1983)
13. J. Léger, M. Schlumberger, P. Czernichow, Diagnostic étiologique de l'hypothyroidie congénitale et thyroglobuline plasmatique, Arch Fr Pediatr 43 : 179 (1986)
14. A. Muir, D. Daneman, A. Daneman, R. Ehrlich, Thyroid scanning, ultrasound and serum thyroglobulin in determining the origin of congenital hypothyroidism, Am J Dis Child 142 : 214 (1988)
15. R. Penny, C. A. Spencer, D. Frasier, J. F. Nicoloff, Thyroid-stimulating hormone and thyroglobulin levels decrease with chronological age in children and adolescents, J Clin Endocrinol Metab 56 : 177 (1983)
16. U. Feldt-Rasmussen, P. H. Petersen, J. Date, C. M. Madsen, Serum thyroglobulin in patients undergoing subtotal thyroidectomy for toxic and non toxic goiter, J Endocrinol Invest 5 : 161 (1982)
17. R. C. Smallridge, F. M. de Keyser, A. J. Vanherle, N. E. Butkus, L. Wartofsky, Thyroid iodine content and serum thyroglobulin : cues to the natural history of destruction induced thyroiditis, J Clin Endocrinol Metab 62 : 1213 (1986)
18. M. Schlumberger, P. Charbord, P. Fragu, J. Lumbroso, C. Parmentier, M. Tubiana, Circulating thyroglobulin and thyroid hormones in patients with metastases of differentiated thyroid carcinoma : relationship to serum TSH levels, J Clin Endocrinol Metab 51 : 513 (1980)
19. J. J. Sauk, Ectopic lingual thyroid. J Pathol 102 : 239 (1970)
20. A. L. Strickland, J. A. MacFie, J. J. Van Wyk, F. S. French, Ectopic thyroid glands stimulating thyroglossal duct cysts : hypothyroidism following surgical excision. JAMA 208 : 307 (1967)
21. W. T. Conklin, R. M. Davis, R. W. Dabb, C. M. Reilly, Hypothyroidism following removal of a "thyroglossal duct cyst", Plast Reconst Surg 68 : 930 (1981)

DISCUSSION

DELANGE (Brussels) : After withdrawal of therapy, I noticed that total T4 remained in the normal range while free T4 levels decreased. What is your explanation ?

CZERNICHOW (Paris) : The total T4 was measured at the time of diagnosis and the free T4 after withdrawal of therapy. I wanted to show that there was no correlation between T4 and free T4. In other words, it is impossible to predict from the initial T4 level which glands will still be "productive" 5 years later.

CHANOINE (Brussels) : Could a thyroglobulin measurement be an useful indicator to follow substitutive therapy ?

CZERNICHOW (Paris) : I don't think so because we find cases with undetectable TG levels and measurable TSH values, in the 4-5 μU/ml range.

MITCHELL (Jamaica Plain) : I have two questions. The first one concerns the possibility of using TG determinations to discriminate between agenesis patients, who sometimes have measurable TG levels, and ectopia. Is it possible to see a difference in the rise of TG when stimulated by TSH, (for instance after withdrawal of therapy) between ectopia and athyreosis ? My second question concerns the possible usefulness of TG determinations to discriminate between transient and permanent forms of CH.

CZERNICHOW (Paris) : I have no answer to your first question because we have avoided stopping therapy in those babies. For the second question, I have seen 2 children with transient hypothyroidism. Both of them had very high TG levels, which you would expect in normal glands actively stimulated by TSH.

DREXHAGE (Amsterdam) : Do you know if your TG RIA measures only intact TG molecules or perhaps also fragments or denatured molecules ?

CZERNICHOW (Paris) : We have no data on this.

SIGNIFICANCE OF EARLY DIAGNOSTIC DATA IN CONGENITAL HYPOTHYROIDISM: REPORT OF THE SUBCOMMITTEE ON NEONATAL HYPOTHYROIDISM OF THE EUROPEAN THYROID ASSOCIATION*

F. Delange, Brussels, J. De Vijlder, Amsterdam, G. Morreale de Escobar, Madrid, P. Rochiccioli, Toulouse, and S. Varrone, Naples

SUMMARY

The objective of this work has been to evaluate the natural history of thyroid dysgenesis during early life in order to explain why infants with congenital hypothyroidism and with similarly low serum T_4 at diagnosis frequently have markedly different clinical signs of hypothyroidism. Standardized data were collected in an international collaborative study from 181 infants (132 females and 49 males) with congenital hypothyroidism detected by neonatal screening and in whom thyroid scanning performed before therapy allowed a precise etiological diagnosis i.e. thyroid agenesis, ectopic thyroid or dyshormonogenesis. In thyroid agenesis, serum thyroxine (T_4) at screening was low but not necessarily undetectable and rapidly decreased thereafter. In ectopic thyroids, serum T_4 decreased by 40 % between screening and diagnosis in approximately one third of the infants. Serum thyroglobulin (Tg) was not necessarily undetectable in thyroid agenesis nor detectable in ectopic thyroids. There was only a poor relationship between serum T_4 and bone maturation at diagnosis in congenital hypothyroidism. The data suggest that loss of functional capacity of the dysgenesic thyroid tissue occurs during early life in some infants with congenital hypothyroidism and that serum Tg cannot replace thyroid scintigraphy in the evaluation of infants with congenital hypothyroidism.

INTRODUCTION

Infants with congenital hypothyroidism and with similarly low serum thyroxine at diagnosis frequently have markedly different clinical signs of hypothyroidism (1).

This discrepancy can be partly explained by the lack of precision in the assessment of clinical data in young infants with congenital hypothyroidism. It also could be explained by differences between individuals in the degree of adaptation to thyroid failure. These mechanisms could be

* Work performed with the acknowledged collaboration of P. Bourdoux and A.M. Ermans, Brussels; J.P. Bourguignon and C. Ernould, Liège; J.H. Dussault, Québec; J.P. Farriaux, Lille; K. Von Reuss and W. Rabl, Munich; R. John, I. Hughes and J. Lazarus, Cardiff.

involved. 1) Differences between individuals in the transfer of maternal thyroid hormones across the placenta during fetal life. 2) Differences in the preferential secretion of T_3 by the dysgenesic thyroid tissue which could maintain a state of euthyroidism in spite of low serum T_4. 3) Differences in the duration of the hypothyroid state at diagnosis in spite of similar serum T_4. This difference could result from a loss of the functional capacity of the dysgenesic thyroid tissue occurring during late fetal or early postnatal life only in some infants.

The first mechanism is particularly difficult to investigate in humans but is possible on the basis of recent experimental data (2-4). The second mechanism is at least partly established (5).

The objective of the present work has been to investigate the third possibility.

PATIENTS AND METHODS

Patients

The Subcommittee on Neonatal Hypothyroidism of the European Thyroid Association has conducted an international collaborative study on infants with congenital hypothyroidism detected by screening thanks to the acknowledged collaboration of the screening centers of Liège and Brussels (Belgium), Quebec (Canada), Toulouse and Lille (France), Munich (Germany) and Cardiff (Great Britain).

The study included 181 infants (132 females and 49 males) with congenital hypothyroidism in whom thyroid scanning data were available before therapy. The age at diagnosis varied from 1 to 93 days (mean $\pm$ SEM : 22 $\pm$ 2 days). Thyroid scintigraphy evidenced thyroid agenesis in 44 of the 181 infants (24.3 %) and ectopic thyroids in 112 (61.9 %). The last 25 infants (13.8 %) had permanent primary hypothyroidism with the thyroid in normal position, strongly suggesting dyshormonogenesis.

Thirty-three infants with elevated serum concentrations of thyrotropin (TSH) but normal thyroxine (T_4) at screening and with normal control examinations were used as controls. For obvious ethical reasons, perfectly normal newborns could not be investigated.

Methods

The serum concentrations of TSH and T_4 were determined locally by the participating centers on spot samples at screening and on serum at diagnosis. The serum concentration of thyroglobulin (Tg) at diagnosis was also determined. A large majority of the Tg assays were performed or at least repeated in Brussels in order to avoid differences between laboratories due to methodological reasons : the antigen hTg was purified according to the method of Van Herle et al. (6). The antiserum was produced by the immunization of New Zealand rabbits with a mixture of equal parts of hTg (1 mg/ml) and complete Freund's adjuvent. Sera were harvested and stored at - 20° C. ^{125}I hTg was prepared freshly with the chloramine-T method and chromatographied on a Sephadex G-200 column. Specific activity was approximately 40 µc/µg. The radioimmunoassay of hTg was made by using a PBS buffer ph 7.3, 0.01 M, containing EDTA (0.01 M), BSA (0.5%) and azide (0.1 %). 100 µl of serum or standard were incubated with 100 µl of buffer and 100 µl of antiserum (dilution 1 : 90.000) for 24 h at 4° C. Subsequently, 100 µl of ^{125}I-hTg were added and the mixture was further incubated for 72 h at 4° C. Finally, 100 µl of normal rabbit serum (1 %) and 100 µl of a 4 % solution of antirabbit gamma-globulins were added and incubated for 24 h at 4° C. All tubes were centrifuged at 4° C for 30 minutes and the supernate was discarded.

The interassay coefficient of variation was 6.7 %.

Thyroid scintigrams were performed in the hypothyroid infants only. Bone maturation was evaluated by using the method of Senecal et al. (7).

The investigation consisted in three parts : 1) a comparison of the serum concentration of T_4 at screening and diagnosis in the same infants with thyroid agenesis or ectopic thyroids.
2) a comparison of the serum concentrations of Tg in infants with thyroid aganesis, ectopic thyroids, dyshormonogenesis and in the controls.
3) an evaluation of the relationship between the serum concentration of T_4 at diagnosis, used as an index of the degree of thyroid failure at this time and bone maturation, used as an index of the supply of thyroid hormones to the tissues on a longterm basis from early fetal life up to the time of disgnosis.

RESULTS

Figure 1 compares serum T_4 at screening and diagnosis in 30 infants with thyroid agenesis and 47 infants with ectopic thyroids. The mean ($\pm$ SEM) time interval between screening and diagnosis was 22.8 $\pm$ 2.8 days.

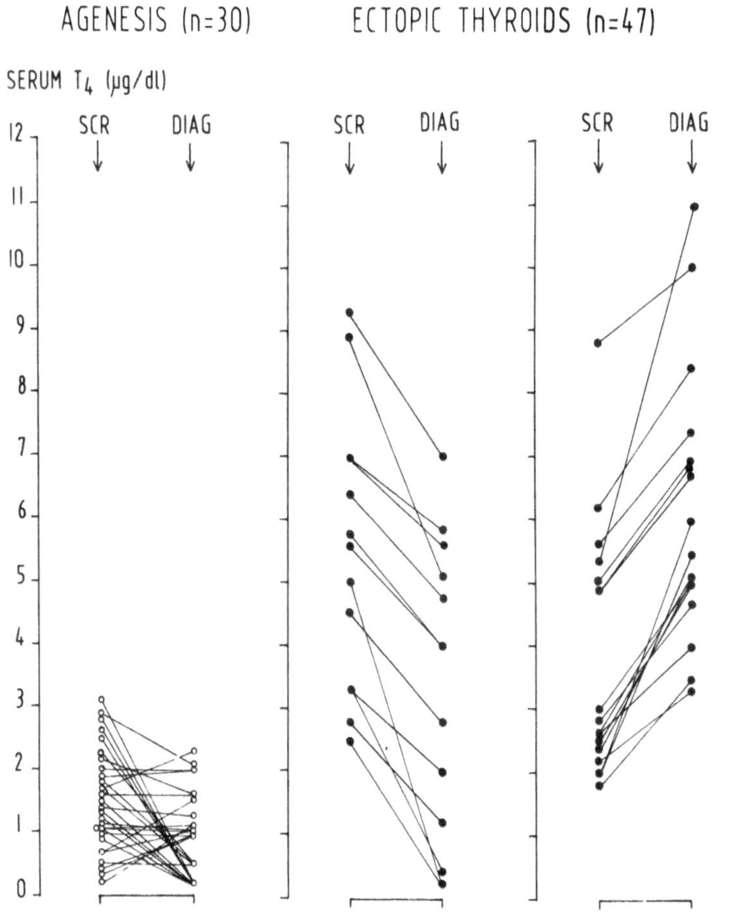

Figure 1 . Comparison of the serum concentrations of thyroxine (T_4) at screening (SCR) and diagnosis (DIAG) in infants with congenital hypothyroidism due to thyroid agenesis or ectopic thyroids. Serum T_4 was unmodified in the last 17 infants with ectopic thyroids.

In thyroid agenesis, T_4 at screening was low but not necessarily undetectable. It did not change between screening and diagnosis in 21 infants and further decreased in the 9 others. The mean (± SEM) values were 1.7 ± 0.2 μg/dl at screening and 1.0 ± 0.1 μg/dl at diagnosis ($p < 0.005$).

In ectopic thyroids, T_4 decreased between screening and diagnosis in 13 infants, i.e. in 28 % of them. The mean decrease was 40 % of the initial value and the difference was significant ($p < 0.001$). At diagnosis, five infants with ectopic thyroids, i.e. 10 % of them, had T_4 values situated within the range observed in thyroid agenesis. On the contrary, T_4 increased in 17 other infants. It was unmodified in the last 17 infants.

Figure 2 compares the serum concentrations of Tg observed in the controls and in 105 infants with congenital hypothyroidism, 18 with thyroid agenesis, 67 with ectopic or hypoplastic glands and 20 with dyshormonogenesis.

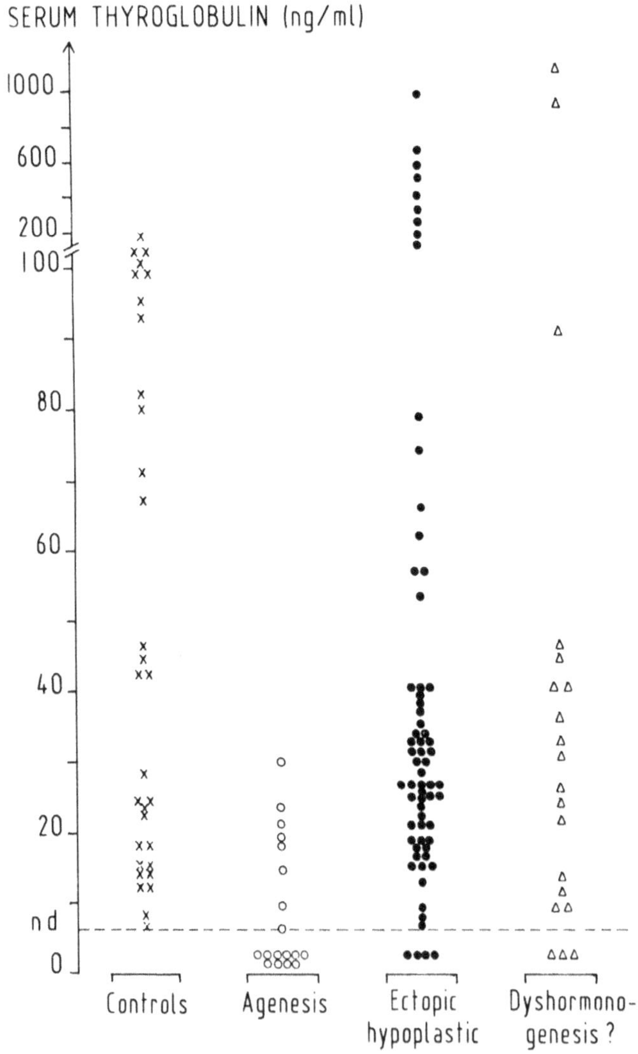

Figure 2 . Comparison of the serum concentrations of thyroglobulin at the time of diagnosis in the controls and in infants with congenital hypothyroidism due to thyroid agenesis, ectopic or hypoplastic thyroids and dyshormonogenesis. nd = non detectable.

Serum Tg was detectable in all the controls. It showed a very important variability between individuals, from 6 to 158 ng/dl. Among the 18 infants in whom thyroid scintigraphy did not reveal any functioning thyroid tissue, serum Tg was not detectable in only ten and was barely or clearly detectable in the last eight infants. Conversely, among the 67 infants in whom thyroid scintigraphy evidenced a limited quantity of functioning thyroid tissue, usually in ectopic position, serum Tg was detectable in only 63 and was undetectable in the last four infants. When detectable, serum Tg varied markedly between individuals, from barely detectable values up to extremely elevated values. An almost similar picture was observed in the 20 infants with dyshormonogenesis.

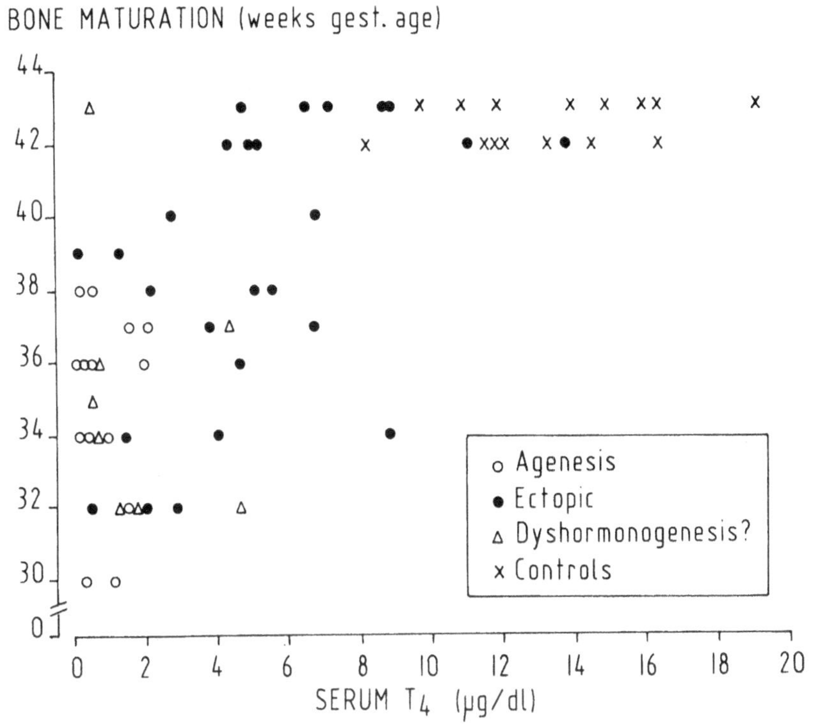

Figure 3 . Relationship between serum T_4 and bone maturation at the time of diagnosis in infants with congenital hypothyroidism and in controls.

Figure 3 shows the relationship observed in the infants investigated in Brussels between serum T_4 and bone maturation at the time of diagnosis. There was a relationship between both variables but again, there was an important variability between individuals. Bone maturation was normal in all infants when serum T_4 was above 9 μg/dl. It was occasionally retarded when T_4 was below 7 μg/dl and was almost systematically retarded only when serum T_4 was below 3 μg/dl. However, even for undetectable serum T_4 at diagnosis, bone maturation varied markedly between individuals from a normal value of 43 weeks to values as low as 30 weeks.

DISCUSSION

This work indicates that in sporadic congenital hypothyroidism due to thyroid dysgenesis, serum T_4 decreases by 40 % between screening and diagnosis in approximately one third of the infants with ectopic thyroids, but in contrast, it increases in another third of them. It also shows that serum Tg is usually but not necessarily undetectable in thyroid agenesis and conversely, usually but not necessarily detectable in ectopic thyroids. In other words, the determination of serum Tg is neither a sensitive nor a specific index of the presence or absence of functioning thyroid tissue in congenital hypothyroidism. Finally, this work shows that there is a poor relationship between serum T_4 and bone maturation at diagnosis.

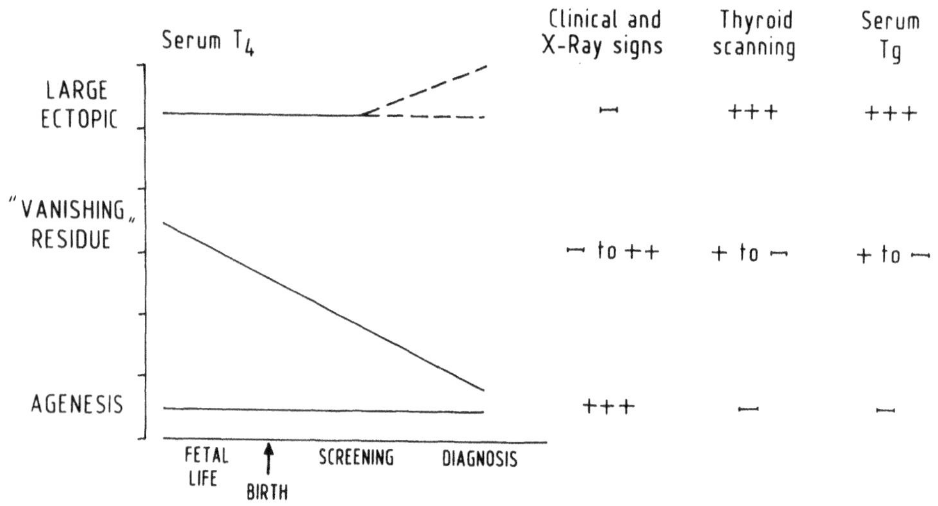

Figure 4 . Hypothetic scheme of the natural history of thyroid dysgenesis. Changes in serum T_4 from fetal life up to the time of diagnosis and consequences on the clinical, radiological and scanning data and on serum Tg. + : Present. ├──┤ : Absent.

Some of the apparently contradictory findings reported in this work could be explained by methodological errors, for example in the scanning data, Tg assays or evaluation of bone maturation. However, such errors cannot entirely explain the overlap observed between the different groups of patients, especially considering that the determination of serum Tg and bone maturation were centralized in order to overcome the problem of variability between laboratories.

Rather, the findings could be reconciled if a dysgenesic thyroid tissue was considered not as a static phenomenon of all or nothing but rather as a dynamic, progressively changing system. Figure 4 summarizes the data and their possible interpretation along this line.

Primary TSH screening identifies infants with large ectopic thyroids and normal or subnormal T_4. They usually have no clinical nor radiological signs, positive scans and elevated serum Tg. A progressive increase of serum T_4 observed in some of them suggests that persistent hyperstimulation by TSH could even improve the functional capacity of their thyroid residue. At the other extreme of the range, are infants who never had any thyroid tissue. They have severe clinical and X-rays signs, no detectable thyroid tissue nor serum Tg. In between are infants with what we would like to call "vanishing" thyroid tissue, explaining the progressive decrease of serum T_4 in spite of persistent TSH hyperstimulation. Depending on the timing and degree of the process and on the possible dissociation in the time course of the trapping and secretory capacities of the dysgenesic thyroid tissue, infants with similarly low T_4 at diagnosis will have variable clinical and radiological signs; thyroid tissue and serum Tg will be detectable or not. This would explain the possible association of undetectable thyroid tissue and detectable serum Tg and vice versa.

Such a concept of congenital hypothyroidism is not purely speculative because progressive loss of functional capacity of ectopic thyroid tissue has been documented in older infants with thyroid dysgenesis several years before the introduction of neonatal screening (8-11). The issue is important because it could indicate that, depending on the time course of thyroid function of the dysgenesic tissue before diagnosis, affected infants of the same age with the same serum T_4 at diagnosis and the same adequate therapy may have a different neurointellectual prognosis (1, 12, 13).

Finally, the observation of clearly detectable but rapidly decreasing serum T_4 and of barely retarded bone maturation in some infants in whom thyroid scintigraphy did not reveal any functioning thyroid tissue (agenesis) could also be explained by a certain degree of transfer of maternal thyroid hormones across the placenta.

In conclusion, our data suggest that loss of functional capacity occurs during early life in some infants with congenital hypothyroidism due to thyroid dysgenesis. Such a process could account at least partly for the variability of the clinical and radiological findings in CH infants with similarly low levels of serum T_4 at diagnosis. The cause of such a process of atrophy is unknown. The role of autoimmune factors would be an attractive possibility (14, 15).

This investigation confirms that the determination of serum Tg contributes to the evaluation of the cause of congenital hypothyroidism (16-21). However, because of its lack of specificity, this determination cannot replace thyroid scintigraphy. Rather, the two variables provide complementary data by exploring different aspects of the function of the dysgenesic thyroid tissue.

ACKNOWLEDGEMENT

The Subcommitte gratefully acknowledges P. Bourdoux, Ph.D., Brussels for Tg assays and Dr. D.C. Evered, London, in the preparation of this report.

Correspondance should be addressed to Dr. F. DELANGE, Department of Radioisotopes, University Hospital Saint-Pierre, 322, rue Haute, B - 1000 - Brussels, Belgium.

REFERENCES

1. J. Letarte, and S. La Franchi, Clinical features of congenital hypothyroidism, in : "Congenital hypothyroidism" J.H. Dussault and P. Walker eds., M. Dekker Publ., New York (1983) : 351-383.
2. G. Morreale de Escobar, R. Pastor, M.J. Obregon and F. Escobar del Rey, Effect of maternal hypothyroidism on the weight and thyroid hormone content of rat embryonic tissues, before and after onset of fetal thyroid function. Endocrinology 117 : 1890-1900 (1985).
3. R.J. Wood, A.K. Sinha and R. Ekins, Uptake and metabolism of thyroid hormones by the rat fetus in early pregnancy. Clin. Sci. 67 : 359-363 (1984).
4. R. Ekins, Roles of serum thyroxine-binding proteins and maternal thyroid hormones in fetal development. Lancet i : 1129-1132 (1985).
5. J. Verelst, F. Delange, P. Bourdoux, Q. Vu and A.M. Ermans, Circulating thyroid hormones in thyroid dysgenesis, in : "Neonatal screening" H. Naruse and M. Irie eds., Excerpta Medica publ., Amsterdam (1983) : 138-139.
6. A.J. Van Herle, R.P. Uller, N.L. Matthews and J. Brown, Radioimmunoassay for measurement of thyroglobulin in human serum. J. Clin. Invest. 52 : 1320-1327 (1983).
7. J. Senecal, M.C. Grosse, A. Vincent, J. Simon and J.N. Lefrèche, Maturation osseuse du foetus et du nouveau-né. Arch. Franc. Ped. 34 : 424-438 (1977).
8. N. Neimann, M. Pierson and J. Martin, Hypothyroïdies par troubles du développement et lésions inflammatoires du corps thyroïde, in : "Les hypothyroïdies."Rapport 18ème Congrès de l'association des pédiatres de langue française. Genève 1961. S. Karger publ., Bâle (1961); 1 : 89-157.
9. J.C. Job, P. Canlorbe and M. Tubiana, Decreasing radioiodine uptake during the course of congenital hypothyroidism, in : "Current topics in thyroid research" C. Cassano and A. Andreoli eds. Academic Press New York (1965) : 828-831.
10. G. Dutau and P. Rochiccioli, Les dysgénésies thyroïdiennes. A propos de 63 observations. Ann. Pediat. 22 : 315-325 (1975).
11. M. Kaplan, R. Kauli, E. Lubin, M. Grunebaum and Z. Laron, Ectopic thyroid gland. A clinical study of 30 children and review. J. Pediatr. 92 : 205-209 (1978).
12. J. Glorieux and F.A. Lavecchio, Psychological and neurological development in congenital hypothyroidism, in : "Congenital hypothyroidism" J.H. Dussault and P. Walker eds. M. Dekker publ. New York (1983) : 411-430.
13. J. Rovet, R. Ehrlich and D. Sorbara, Intellectual outcome in children with fetal hypothyroidism. J. Pediatr. 110 : 700-704 (1987).
14. R.D. Van der Gaag, H.A. Drexhage and J.H. Dussault, Role of maternal immunoglobulins blocking TSH-induced thyroid growth in sporadic forms of congenital hypothyroidism. Lancet i : 246-250 (1980).
15. J.H. Dussault and D. Bernier, ^{125}I uptake by FRTL5 cells : a screening test to detect pregnant women at risk of giving birth to hypothyroid infants. Lancet ii : 1029-1031 (1985).
16. J.L. Ket, J.J.M. De Vijlder, H. Bikker, M.H. Gons and W.H.H. Tegelaers, Serum thyroglobulin levels : the physiological decrease in infancy and the absence in athyroidism. J. Clin. Endocrinol. Metab. 58 : 1301-1303 (1981).

17. E.G. Black, S.J. Bodden, J.A. Hulse and R. Hoffenberg, Serum thyroglobulin in normal and hypothyroid neonates. Clin. Endocrinol. 16 : 267-274 (1982).

18. P. Czernichow, M. Schlumberger, R. Pomarede and P. Fragu, Plasma thyroglobulin measurements help determine the type of thyroid defect in congenital hypothyroidism. J. Clin. Endocrinol. Metab. 56 : 242-245 (1983).

19. M.H. Gons, J.H. Kok, W.H.H. Tegelaers and J.J.M. De Vijlder, Concentration of plasma thyroglobulin and urinary excretion of iodinated material in the diagnosis of thyroid disorders in congenital hypothyroidism. Acta Endocrinol. 104 : 27-34 (1983).

20. F. Pacini, R. Lari, P. La Ricca, L. Grasso, F. Di Bartolo, G.F. Fenzi, M. Ciampi, S.G. Gragnani, L. Baschieri and A. Pinchera, Serum thyroglobulin determinations in the differential diagnosis of congenital hypothyroidism. J. Endocrinol. Invest. 7 : 29-33 (1984).

21. F. Dammaco, A. Dammaco, T. Cavallo, S. Sansonna, N. Bafundi, C. Torelli, E. Frezza, F. Vitale and D. Griseta, Serum thyroglobulin and thyroid ultrasound studies in infants with congenital hypothyroidism. J. Pediatr. 106 : 451-453 (1985).

DISCUSSION

FISHER (Torrance) : Would you care to comment on the possibility that there is a significant variation in the timing of thyroid hormone effects which, in turn, could explain why some of these infants with T4 under 2 µg/dl have a bone maturation equivalent to 38 to 40 weeks ?

DELANGE (Brussels) : I agree that it is that additional possibility that we have to consider, especially after the presentation of the animal models you reviewed for us yesterday.

KLEIN (Hanover) : I think that you need not limit those patients whose thyroid function "vanishes" to those who have ectopic tissue with absence of eutopic tissue. Many years ago, it was shown that a loss in the ability to take up radioactive iodine occurred over a period of 9 to 10 months postnatally in patients with hypothyroidism. After all, it is plausible that the bulk of these patients represent the same disease. The ectopic function could represent a compensatory mechanism. The cells left behind at the original thyroid location are variably compensated so that we see a large variation in the extent of thyroid degeneration at the time of birth.

DELANGE (Brussels) : Thank you for your comments. I agree that my theory is speculative because we have no histological proof in these babies. It is also possible that maternal transfer of thyroid hormones explains the measurable T4 levels in some of these infants with undetectable thyroid tissue.

MORREALE (Madrid) : I am happy that I did not have to make this comment. However, measurements are usually performed 5 days after birth. In the newborn the half-life of T4 is very rapid. Therefore, the serum T4 level could have been higher if measured in cord blood. What we need is a collaborative international project to get cord samples from large numbers of screened babies.

DELANGE (Brussels) : It requires, indeed, a well-planned collaborative program. In our own series, we have, however, seen no change between cord levels and day 5.

VULSMA (Amsterdam) : In 3/15 children with organification defects we observed that 3 days after birth T4 levels were 40-50 mM/l. Later on, in those infants, after withdrawal of therapy, the T4 fell to 5 mM/l. So, we have to explain where this T4 was coming from if the inborn metabolic error was the same at birth and 5 years later. Could it have come from the mother ?

GLINOER (Brussels) : There are alternative explanations : you could have an organification block which is not complete, so you may still have some T4 formed in those patients.

SACK (Tel Aviv) : I do not understand how ectopic tissue may still be functioning after several years, and be "spared" from auto-immune destruction.

DELANGE (Brussels) : It is possible by analogy with the testes for example, that for anatomical reasons, thyroid tissue located in an abnormal situation does function for a given period of time and then stops producing hormones. This need not necessarily imply auto-immune destruction.

DUSSAULT (Quebec) : Two comments. First, Paul Walfish did measure T4 both in cord serum and after 5 days. Second, a T4 level of 4 µg/dl in a neonate (knowing the T4 clearance rate and the secretion rate of the mother) would represent an almost free clearance of T4 from mother to fetus, which doesn't make sense to me. So, perhaps, 1-2 µg/dl might be coming from the mother but not higher levels.

DELANGE (Brussels) : I am personally not ready to accept that athyreotic babies have T4 levels of 4 µg/dl. I would rather suspect that the thyroid scan was not properly performed.

DUSSAULT (Quebec) : However, if antibodies block TSH and thyroid uptake, you can see a negative scan.

DREXHAGE (Amsterdam) : I learned today that in so-called thyroid agenesis, some thyroid cells could still be there for years which, in turn, gives an argument to explain why we still detect antibodies after several years. Also, destruction of cells by autoimmune processes is not a simplistic direct mechanism. In autoimmunity, destruction is always the result of multiple factors and of multiple cells associated with the process, sometimes for years. There could be a balance between regenerating forces and destructive forces or between blocking and repair.

PART 4. THERAPY AND NEUROPSYCHOINTELLECTUAL DEVELOPMENT

PROBLEMS IN THE MANAGEMENT OF PATIENTS WITH INFANTILE HYPOTHYROIDISM

Marvin L. Mitchell, Rosalie J. Hermos, Deborah L. Frederick, and Robert Z. Klein

New England Regional Screening Program, State Laboratory Institute, 305 South Street, Jamaica Plain, MA 02130, USA

There is general agreement that treatment of hypothyroid infants diagnosed by screening has proved to be remarkably effective in preventing the intellectual deficit commonly seen in infantile hypothyroidism of an earlier era. However, not all programs have experienced the same degree of success with respect to neuropsychological outcome among their patients. Both the Toronto and Quebec investigators have reported that children with delayed skeletal maturation had lower global IQ scores than those of the children with normal bone maturation at birth.[1,2]

These findings differ from observations made by us, in which the only consistent variable associated with outcome was adequacy of treatment.[3] We found that failure to maintain the circulating level of T4 in the upper half of the normal range during the first year of life was the only significant factor related to low IQ on stepwise regression analysis. In view of this, we wondered whether the difference in results between us and the Canadians might have been related to suboptimum levels of circulating T4 early in treatment and/or during the first year of life in those patients of the Canadian cohorts who were more severely affected. If this were the case, the consequences would have been especially harmful to those children who had little or no endogenous circulating thyroid hormone. The inadequate therapy in these instances could have resulted from lack of compliance, insufficient medication, poor absorption of the hormone, or other causes not yet defined.

Considering the circumstances, it is impossible at this juncture to obtain direct laboratory evidence to support this proposition. However, we reasoned that if there were any basis in fact to the notion that variability in response played an important role in outcome, we might find some hint of this in data from our patients. Therefore, we reviewed the overall pattern of response to thyroid medication, as reflected by changes in T4 and TSH, in the cohort of patients followed since 1976.

Before turning to the results, I would like to remind you of the manner in which these patients were managed clinically when we began our program. Approximately, three quarters of the cohort are treated by members of the New England Congenital Hypothyroidism Collaborative, and the remainder by pediatricians--all of whom are guided by a standardized protocol. Our goal was to achieve and maintain serum T4 concentrations above 10 μg/dl during the first year of life, this to be done as expeditiously as possible. To accomplish this end, 10 μg/kg of ℓ-thyroxine was recommended at the start of treatment, to be adjusted according to the blood level of T4. The initial dose is usually adequate until the child reaches 6 kg, at which time the dosage is changed to 6 μg/kg. Above 10 kg of weight additional 2-3 μg are added for each kg, and at 20 kg of weight the dosage is 90 μg plus 1 μg for each kg above 20. The purpose of these recommendations was to aid the physician in gauging whether the infant was receiving too little or too much medication and thus to assess adequacy of treatment as well as compliance.

In order to monitor the response to this regimen, measurements of T4 and TSH were requested two and four weeks after treatment began, two to four weeks after a change in dosage, and routinely at 3, 6, 9, 12, and 18 months of age and at each birthday. Thus the protocol called for a minimum number of seven determinations during the first year of treatment.

These guidelines were provided to all pediatricians in the New England network, and it has been our experience that with few exceptions the physicians have made a concerted effort to adhere to the instructions. Despite this we found considerable variability in response both to identical doses of ℓ-thyroxine in the same infants and among different infants. Variation within the same patient was characterized by fluctuating T4 and TSH values while on a constant dose of T4, as seen in Table 1.

However, management of a majority of the patients was satisfactory, as reflected by a TSH level below 20 μU/ml in approximately 75 percent of the cohort four to eight weeks after the start of treatment. Comparisons of characteristics between the infants whose TSH levels were and were not within the normal range after eight weeks of treatment revealed several points of interest (Table 2). The mean T4 concentration of 13 μg/dl of the infants whose TSH value was below 20 μU/ml was significantly higher ($p<0.01$) than the mean T4 of 9.7 μg/dl of the group with elevated TSH levels. In addition, the mean ℓ-thyroxine treatment

Table 1. Case #1986

Age	Rx Dose (μg)	T4 (μg/dl)	TSH (μU/ml)
2 Wks	- 25	<1.0	>150
1 Mo	25	17.4	27
2 Mos	25	11.9	7.3
3 Mos	25 37.5	6.3	44
6 Mos	37.5 50	10.7	20
9 Mos	50	9.9	9.8

dose of 37.3 μg of the normal TSH group also was significantly higher ($p<0.01$) than the mean dose of ℓ-thyroxine of 31.3 μg in the infants with abnormal TSH values. Interestingly, the mean birth weights of the two groups were identical. We were surprised to find that even after one year of treatment, approximately 10 percent of our cohort (9/83) had TSH values that never fell below 20 μU/ml.

These data, taken collectively, confirm the observations of Sack[4] that despite the diminished sensitivity of the hypothalamic feedback control system to thyroxine,[5] inhibition of TSH can be attained with the appropriate level of T4. To accomplish this, in our experience, usually requires only modest elevations in circulating T4, as exemplified by the case in Table 3a. During the first few months of treatment, the infant's TSH remained consistently above 20 μU/ml until the level of T4 rose from 10 μg/dl to 13 μg/dl.

We expected that as the children grew older, the fluctuating levels of T4 and TSH would diminish and thus eliminate the need for frequent monitoring. In practice this has not proved to be the case, as based on findings in the children through the ages of nine years. The results of

Table 2. Characteristics of Infants with TSH Above and Below 20 μU/ml after 8 Weeks of Treatment

% Patients	Mean Treatment Dose (μg)	TSH (μU/ml)	Mean T4 + SD (μg/dl)		T4 Range (μg/dl)	Body Weight (kg)
75	37.3*	<20	13.1*	4.0	6-29	5.4
25	31.3*	>20	9.7*	3.4	3-21	5.4

* $p<0.01$

Table 3a. Case #2102

Age	Rx Dose (μg)	T4 (μg/dl)	TSH (μU/ml)
2 Wks	-	2	445
	37.5		
5 Wks	37.5	10	>30
	25		
10 Wks	25	10	>50
	50		
6 Mos	50	13	10
1 Yr	50	12	12

two of these children, detailed in Tables 3b and 4, graphically illustrate the problem of maintaining the circulating T4 at a designated level. Sporadic elevations of TSH concentrations in the face of normal T4 values in Table 3b suggest serious gaps in the management of the patient. Such a laboratory profile can be explained by temporary periods of noncompliance until shortly before the scheduled visit to the physician's office when a large dose of thyroxine is administered.

These are far from isolated instances, as documented by the fact that approximately 30 percent of the patients had abnormally elevated TSH values on one or more occasions during a period of two or more years. These TSH elevations were not trivial, ranging in concentration from 10 to 280 μU/ml with T4 concentrations as low as 4 μg/dl. The cause of this unhappy state of affairs is probably related to the cavalier attitude of both the physician and family toward treatment with the passage of time. In addition, as the older children assume more responsibility for medicating themselves, there is a greater risk of noncompliance.

The uneven treatment and/or response of some of our patients, as characterized by low T4 and elevated TSH levels, strengthened our belief that the poorer outcome of some of the patients in the Canadian cohort was in all probability the result of inadequate therapy. Because of the dearth of published laboratory values, we can only speculate that this was the underlying cause in the Quebec program. However, there seems to be irrefutable evidence that some of the patients in the Toronto cohort were treated unsatisfactorily. After the start of treatment, the mean time required for serum T4 to reach 10 μg/dl or more ranged from 79 to 92 days in their various subgroups with standard deviations of 56 to 92 days. The mean serum T4 in approximately half the Toronto cohort was 8.8 $\pm$ 1.16 μg/dl after one month and 9.9 $\pm$ 0.84 μg/dl at two months.[6]

None of this is meant as a criticism but is intended to underscore the difficulty inherent in the management of such patients. The one overriding message that comes through from the review of our data is the uncertainty that a prescribed dose of ℓ-thyroxine will achieve the desired results in a particular child. Whether this can be explained by

Table 3b. Case #2102

Age	Rx Dose (μg)	T4 (μg/dl)	TSH (μU/ml)
1 Yr	50	12	12
2 Yrs	50 75	12	44
3 Yrs	75 87.5	13	14
4 Yrs	87.5	17	1
5 Yrs	87.5 100	13	18
6 Yrs	100	14	2

Table 4. Case #3152

Age	Rx Dose (μg)	T4 (μg/dl)	TSH (μU/ml)
1 Wk	-	1.0	500
	50		
3 Wks	50	26.0	2
	25		
5 Wks	25	13.0	-
9 Wks	25	9.2	14
	37.5		
4 Mos	37.5	17.0	1
	25		
8 Mos	25	10.0	54
	37.5		
12 Mos	37.5	8.8	83
	50		
18 Mos	50	10.0	3
24 Mos	50	11.0	8
	62.5		
36 Mos	62.5	13.0	10
6 Yrs	62.5	5.2	280
	75		
9 Yrs	75	10.0	11

the failure of the physician to follow through aggressively in any given case, the lack of compliance, erratic absorption of the medication, or yet some other factor is unclear. Because of several incidents in which instructions for administration of the thyroxine tablets were misunderstood, coupled with reports of deliberate refusal to follow instructions by some families, our inclination is to suspect noncompliance as the major issue.

Regardless of the cause, however, it seems reasonable to believe that our experience is not unique and that similar problems to a greater or lesser degree, with regard to patient management, also beset other programs. Thus it may be time to rethink our approach to management, with the goal of ensuring optimal treatment for all patients. To accomplish this end will require a somewhat more radical approach than some physicians may be willing to take. For example, after initiating treatment, measurements of T4 and TSH might be performed every 10 to 14 days until the T4 is above 10 μg/dl and the TSH is below 20 μU/ml. To suppress TSH concentrations to the normal range frequently requires raising the serum T4 between 14-16 μg/dl. The fact that the threshold for feedback inhibition of TSH release by thyroid hormones is higher in hypothyroid children[7] should not minimize the importance of TSH as a marker of adequacy of treatment. This receives some support from studies in which comparison of thyroid function tests were made between two groups of treated hypothyroid children, one of which had a high level of TSH. Although the mean total T4 concentrations were essentially the same in both groups, the mean free T4 values of the children with the elevated TSH levels were significantly lower ($p<0.01$) than that of the children with the normal TSH concentrations.[8]

After the circulating T4 and TSH have reached their appropriate levels, repeat determinations would be performed three weeks after any change in dosage and routinely at two-month intervals until 12 months of age and then at 18 and 24 months. Blood specimens would be obtained more often whenever there was any question of compliance or inconsistency of response. After two years measurements of T4 and TSH could be made every four months in order to avoid the possibility of protracted intervals of suboptimal therapy. We presume that transient periods of hypothyroidism in the older children produce no permanent damage. However, there is no available information concerning the possible influence of prolonged periods of transient hypothyroidism on learning behavior in school or on the timing or completeness of sexual maturation. Therefore, it behooves us not to relax our efforts to achieve the most favorable outcome possible for these patients.

In summary, it is clear from the data presented here that some of the patients in the New England cohort were treated or responded unevenly. This group of patients was characterized by a mean T4 concentration below 10 µg/dl and a mean TSH above 20 µU/ml after eight weeks of therapy. In addition, there were numerous episodes of low T4 and elevated TSH levels in some of these children after the first year of life.

We believe that these findings are not atypical of screening programs but mirror the pattern of practice as applied to the clinical management of patients with infantile hypothyroidism. In order to remedy this situation, physicians will have to monitor patients more closely, more often, and more aggressively. In view of this, we speculate that the difference in outcome of the patients between the New England and Canadian programs was the result of less than adequate therapy in those Canadian patients who were more seriously affected. This does not imply better management of the patients in New England; this means only that we have had the good fortune of caring for families who were probably more compliant.

REFERENCES

1. J. Glorieux, J. H. Dussault, M. Morissette, M. Desjardins, J. Letarte, and H. Guyda, Follow-up at ages 5 and 7 years on mental development in children with hypothyroidism detected by Quebec screening program, J Pediatr, 107:913 (1985).
2. J. Rovet, R. Ehrlich, and D. Sorbara, Intellectual outcome in children with fetal hypothyroidism, J Pediatr, 110:700 (1987).
3. New England Congenital Hypothyroidism Collaborative, Effects of neonatal screening for hypothyroidism: prevention of mental retardation by treatment before clinical manifestations, Lancet, 1095 (1981).
4. J. Sack, The pituitary response in neonatal hypothyroid (nh) children treated with thyroxine, in: "Neonatal Screening," H. Naruse and M. Irie, eds., Excerpta Medica, Amsterdam-Oxford-Princeton, (1983).

5. G. P. Redmond and L. F. Soyka, Abnormal TSH secretory dynamics in congenital hypothyroidism, J Pediatr, 98:83 (1981).
6. R. Klein, Screening for congenital hypothyroidism, Lancet, 403 (1986).
7. T. Sato, Y. Suzuki, T. Taketani, K. Ishiguro, and H. Nakajima, Age-related change in pituitary threshold for TSH release during thyroxine replacement therapy for cretinism, JCE&M, 44:553 (1977).
8. F. Focarile, G. F. Rondanini, A. Bollati, A. Bartolucci, and G. Chiumello, Free thyroid hormones in evaluating persistently elevated thyrotropin levels in children with congenital hypothyroidism on replacement therapy, JCE&M, 59:1211 (1984).

DISCUSSION

DUSSAULT (Quebec) : We have data to indicate that in CH children "affected" (that is, with minor neurological disturbances) and "non affected" (that is, normal based on our evaluation techniques), there was no difference in T4 serum levels 6 weeks after birth and onwards up to 7 years of age. Similarly, there was no difference in serum for TSH values between the 2 subgroups. If anything, the highest TSH levels were found in "non affected" infants. I feel comfortable in the way we treat our patients and I believe that what is involved here is the severity of the disease during fetal life and the timing of onset of the hypothyroidism during development.

MITCHELL (Jamaica Plain) : It has been our experience that unless you can take a series of samples at close intervals, you don't know what goes on in-between. Random samples in some of the children may not really reflect what is going on. The second point is that we may be dealing with a different subset when we talk about the more affected children. In that subset, despite the fact that the serum T4 is 10 ug/dl, it may not be sufficiently high because these infants have no endogenous circulating thyroid hormones at birth. They've been deprived of hormone in utero for a long time. It may be necessary and essential, instead of raising their T4 up to 10 or 11 ug/dl, to increase the level to 14 ug/dl or higher in order to compensate for that.

ROVET (Toronto) : Instead of using a general IQ which is based on many skills and abilities, reflecting many different areas of the brain are developing at different times, we, and I presume the Quebec group, have looked at individual skills and abilities. We found that different factors predict different outcomes : prenatal factors seem to predict more visual spatial abilities whereas postnatal factors appear to reflect language abilities. I suspect that if we were to group all the abilities together , we would probably come up with something similar to what you observed. Have you noticed any effects on behaviour of the children as a response to having higher circulating T4 levels ? It seems that children with very high levels are more difficult temperamently and seem to have a more "sensitive" nervous system.

MITCHELL (Jamaica Plain) : Not really. I think this is very difficult to evaluate because as controls you need parents who don't know whether their kids receive thyroid hormone or not.

BRAVERMAN (Worcester) : One should be cautious in the United States about the use of generic T4 which, as you know, may contain variable amounts of levo-thyroxine. My second comment concerns compliance to therapy. In adults, we have sometimes required a visiting nurse once a week to give the patient a full weekly dose of thyroxine at one time. I would suggest that it is probably the best way to treat children. There are no side effects.

TREATMENT STRATEGY AND LONG TERM FOLLOW UP OF CONGENITAL HYPOTHYROIDISM

G. Van Vliet[1], Th. Barboni[1], M. Klees[1], F. Cantraine[2], and R. Wolter[1]

[1]Département de Pédiatrie, Hôpital Universitaire des Enfants Reine Fabiola and [2]Département de Calcul Scientifique, Faculté de Médecine, Université Libre de Bruxelles, Belgium

Congenital hypothyroidism is one of the most common endocrine disorders in childhood (1). Since it became possible to establish mass screening programs for this condition, its outlook has changed dramatically : early diagnosis and treatment have led to the disappearance of mental retardation (2), which was the most dramatic long-term sequella of congenital hypothyroidism (3). Whether the physical and neuropsychological development of children diagnosed as having congenital hypothyroidism by neonatal screening is entirely normal remains controversial (2, 4-11).
The present paper describes our findings in a group of children with congenital hypothyroidism detected by neonatal screening. Our analysis attempted to answer the following questions : 1) is postnatal physical growth and bone maturation normal ? 2) what are the appropriate biological parameters to assess the adequacy of treatment ? 3) is neuropsychological development normal ?

SUBJECTS AND METHODS

Between June, 1975 and March, 1986, thirty-seven children with permanent primary congenital hypothyroidism detected by a screening program were referred to our Department. The details of the screening and diagnostic procedures will be published elsewhere (12). Briefly, screening was performed on the 5th postnatal day by using TSH as the primary screening test. The diagnostic evaluation, aside from clinical assessment and confirmation of laboratory findings, consisted of : 1) determination of skeletal maturation by the method of SENECAL et al (13); 2) a thyroid scintigraphy whenever no thyroid gland was palpated.

Treatment was initiated on the day of the diagnostic examination. It consisted of tablets of levothyroxine given daily by mouth. The initial dose was 25 or 50 mcg/day and was adjusted by biweekly determinations of plasma free thyroxine. The frequency of blood sampling was decreased as soon as the level of free thyroxine was stabilized at about 2 ng/dl. Growth was assessed at 3, 6, 9, 12, 18 and 24 months and yearly thereafter. Bone age was assessed yearly. Neuropsychological evaluation was performed by a single examiner (Th.B.), using the BRUNET-LEZINE test. For practical reasons, the children were evaluated at somewhat irregular intervals. Some older children were also examined

by use of the McCARTHY test but their numbers were too small for statistical evaluation.
For the purpose of neuropsychological evaluation, we felt it was necessary to validate the BRUNET-LEZINE test in a contemporary group of Belgian children. A day-care center in a lower-middle class neighborhood was selected in the southern part of Belgium. Thirty-four children, aged 7 to 34 months, whose parents had educational and professional backgrounds similar to those of the hypothyroid group, were tested once; they constitute the control group. The parents of both the hypothyroid and control children gave their consent to the study.

STATISTICAL ANALYSIS

The clinical data are summarized as mean + SEM (or as median and range when the distribution was markedly skewed). Bone ages and growth parameters are given as standard deviation scores (SDS); the formula used is SDS = $x-\bar{x}/SD$, where x is the observed value, $\bar{x}$ the mean and SD the standard deviation taken from reference tables. The standards of GREULICH and PYLE (14) have been used for bone age, those of TANNER for growth parameters (15).
Comparisons between subgroups of patients were performed by use of the MANN-WHITNEY U test. The results of neuropsychological evaluation were analysed by regression analysis of developmental age over chronological age.

RESULTS

Findings at diagnosis

There were 28 girls and 9 boys. Twenty-five had ectopic glands, 5 had athyreosis, 6 had goiters; in one case, the result of the thyroid scintigraphy was never sent by the referring physician.
Median age at diagnosis (and of beginning treatment) was 14 days (range : 9-100 days). Median bone age at that time was 38 weeks (range : 30 weeks of gestational age - 3 months of postnatal age).

Growth during follow-up

Growth in height and weight was normal (data not shown); the mean head circumference, which was large at birth, increased further during the first three years of life, and remained higher than normal up to 8 years of age (Table 1).

TABLE 1 . Head circumference (standard deviation score) in early-treated congenital hypothyroidism.

	$\bar{x}$	SEM	N
Birth	+ 0.9	0.2	17
3 months	+ 0.8	0.2	30
6 months	+ 0.9	0.2	27
9 months	+ 0.7	0.2	27
12 months	+ 1.1	0.2	30
18 months	+ 1.0	0.2	29
24 months	+ 1.2	0.2	26
36 months	+ 1.3	0.3	21
4 years	+ 0.8	0.3	19
5 years	+ 0.9	0.3	17
6 years	+ 0.7	0.4	12
7 years	+ 0.8	0.4	12
8 years	+ 0.9	0.7	6

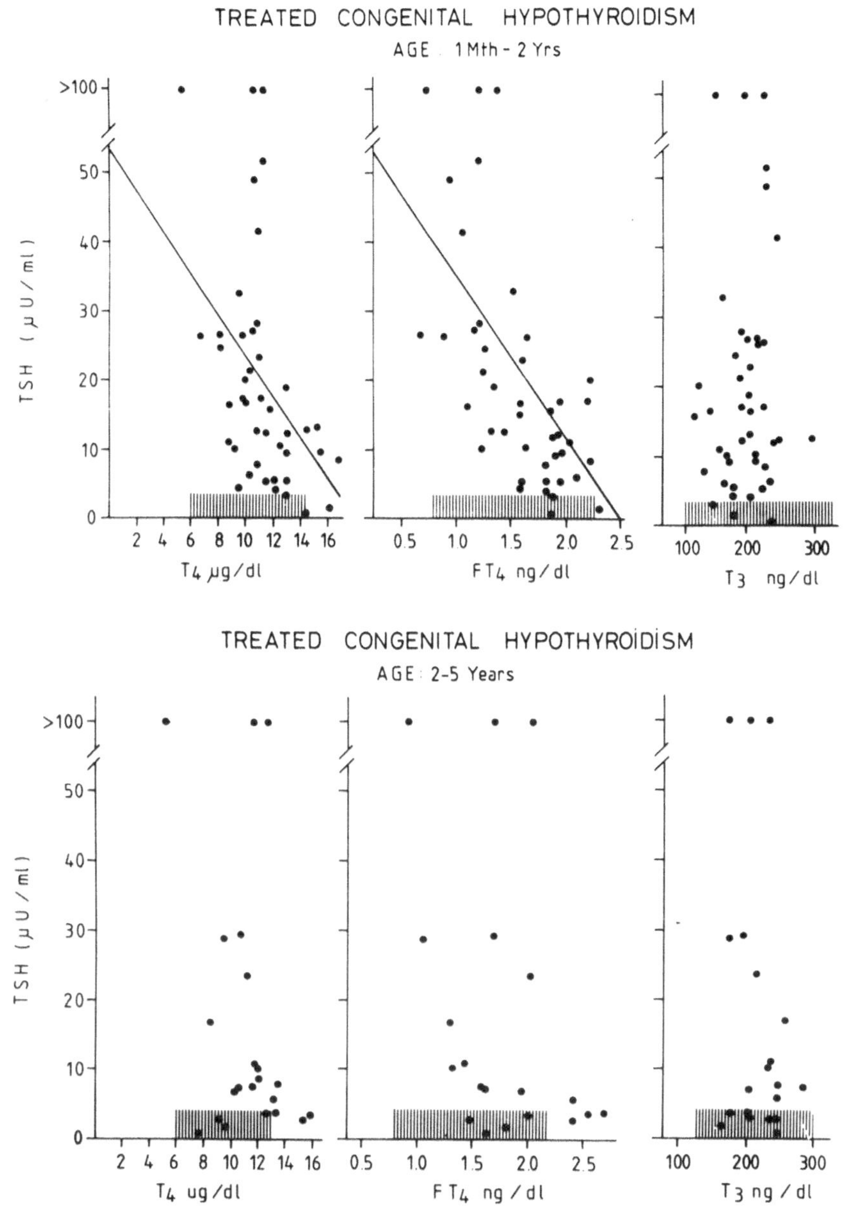

Fig. 1 . Relationship of plasma TSH (mcU/ml) and total T_4 (left), free T_4 (middle) and total T_3 (right) in children with early-treated congenital hypothyroidism aged 1 month to 2 years (top panel) and 2 to 5 years (bottom panel).

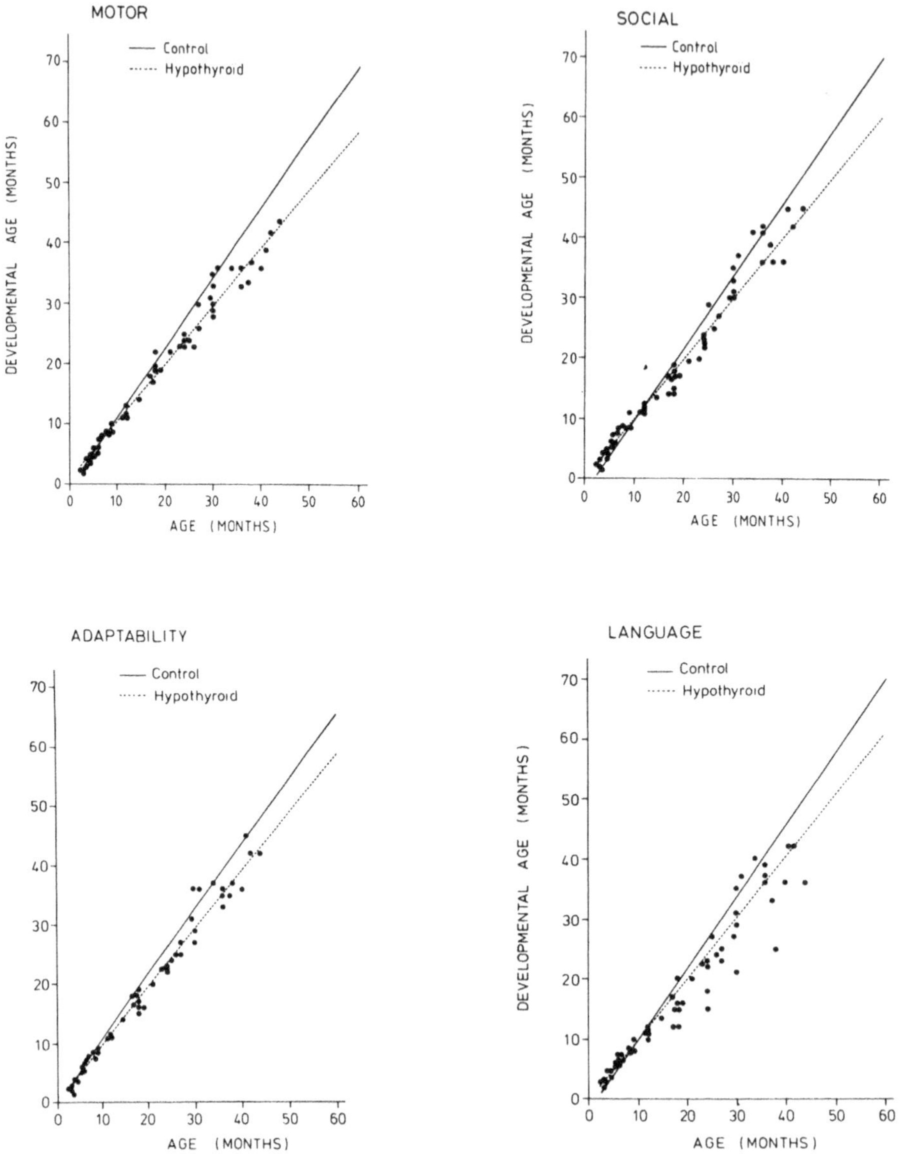

Fig. 2 . Linear regression of developmental age (assessed by the BRUNET-LEZINE test) over chronological age in control children and in children with early-treated congenital hypothyroidism. The individual values for children with hypothyroidism are shown, whereas only the regression line of the control group is shown.

Bone age during follow-up

Table 2 shows the bone age standard deviation score as a function of chronological age. Up to three years of chronological age, bone age remained on average about 1 standard deviation below the mean; it is only after a chronological age of four years that the mean bone age became close to normal.

TABLE 2 . Bone age (standard deviation score) in early-treated congenital hypothyroidism

Bone age (SDS) / Chronological age	$\overline{x}$	SEM	N
1 year	- 1.3	0.2	28
2 years	- 0.8	0.2	27
3 years	- 1.0	0.2	9
>4 years	- 0.1	0.2	17

Biological parameters during follow-up

Figure 1 shows the relationship between plasma TSH and T_4, free T_4 and T_3 in infants aged 1 month to 2 years (top panel) and in children aged 2 to 5 years (bottom panel). In the infants, high plasma TSH values were observed even in those patients with T_4 and free T_4 above the age-related normal range; in the children, TSH values came down while the T_4 and free T_4 remained similar to those of the infants.

Psychomotor development

Figure 2 displays the results of the BRUNET-LEZINE test in the children with congenital hypothyroidism and in the controls. The linear regression line of developmental age versus chronological age was for the four scales higher in the control group than in the hypothyroid group : the slopes of the regression lines for the controls ranged from 1.11 to 1.20, whereas those of the hypothyroid group ranged from 0.97 to 1.02. It should be pointed out, however, that the global developmental quotient was above 85 in all the hypothyroid children tested.

Influence of bone age at diagnosis on subsequent growth and psychomotor development

The mean head circumferences were not significantly different in the subgroup of children with hypothyroidism of prenatal onset (as estimated by a bone age of 37 weeks or less at diagnosis) than in the remainder of the patients. Mean bone age at 1 year was slightly more retarded in children with hypothyroidism of prenatal onset (-1.8 ± 0.3 vs -0.7 ± 0.3, $p < 0.01$).
The slopes of the regression lines shown on Figure 2 were not different when calculated for the two subgroups according to bone age at diagnosis.

DISCUSSION

Our series is very similar to most other recently published studies of children born in the era of neonatal screening for congenital hypothyroidism in terms of sex distribution, etiological diagnosis, chronological and bone age at diagnosis.

Our findings of normal postnatal growth in length and weight is also similar to those reported by others. The evolution of head circumference however differs from that in two previous studies : BUCHER et al (9) found a large head circumference at birth (+ 0.83 SDS, compared to + 0.9 in our series) but showed that the head grew less during the first year of life so that from one year onwards, mean head circumference was no longer different from normal; in contrast, THOMPSON et al (6) reported normal head size at birth with accelerated postnatal head growth so that the mean head circumference was around the 90th centile from 3 months to 4 years. In our series, we found a large head size at birth with a further increase compared to the reference values up to 3 years of age and later a stabilization of mean head circumference SDS at about + 0.8 until 8 years of age, the end of our observation period.
The mechanisms whereby thyroid hormone deficiency and its treatment may influence head growth have been discussed by BUCHER et al (9). Our data as they suggest, are consistent with the concept that hypothyroidism of perinatal onset leads to a shortened cranial base : increased growth of the calvarium then ensues, irrespective of the size of the brain or of the cerebral ventricles. It is noteworthy, however, that in older children with primary hypothyroidism, initiation of levothyroxine therapy may lead to pseudotumor cerebri (16).

Our finding of a delay in bone age that persists up to a chronological age of 3 years is also somewhat at variance with published data : BUCHER et al (9) report a normal bone age at 1 year, whereas THOMPSON et al (6), who found a marked delay in bone age at one year, report that this delay disappears between 2 and 3 years. A delay in bone age that persists during the first three years of life, as we found, could suggest that the treatment is in fact insufficient. Indeed, even with plasma free T_4 levels around 2 ng/dl, elevated plasma TSH levels are seen in infants with congenital hypothyroidism (Figure 1). It is generally considered that such a phenomenon results from a higher threshold for feedback regulation of TSH release in hypothyroid children (17). An alternative explanation would be that hypothyroid children need higher plasma free T_4 levels not only to inhibit TSH release but also to ensure a catch-up of bone maturation.
Our preliminary results on the developmental evolution of hypothyroid infants (Figure 2) suggest a slight delay in maturation. Our impression and preliminary reports in the literature (11) indicate that, by 9 years of age, children with congenital hypothyroidism become completely indistinguishable from controls in their neuropsychological status and school performance. The fact that they appear to lag slightly behind their coevals during the first three years of life, a lag also observed by GLORIEUX et al (2),could also be viewed as a result of suboptimal treatment with thyroid hormone.

In the present sample, the psychomotor development was not worse in children with hypothyroidism of prenatal onset. In our previous neuropsychological study of children with thyroid dysgenesis (born before the screening era), bone age was an important prognostic factor (3). Median age at start of treatment was about 6 months in that earlier series, in contrast to 14 days in the present one, and this may account for the fact that bone age at diagnosis is no longer essential in determining long term outcome. It should be noted, however, that in a recent study by ROVET et al (5), a bone age of less than 37 weeks at diagnosis was associated with a slightly poorer neuropsychological performance at 6 years, in spite of treatment at 16.7 days.

In summary, we found that children with congenital hypothyroidism treated early thanks to a screening program nevertheless had some slight abnormalities in their postnatal somatic and psychomotor development;

these were : 1) an accelerated growth of cranium with a tendency toward macrocephaly that persisted until at least 8 years of age; 2) a delay in bone age that persisted up to age 3 years; 3) a sligthly delayed psychomotor development during the first three years of life. Taken together, these data suggest either that hypothyroidism of perinatal onset leads to some abnormalities which cannot be prevented by early treatment and/or that our current treatment regimen, in spite of the monitoring by measurement of plasma free T_4 levels, is suboptimal.

REFERENCES

1. L. Wilkins, The diagnosis and treatment of endocrine disorders in childhood and adolescence, Springfield, Ill : Charles, C. Thomas, (1965).
2. J. Glorieux, J.H. Dussault, J. Morissette, M. Desjardins, J. Letarte, H. Guyda, Follow-up at ages 5 and 7 years on mental development in children with hypothyroidism detected by Quebec screening program, J. Pediatr. 107 : 913-915 (1985).
3. R. Wolter, P. Noel, P. De Cock, M. Craen, Ch. Ernould, P. Malvaux, F. Verstraeten, J. Simons, S. Mertens, N. Vanderbroeck, M. Vanderschueren-Lodeweyckx, Neuropsychological study in treated thyroid dysgenesis, Acta Paediatr. Scand., suppl 277 : 41-46 (1980).
4. New England Congenital Hypothyroidism Collaborative. Neonatal hypothyroidism screening : status of patients at 6 years of age. J. Pediatr. 107 : 915-919 (1985).
5. J. Rovet, R. Ehrlich, D. Sorbara, Intellectual outcome in children with fetal hypothyroidism. J. Pediatr. 110 : 700-704 (1987).
6. G.N. Thompson, R.B. McCrossin, J.L. Penfold, Ph. Woodroffe, W.A. Rose, E.F. Robertson, Management and outcome of children with congenital hypothyroidism detected on neonatal screening in South Australia, The Medical Journal of Australia 145 : 18-22, 1986.
7. R. Illig, R.H. Largo, Q. Qin, T. Torresani, P. Rochiccioli, A. Larsson, Mental development in congenital hypothyroidism after neonatal screening, Archives Dis. Childh. 62 : 1050-1055 (1987).
8. A. Ilicki, A. Larsson, Psychomotor development of children with congenital hypothyroidism diagnosed by neonatal screening. Acta Pediatr. Scand. 77 : 142-147 (1988).
9. H. Bucher, A. Prader, R. Illig, Head circumference, height, bone age and weight in 103 children with congenital hypothyroidism before and during thyroid hormone replacement, Helv. Paediatr. Acta 40 : 305-316 (1985).
10. J.A. Hulse, D.B. Grant, D. Jackson, B.E. Clayton, Growth, development and reassessment of hypothyroid infants diagnosed by screening, Brit. Med. J. 284 : 1435-1437 (1982).
11. New England Congenital Hypothyroidism Collaborative, Preliminary report on patients with infantile hypothyroidism after the third grade, Pediatr. Res. 21 : 243A (1987)(Abstract).
12. J.P. Chanoine, V. Toppet, G. Van Vliet, R. Lagasse, P. Bourdoux, M. Spehl, F. Delange, Can thyroid echography replace thyroid scintigraphy in congenital hypothyroidism ? (submitted for publication).
13. J. Senegal, M.C. Grosse, A. Vincent, J. Simon, J.N. Lefreche, Maturation osseuse du foetus et du nouveau-né, Arch. Fr. Pédiatr. 34 : 424-438 (1977).
14. W.W. Greulich, S.I. Pyle, Radiographic atlas of skeletal development of the hand and wrist, 2d ed., Stanford, Calif : Stanford University Press (1959).
15. J.M. Tanner, Means and standard deviations of height, weight and head circumference, in Smith, D.W., Growth and its disorders, W.B. Saunders Company (1977).

16. C. Van Dop, F.A. Conte, T.K. Koch, S.J. Clark, S.L. Wilson-Davis, M.M. Grumbach, Pseudotumor cerebri associated with initiation of levothyroxine therapy for juvenile hypothyroidism, New Engl. J. Med. 308 : 1076-1080 (1983).
17. F. Focarile, G.F. Rondanini, A. Bollati, A. Bartolucci, G. Chiumello, Free thyroid hormones in evaluating persistently elevated thyrotropin levels in children with congenital hypothyroidism on replacement therapy, J. Clin. Endocrinol. Metab. 59 : 1211-1214 (1984).

DISCUSSION

KLEIN (Hanover) : You cannot use "weight" in your statistical analyses because weight is not normally distributed. Was cranial size a Gaussian distribution ?

VAN VLIET (Brussels) : Strictly speaking you are right for the use of "weight". However, the error made by converting weight in standard deviations scores is in fact minimal and there are tables of standard deviations for weight in the book by D.W. Smith on growth and its disorders. The tables I've used are those tables where everything, including head circumference and weight, is expressed in standard deviation.

INTELLECTUAL OUTCOME IN HYPOTHYROID CHILDREN SCREENED AT BIRTH

J.P. Farriaux, J.L. Dhondt, and M.F. Lebecq

Service de Génétique et Maladies Héréditaires du Métabolisme
Centre Hospitalo-Universitaire
Hôpital C. Huriez
place de Verdun
F-59037 Lille
France

INTRODUCTION

Screening newborn infants for congenital hypothyroidism is effective in preventing mental retardation. Children who are identified and treated within the first month of life are normal in most aspects of physical and global intellectual development. However, longitudinal studies have indicated the existence of persistent sequeales. "We are beginning to realize that in spite of early recognition and onset of therapy, children with congenital hypothyroidism exhibit mild abnormalities of perception and coordination at later ages."(Jean H. Dussault, Peter Walker, 1983). The long term prognosis can be correlated with the severity and duration of congenital hypothyroidism at the time of initial diagnosis, that signify in utero deleterous effect of fetal hypothyroidism. Many studies have shown that factors such as etiology of the hypothyroidism or levels of residual hormone synthesis also influence the degree of neurological impairement. Recent investigations (5,8) have also focused on significant correlations between skeletal maturation at the time of diagnosis and severity of congenital hypothyroidism.

This report will present the intellectual outcome of patients with congenital hypothyroidism and an attempt to define which parameter can be of a prognosis value to detect patients who are at particular risk.

MATERIAL AND METHOD

The sample consisted of a total of 94 children (26 boys and 68 girls) who were identified through the neonatal screening program set up in Nord Pas-de-Calais area in 1976. This sample represents a part of the 188 patients screened by the Regional Centre and participating in a longitudinal prospective study . The cause of congenital hypothyroidism was established by 123-iodine scan: 19 had athyrosis, 65 ectopic gland and 10 a normally located gland.

Skeletal maturity was assessed at the time of confirmatory diagnosis. Knee radiographs were performed by a standardized protocole and a bone surface score (BSS) was defined as the sum of surfaces ([D x d] x /4) of distal femoral and proximal tibial bone centers.

The therapeutic protocole was based on L-thyroxine administration (liquid solution). The dose was adapted to keep TSH below 7 µU/ml and T4 at upper limit of the normal range.

The psychological test protocole involved assessments at 1 year, 2 years (Brunet Lézine), 5 years (WPPSI, Wechsler Preschool and Primary Scale of Intelligence) and 7 years (WISC-R, Revised Wechsler Intelligence test for Children). These scales provide an estimate of overall development (global DQ or IQ) and scores on specific intellectual and behavioral functioning : posture, eye-hand coordination, speech, sociability (Brunet Lézine test), verbal and non-verbal performance (Wechsler scales). All the patients were evaluated by the same psychologist. Because the majority of children are still young, analysis of school performances was limited to their entrance at the primary school.

RESULTS

At 1 and 2 years of age, no significant difference in the overall development was found between hypothyroid children and controls. A decrease of global IQ/DQ was observed at age 5 and 7 years, which was more significant when median values were considered. The scatter of IQ values was more important at these ages than at 1 and 2 years. Considering subscale developmental quotients, lower scores of tests involving speech and verbal performances were noted at each age (Fig.1).

Intellectual development and etiology : Distribution of patients in respect to the etiology of congenital hypothyroidism is shown in table I. Except for initial T4 level at time of screening, the different etiologic groups did not differ in age at treatment onset, in T4 levels obtained on L-T4 therapy or in age at which TSH suppression was obtained.

Global IQ/DQ and subscores analysis did not reveal specific abnormalities except those already found in the whole group (Fig.2).

Intellectual development and indexes of severity of congenital hypothyroidism: Among parameters, which may reflect the severity of congenital hypothyroidism, only initial T4 and bone surface area have shown a significant relationship with developmental scores at 1 and 2 years.

The greatest proportion of BSS different from zero was observed in the ectopy group (71%), and a significant correlation was found between BSS and initial T4 levels ($r=0.493$, $n=93$, $p< 0.001$). According to previous studies (5), patients were subdivided into two groups on the basis of degree of skeletal maturity at the time of diagnosis. Children with BSS below 5 mm2 were assigned to delayed bone maturity group, children with BSS higher than

Table I. Etiologic distribution of patients

	Athyreosis	Ectopic gland	Normally located gland
male	7	14	5
female	12	51	5
total	19	65	10

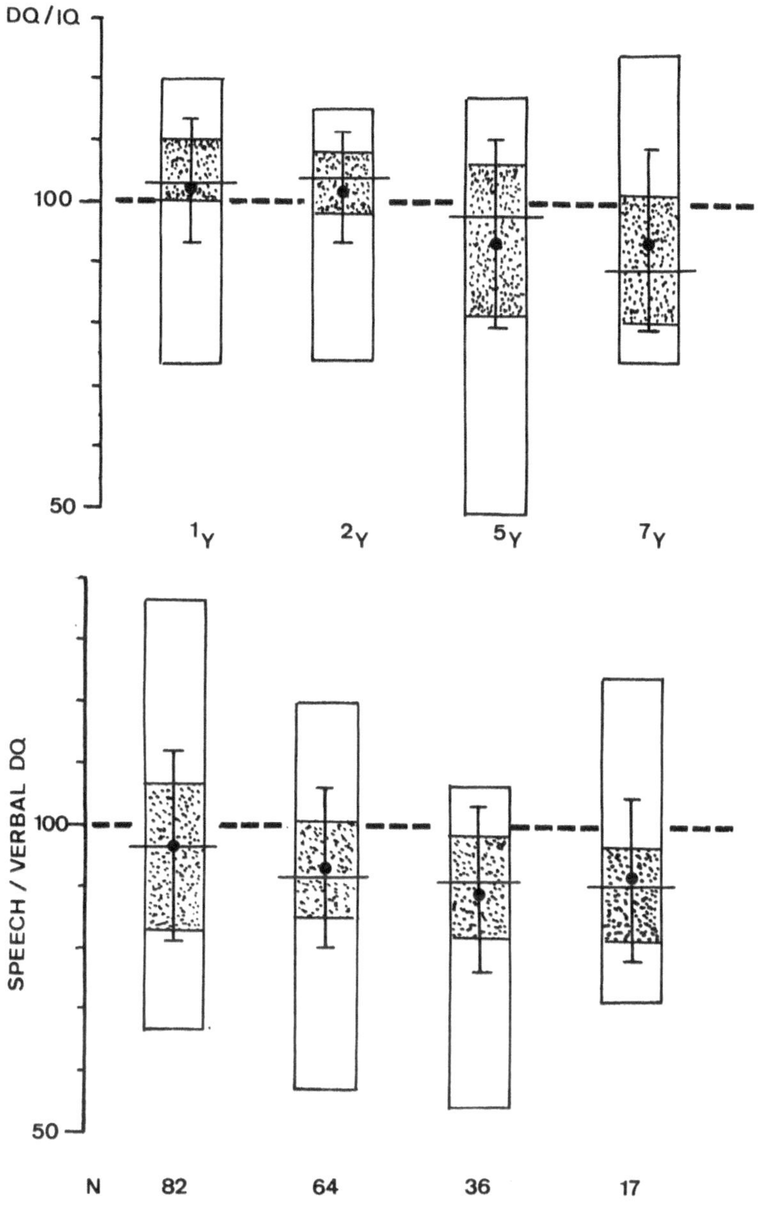

Figure 1. Global and verbal scores at age 1, 2, 5 and 7 years.
mean ± 1 sd, range, interquartile range,
median
N: number of patients.

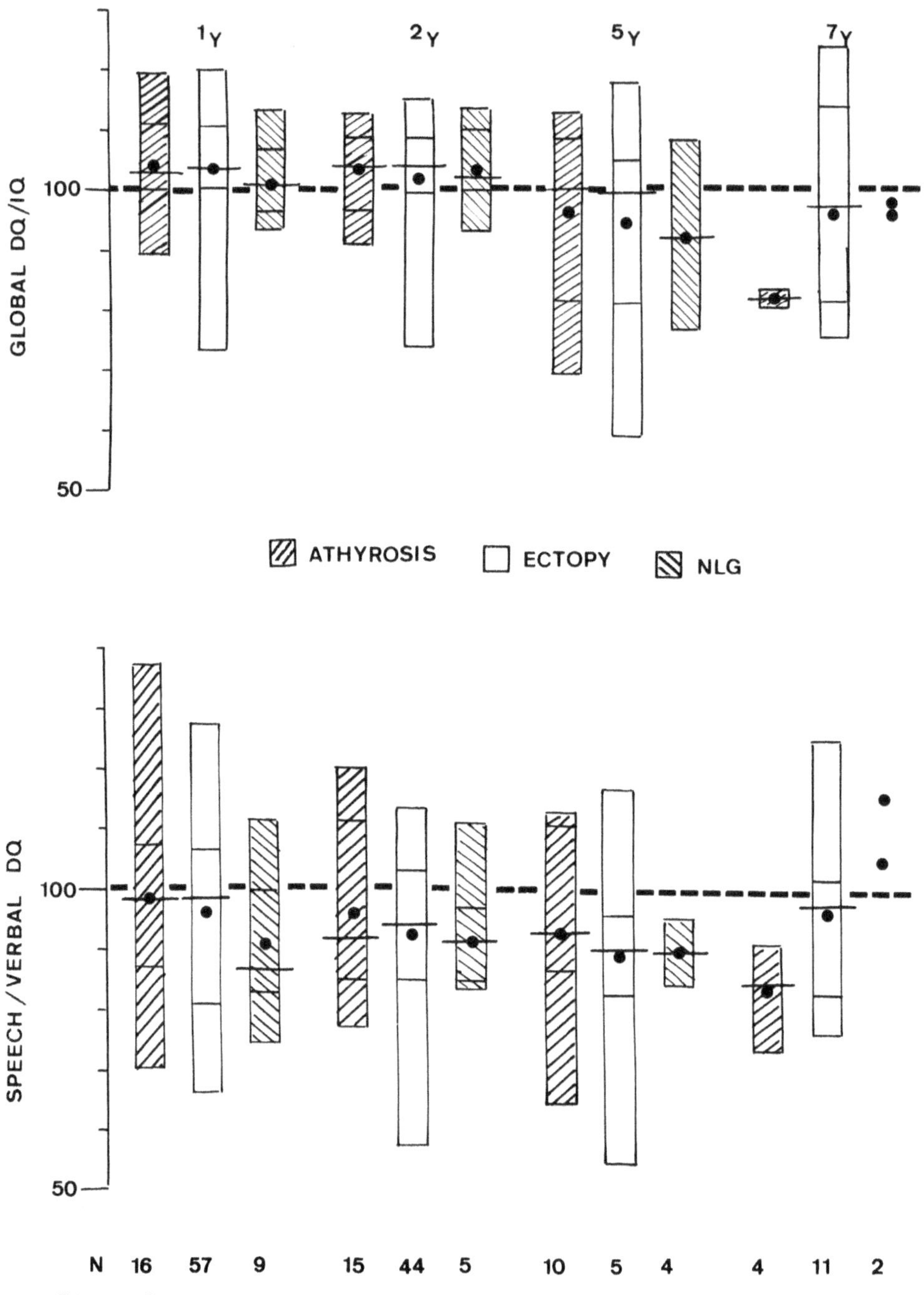

Figure 2. Global and verbal scores at age 1, 2, 5 and 7 years in the three etiologic groups.
mean ± 1 sd, range, interquartile range, median
N: number of patients
NLG: normally located gland.

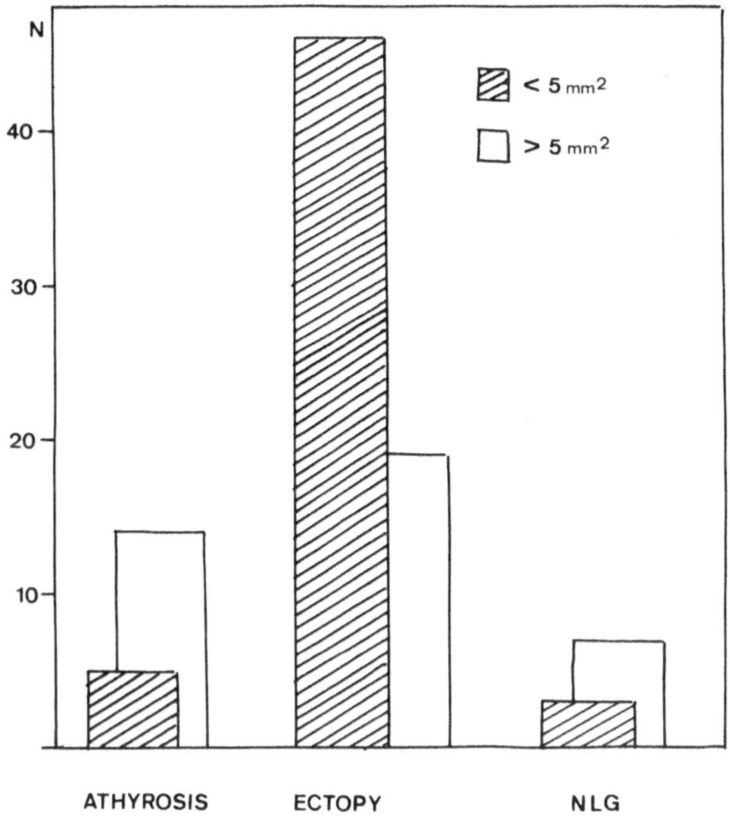

Figure 3. Etiologic distribution in the patients with bone surface score below and upper 5 mm2.
NLG: normally located gland.

Table II. IQ/DQ scores in the two subgroups according to the bone surface score

Bone surface	> 5 mm2			< 5 mm2		
age	N	IQ/DQ mean ± sd	< 90	N	IQ/DQ mean ± sd	< 90
1 year	49	106 ± 8	2%	33	100 ± 11 ***	18%
2 years	35	104 ± 9	6%	29	100 ± 8 **	10%
5 years	18	100 ± 11	17%	18	89 ± 18 *	50%
7 years	9	100 ± 16	33%	8	86 ± 9 **	75%

N : number of patients
*** $p < 0.005$
** $p < 0.05$
* $p < 0.025$

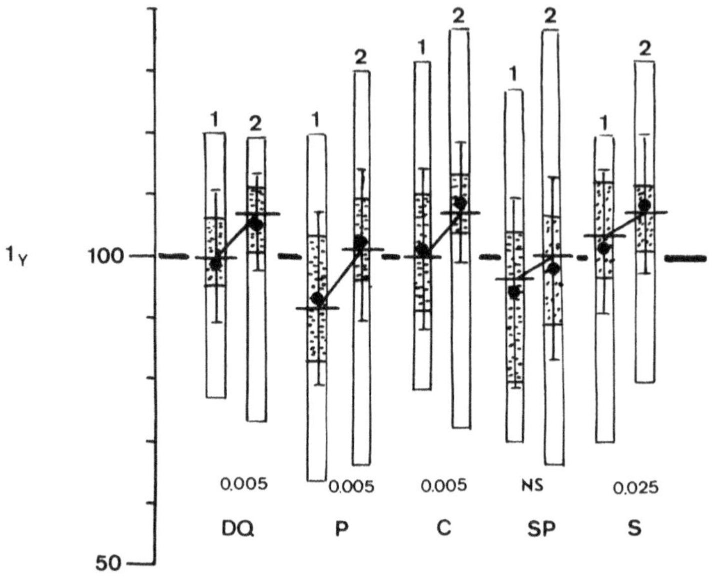
1Y
100
50
1
2
0.005
0.005
0.005
NS
0.025
DQ
P
C
SP
S

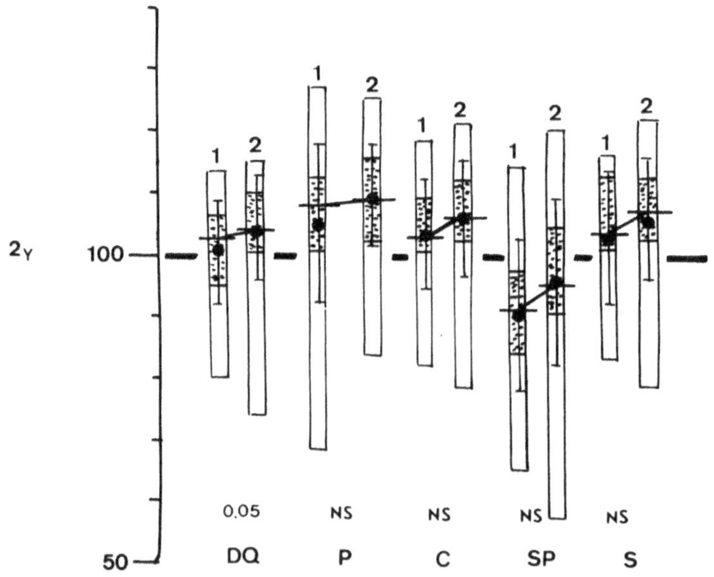
2Y
100
50
1
2
0.05
NS
NS
NS
NS
DQ
P
C
SP
S

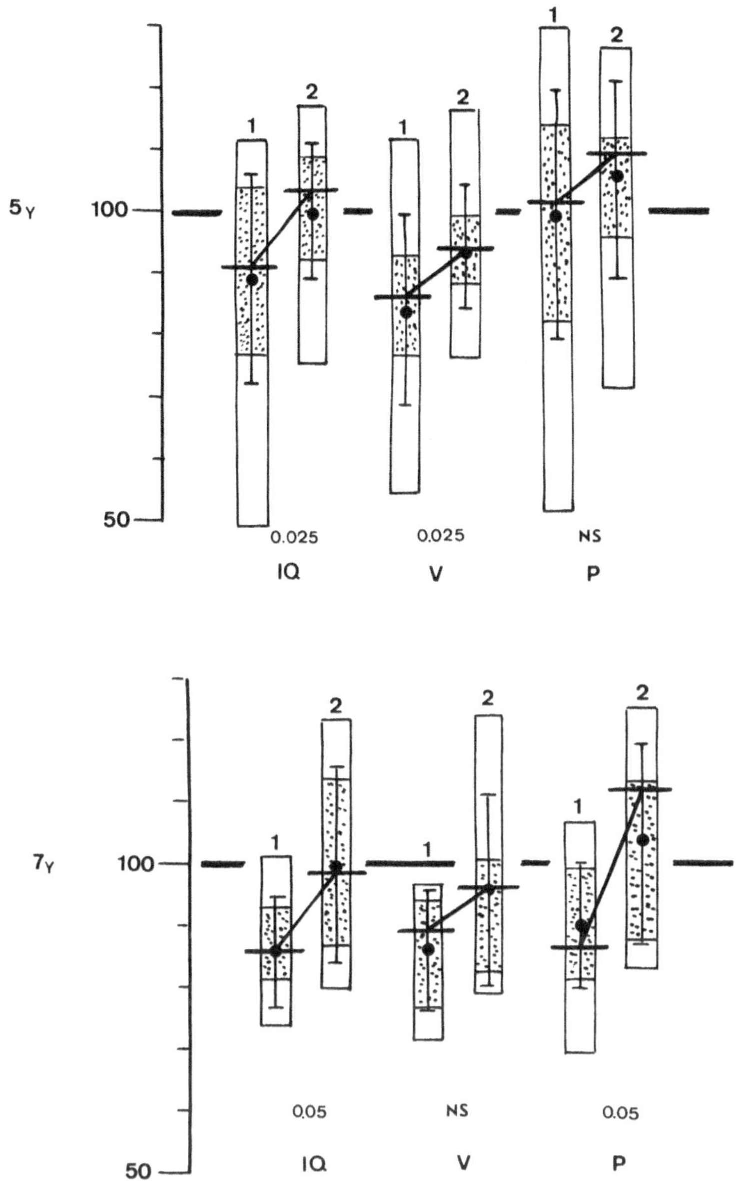

Figure 4. Global DQ and IQ scores at age 1, 2, 5 and 7 years in regard to the "bone surface score" group.
1. BBS < 5 mm2
2. BBS > 5 mm2
mean ± 1 sd, range, interquartile range,
median
DQ: global score, P: posture, C: eye-hand coordination, SP: speech, S: sociability (Brunet Lézine test),
IQ: global score, V: verbal, P: non-verbal performance (Wechsler scales).
t-test: p value, NS: not significant

5 mm2 to nondelayed bone maturity group. Etiologic distribution among these two groups is shown in figure 3.

Global IQ/DQ and subscores distribution at 1, 2, 5 and 7 years showed that mean and median values are lower in the delayed bone maturity group (Fig.4). In this group, IQ/DQ scores below 90 were more frequent (Table II). Five patients, out of 20, entered primary school at a later age, these 5 patients belonged to the delayed bone maturity group.

COMMENTS

The success of congenital hypothyroidism screening in preventing for severe mental retardation is undoubtly. However, it has been shown that some patients have some degree of impaired psychologic and neuromuscular function at later ages, even when therapy has been started early (3). Because this might be related to screening and management, it is important to identify the children most at risk for subsequent impairement and to determine the predisposing factors.

Like in a normal population, scores obtained during the first 24 months of life do not predict later IQ. A tendency to lower IQ values and scatered range with age was observed, and agreed with data previously reported (4,8). Differences with other reports about results of IQ testing and subscales scores might be due to the various tests used in these studies (1,6,7). Nevertheless, declined verbal scores have been reported by Glorieux (5), Rocchichioli (9), David (2) and Rovet (11).

To evaluate the various factors that could interfere with the future developmental progress of our children, we have tried to establish correlations and group differences according to parameters that could be objectively measured at the first diagnostic visit. These include hormonal values, skeletal maturation and thyroid scan.

Etiology

The presence or absence of thyroid tissue did not alter the psychological performance as previously reported by Letarte (8).

Age at start of treatment

In our population, the age of the patients at the onset of therapy was not correlated with their later mental performance. That confirms previous observations (6,8). However, most of the patients have been treated early in life and the range of age at initiation of therapy is narrow.

Initial T4

A correlation between initial T4 and mental development was observed only at 1 and 2 years, as reported by Dussault (3), but not found at later age.

Bone maturation

Letartre and LaFranchi (8) first reported that the indice of bone surface area correlated the intensity and duration of thyroid deficiency to lower developmental scroes. In deed, the duration of disease can be estimated by the mesure of bone maturation (10). Although there is a significant (and expected (3)) correlation between BBS and initial T4 levels, the high percentage of ectopia in the "delayed bone maturity group" remains surprising.

The present results confirmed previous reports (3,4,5,8,9) that children with significantly retarded skeletal maturation at the time of diagnosis obtained lower IQ scores and subscores at every assessment.

In the present study, many factors have not been taking in account (compliance, IQ of relatives, associated psychological factors, socioeconomic status of families), that can explain the scatter of data.

The variable correlations between these parameters and mental development reported in the literature may be explained by the different methodologies and criteria used in these studies, but emphasize the need of international collaborative programs aimed to a certain degree of standardisation.

CONCLUSION

In general, and according to previous works, it can be say that the intellectual performance of hypothyroid children was lower when they had biological or radiological features of a more intense thyroid deficit at the time of diagnosis. However, the scatter of the different parameters or their overlap when different a priori at risk group were compared did not allow a definite prediction of the risk. The BSS appears to give the best indicator of ante or postnatal start of hypothyroidism. Although it remained difficult to identify patients who are at particular risk on a case to case basis.

The most significant observation was low verbal scores. That suggests that the children may benefit from speech and psychologic therapy in addition to close psychological survey, because of the risk for future learning disabilities.

ACKNOWLEDGMENT

This study was supported by grants from the University of Lille II and from the "Conseil Régional Nord Pas-de-Calais".

REFERENCES

1 New England Congenital Hypothyroidism Collaborative. Neonatal hypothyroidism screening: status of patients at 6 years of age. J.Pediatr. 107: 915 (1985)

2 M.David, C. Dorche, J. Courdier, and M.O. Rolland. Follow up results of congenital hypothyroid infants found by neonatal screening. In: Neonatal screening, H. Naruse and M. Irie eds., Excerpta medica, Amsterdam, pp. 183 (1983)

3 J.H. Dussault, J. Glorieux, J. Letarte, H. Guyda, and J. Morissette. The mental development at 3 years of age of hypothyroid infants detected by the Quebec dcreening program. In: Neonatal screening, H. Naruse and M. Irie eds., Excerpta medica, Amsterdam, pp. 163 (1983)

4 J. Glorieux, J.H. Dussault, J. Morissette, M. Desjardins, J. Letarte, and A.Guyda. Follow up at ages 5 and 7 years on mental development in children with hypothyroidism detected in Quebec screening program. J.Pediatr. 107: 913 (1985)

5 J. Glorieux, M. Desjardins, J. Dussault, J. Letarte, and J. Morissette. Etude longitudinale du développement mental des hypothyroïdiens congénitaux. Dix années de dépistage. Arch.Fr.Pédiatr. 44: 709 (1987)

6 A. Ilicki, and A. Larsson, Psychomotor development of children with congenital hypothyroidism diagnosed by neonatal screening. Acta Paediatr.Scand., 77: 142 (1988)

7 R. Illig, R.H. Largo, M. Weber, Th. Augsburger, A. Lipp, D. Wissler, Perrenoud A.E., and T. Torresani. Sixty children with congenital hypothyroidism detected by neonatal thyroid: mental development at 1, 4, and 7 years: a longitudinal study. Acta Endocrinol. (Suppl.) 279: 346 (1986)

8 J. Letarte, and S. Lafranchi. Clinical features of congenital hypothyroidism. In: Congenital hypothyroidism. J.H. Dussault and P. Walker eds., Dekker, New York, pp.351 (1983)

9 P. Rochiccioli, B. Roge, F. Alexandre, J. Coll, G. Dutau, C. Enjaume, and D. Augier. Résultats du développement psychomoteur des hypothyroïdies dépistées à la naissance. Arch.Fr.Pédiatr. 40: 537 (1983)

10 P. Rochiccioli, G. Dutau, F. Despert. La surface des épiphyses du genou: index d'ancienneté de l'hypothyroïdie néonatale. Arch.Fr.Pédiatr. 41: 329 (1984)

11 J. Rovet, R. Ehrlich, and D. Sorbara. Intellectual outcome in children with fetal hypothyroidism. J.Pediatr. 110: 700 (1987)

DISCUSSION

FISHER (Torrance) : Total T4 values are relatively higher early in infancy and fall later. Corrected total T4 levels at one week are about 15 µg/dl and fall to 11-12 µg/dl by 1 month. The levels approximate the 8 µg/dl adult mean at about 18 years. Normally, endogenous secretion of thyroid hormones, includes about 10 % of T3 in children. To compensate for the absent endogenous T3 during the first year of life, we need a serum corrected T4 in the 10-15 µg/dl range. We do not usually measure T3 values, so we don't know whether they are absolutely normal or not in most infants. Many of the T4 results we have are total T4 levels (i.e. not corrected for TBG). The data we have for free T4 values depend upon the methods used. Furthermore, we have no reliable TSH data to help us control our T4 results in many infants because of the abnormal TSH feedback. I think that the increased head circumference, the decreased bone age, and the decreased developmental score correlations with low IQ values at 5 to 7 years, as well as the correlation in some of the programs of low IQ values with delayed bone maturation at birth, all suggest that the treatment we have been providing is inadequate until proven otherwise. We cannot conclude that the IQ is not correctable in these infants until we are sure we have provided adequate therapy. I suggest that we start all term neonates on 50 µg T4/day to raise their T4 to the 10-16 µg/dl range in the first 1-2 months and then keep them in some range that we can agree on (for a corrected T4) at least for the first year, with rather frequent testing.

DELANGE (Brussels) : Dr. Dhondt, I noticed that the age at the time of screening was higher in the infants affected by hypothyroidism compared to those affected by PKU. Was this due to the fact that some of the affected infants had obvious clinical symptoms for which they were transferred into a neonatal section and because of that, their screening was delayed ?

DHONDT (Lille) : We cannot be sure. The difference was small : only 2 days. However, I think there is something in that delay. We plan to look at this now on a prospective basis.

RONDANINI (Milano) : Eight years ago we started a prospective study in 30 patients (30 screened babies) divided into 2 subgroups. In one group we tried to achieve normal TSH levels while in the second group we did not. The first group had upper normal range FT3 and FT4 levels. Some of these children had signs of hyperthyroidism and the parents complained because of poor sleep, diarrhea, nervousness, etc... Thus we stopped the study and now adjust to normal FT3 and FT4 levels and leave the TSH levels in the upper range until 1-2 years of age. Thereafter, TSH normalized spontaneously. I believe that we have done well. These babies are reported in our poster at this meeting. After 5 years of study, they have normal neurological and psychological development. Also, concerning the suggestion of Dr. Fisher to start therapy with 50 µg l-thyroxine/d, we have tried that and observed frequent clinical signs of hyperthyroidism. Therefore we now start with a lower dose.

ADDISON (Manchester) : I wish to reply to Dr. Delange's point about the delay in diagnosis. In fact, PKU patients are more delayed due to transfer to special intensive care units, since PKU testing is not usually done until these patiens are put on a protein diet. Therefore, the difference in time of diagnosis between PKU and hypothyroidism cannot be due to the fact that these children are in intensive care units. A more general comment is that we must be very careful about the end-point in studies assessing treatment and IQ. IQ can change quite substantially throughout development. The end-point we should be looking for is not the value at age 3-7 years. It is when the children have finished their education and have established themselves in life. Otherwise, we might be doing a lot of unnecessary manipulation of these children and of their treatment, until we actually know what the real end-point is.

CZERNICHOW (Paris) : I want to make a comment about the health care system, the problem of dosage and of normal values in infants and children. By seeing children every 2 weeks during the first 6 months and by explaining to the parents what the situation is, we have not encountered any problem with compliance. Parents are very cooperative and we have not seen any "escape" of TSH that could be related to non-compliance. Concerning the dosage, we give l-T4 in aqueous solution (8 µg/kg body weight) and adapt the dosage for the weight of the infants after 2 weeks. Our mean TSH is 12 µU/ml. Free T4 levels were always maintained at the upper normal range and free T3 close to the mean value for the age.

INTELLECTUAL AND NEUROPSTCHOLOGICAL ASSESSMENT OF CHILDREN WITH CONGENITAL HYPOTHYROIDISM

Marianne Klees

Département de Pédiatrie, Hôpital Universitaire des Enfants Reine Fabiola, Université Libre de Bruxelles, Belgium

The brain at some point in its development becomes a critical target organ for thyroïd hormones, and is therefore affected by hypothyroïdism more profoundly than any other organ.

The main sequella of thyroïd insufficiency occuring during the perinatal period is brain dysfunction which may be so severe as to cause mental retardation.

Mental retardation was extremely frequent, and a degree of brain dysfunction was the rule in infants with congenital hypothyroïdism before the era of systematic neonatal screeming for congenital hypothyroïdism.

This screeming was initiated approximately 14 years ago, and long term neurointellectual follow-up is now available on large series of children.

It is obvious that neonatal screening has reached its main objective, which is the prevention of mental retardation. (4) (5) (11) (12)

With very few exceptions, usually explained by other causes, infants with congenital hypothyroïdism detected in the very first days of life, and rapidly and appropriately treated, are normal in most aspects of intellectual development.

However, some discrepancies exist in the literature on the possible long term consequences of hypothyroïdism of prenatal onset in spite of early recognition. (10)

These discrepancies point to the need for further study of all the correlations between biological and psychological parameters.

The objective of the present paper is :

1. to critically review the different psychometric tests used by the researchers, to delineate their field of application, their strength and their weaknesses;to point out how they can reveal neuropsychological signs ; to briefly describe the way to conduct a neuropsychological examination.
2. to evaluate the possible mechanisms that may explain the results obtained with infants affected by hypothyroïdism of prenatal onset.
3. to speculate on the possible long term consequences of the neurocognitive impairments observed, and especially on risks of the evolution.

TABLE 1

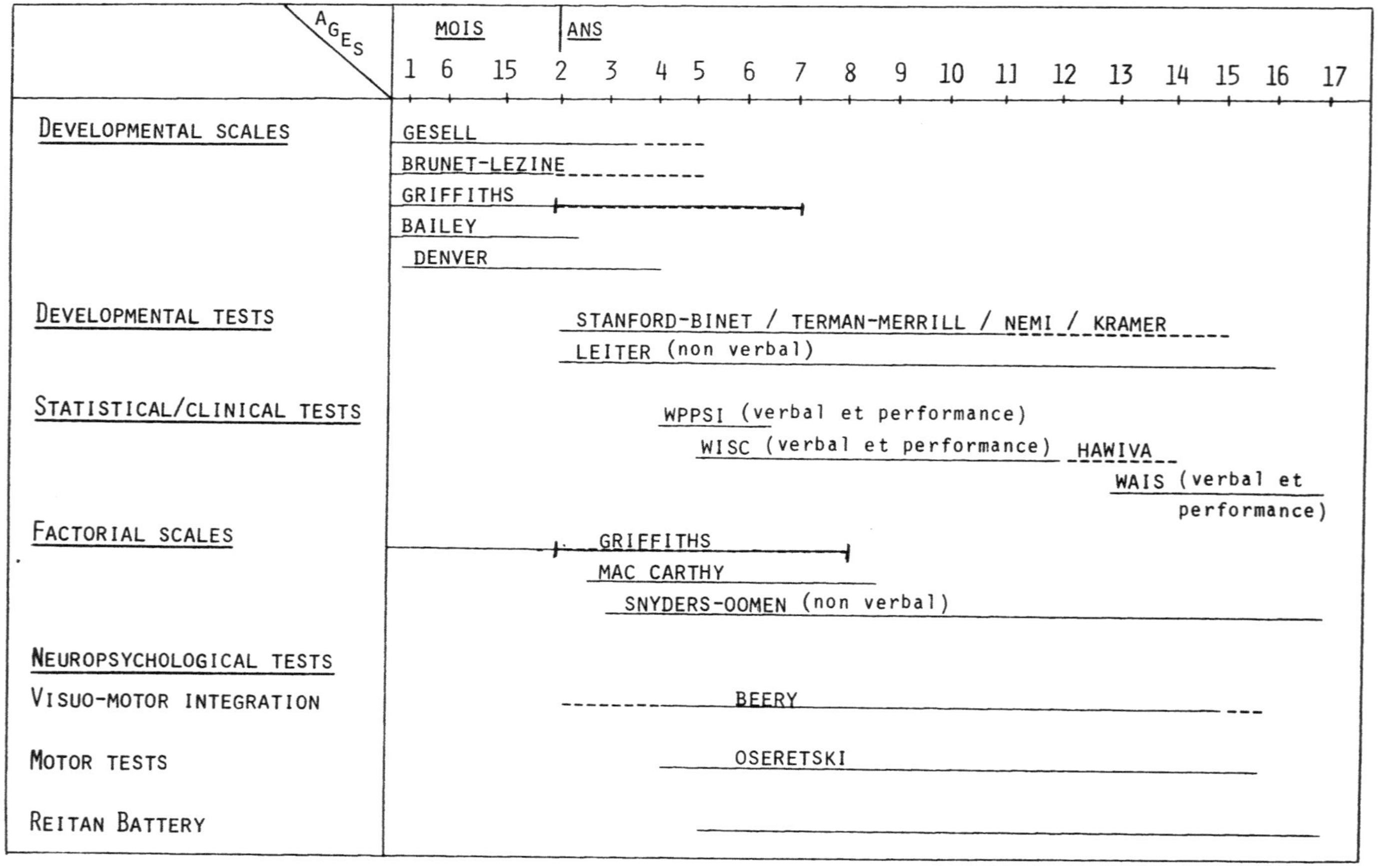

——— DISCRIMINATIVE ----- NOT ENOUGH DISCRIMINATIVE

1. INTELLECTUAL AND NEUROPSYCHOLOGICAL ASSESSMENT (TABLE 1)

Human development proceeds in an orderly predictable manner, with the same variability in behavior for normal infants and children found in all biological measurements.

By observing the children's behavior, by asking some questions to the parents, and by recording and analysing these informations systematically, an estimate of the level of function in various area of behavior can be established.

Social, familial and educational variables have a great influence on cognitive outcome. Environmental factors play a major role in human beings, and are even more significant than inner characteristics of the child predicting "intelligence", except for the most severely damaged children.

Before the age of 2 years, behavioral observation and developmental scales are available.

1.1. The Gesell Developmental Schedules (1919) (25) are ones of the oldest and most established infant "intelligence" measure. From 4 weeks to 5 years of age (not very discriminative after 3 1/2 years). Individual administration. Taking the test lasts 45 minutes. A developmental quotient is determined. Its current use is based on american and high economic level normative data, from a carefully conducted longitudinal study of early development. For certain questions, the parents testimony is the only ressource.

For the well trained examiner it represents an excellent clinical device for the study of mental growth of babies. It points out the sequences over time and the stages of development in the four fields of
- postural and motor development
- oculo-motor coordination
- language
- relational behavior

The number of items is different from age to age, and from sector to sector. This is not very practical and leads to a loss of homogeneity and reliability of the test.

1.2. Brunet et Lezine (1951) (23) in France, have presented a revised version of the Gesell Schedules, based on the same material and the same sectors of development : the presentation of the scale is precise and items are ranked according to difficulty. The examiner's influence is reduced to a maximum. The test consists in 10 items per age group. The total of the points obtained in each sector corresponds to a psychomotor developmental age for these precise sector (postural, oculo-motor, coordination, language, socio-lization development). When the observed level is homogeneous in the sector, a global developmental quotient may be rapidly calculated.

From 1 month to 5 years, has a major interest in research which is to provide quotients of development in 4 different sectors. This allows
- specific analysis for each sector (which is especially interesting when the results are heterogeneous)
- and analysis of the profile of development over time.

1.3. The Griffiths Mental Developmental Scale (1964-1970) (26)
The Griffiths scales (first level) are age scales consisting of a total of 260 items some derived from other infant tests, for assessing babies from birth to two years ; another scale assesses children from 2 to 8 years. Quotients are obtained from the developmental ages.

The author grouped the items for babies into five subscales : locomotor, personal-social, hearing and speach, eye and hand, and performance ; she added practical reasonning for children older than 2 years (see 1.9). The test contains an equal number of items in each subscale for each age period, and therefore it is possible to calculate separate Developmental Age and Developmental Quotients for the subscale.

While the majority of items are tested for, in quite a number of instances, the parent's testimony has to be relied upon, e.g. items regarding the number of different sounds the infant makes whether he enjoys a bath, and whether he listens to music ...
The items were carafully selected and this has resulted in a clear stable progression score with age. It has the advantage of having an unusually large number of items, equal numbers of items for each subscale at each age, and reasonable stability of mean quotients and standard deviation 13 from age to age.

Insufficient evidence is given regarding the internal consistency, and standard deviation of subscales, so that one must advise that profiles be used only with caution, and be regarded as providing pointers to be confirmed by other means. (20)

Dr Griffith sells her testing materials only to person who have been trained by her in their use.

1.4. The Bailey scales of infant development (1969) (21) are also mental and motor scales for early mental and psychomotor development based on normative maturational development data. From 2 months to 30 months of age. Individual administration. Taking the test lasts 45 minutes.

The test is providing a Mental Developmental Index (MDI) and two main scores (mental and motor). Performances as well as the contact with the child are analysed (30 ratings of behavior).

To assess mental functions, the Bailey scale uses measures such as the infant's response to a bell, the ability to follow an object with the eyes, and in older infants, the ability to follow oral instructions.
The heart of the Bailey Scale is the motor scale because a major underlying assumption is that later mental function is dependant on motor development. The standardization of the Bailey Scale is excellent.

1.5. The Denver Developmental Screening Test (1966) (24) consists in an inventory of 142 items of ease administration, clarity and scoring, analysing the gross-motor, the fine-motor, adaptative, the language and the personal-social development. From 3 months to 4 years of age, individually administrated. It is a test of good reliability and validity.

x x x

When the child reaches the age of 18 months to 2 years, a very important step in the mental development is the birth of the symbolic way of thinking : the child begins to talk, to play, to imitate. From that very moment, the psychologists are speaking in terms of intelligence, and tests have been constructed to measure it (even though discussion remains open as to what is intelligence).

These following tests are made of a mixture of developmental, factorial, clinical and statistical data in different proportions.

x x x

Among the most commonly used "developmental tests", we find

1.6. The Stanford-Binet Intelligence Scale (beginning of the 20th Century) (33) is based on two major principles of test construction which are age differenciation and general mental ability. From 2 years to "adult intelligence". Individual administration. Items are verbal and non verbal. Good validity until 11 years. Reliability.88.

Binet restricted his search for measures of intelligence to anything related to the whole or final product of intelligence, rather than to hypothetical individual process underlying intelligence.

Mental age is a unit of measure for expressing results a certain kind of intelligence tests. A subject's mental age is based on his or her performance compared to the average performance of individuals in a specific chronological age group. Compared whith the chronological age, it gives a quotient, the IQ . The standard IQs have a mean of 100 and standard deviation of 16. In the Stanford-Binet, the items are heterogeneous, and irregulary spread, and therefore this scale permits only comparisons based on the concept of mental age.

The Terman-Merrill's (1931) (35) scale is a revision of the Binet scale.

Such is the Nemi (Zazzo) (29) shortened and precise.

1.7. The Leiter International performance Scale (1969) (27) is a non-verbal test, requiring perceptual matching, analogous progression, understanding, memory and other varied items, many of them similar to verbal tests. The subject could neither ask question, nor be told anything about it.
Age : 2 to 18 years of age. Individual administration. Taking the test lasts 30 minutes.

This scale was developed to provide a culture free, non verbal means of assessing general intelligence, based on primiarily abstract concepts.

To answer, the subject must place wooden blocks (6 to 8) representing shapes, color, numbers, objects ... in a sort of wooden frame. In this way, the motor skill of the child may be observed.

Some items, such as reconstructing blok designs, gives information about the visuo-spatial integration ; some other items, like understanding analogous progression allows to estimate what Piaget called the "operative thinking".

Good reliability and validity ; easy to handle ; economical of time and effort on the part of the examiner.

x x x

In another kind of tests, statistical-clinical tests, the same task has to be performed by people of all ages, and the IQ is determined by a mathematical transformation of scores. Means of each age group are established without any reference to mental age score.

1.8. That is the case of the Wechsler Intelligence Scale (1967) (36) which were developed by David Wechsler to fit the clinical concept of "intelligence" rather than in terms of "age differentiation". The IQ is determined by a transformation of standard scores ; the usual mean is 100 and the standard deviation of 15. The mean of each of the 12 subtest's score is 10, and the standard deviation is 3. The IQ consistency would tolerate a variation of 5 points, according to the author.

So, it is not clinically exact to consider as contributive a difference of 5 points, even if it is significative at P = 0.005.
A difference of 15 points (one sigma) may be interpretated as a cue of regression of or progression in the intellectual efficiency, in relation to individual historical factors of the subject.

Separate set of verbal and performances tests were constructed allowing for both a verbal and a performance IQ (VIQ and PIQ). The 12 subtests were selected by combining the test literature for tests that had worked well in practice.

Individual passation in more or less 1 hour. The total IQ is highly reliable, even if marked fluctuations are to be expected in some of the subtest scores.

The Wechsler scales have been adapted in different languages, and for different groups of age :
WPPSI : age 4 - 6 1/2 years of age - Wechsler Preschool and Primary Scale of Intelligence
WISC-R : age 5 - 16 years - Wechsler Intelligence Scale for Children - Revised
WAIS : Wechsler Adult Intelligence Scale
in German : Hamburg Wechsler Test (HAWIVA and HAWIK)

The subtests of the Wechsler may be used in profile analysis to detect different clinical disorders, and signs of neurocognitive dysfunction : object assembly is related to body schema, Kohs block designs are strongly related to spatial abilities ; picture completion to visual perception ; code to speed and attention span ; pictures arrangement to time perception ; digit memory to span and concentration ; vocabulary to audio-verbal perception, ...

Intra subtests comparison in the same subject brings a lot of useful clinical informations (A survey about the use of the WISC for clinicians is given by Kaufman et al. 1986)(6).

As it is a mathematically constructed test, and since its reliability is excellent (about .90), it represents a very useful tool in research, and more precisely in international research.

x x x

Two children with the same IQ can be entirely different persons, at it is well known.

There is nothing wrong about using a mental age, or an IQ, if it is understood as an <u>average of several tests</u>.

The error is interpreting it as measuring some basic <u>unity</u>, when it is known to be nothing more than a composite of <u>many functionnal units</u>.

According to Spearman, individual differences in all tests of abilities can be attributed to <u>group factors</u>, and the common variance among tests of abilities can be explained by one <u>general factor</u> (intelligence ?). The analysis of batteries of tests revealed 6 primary factors : Verbal Comprehension, Word fluency, Number, Space, Memory and Reasonning.
Primary abilities can be identified in children from 4 years of age to adult age.
This makes possible studies on the rates of development of the separate abilities at various age.

Among the most commonly used "factorial tests", we find :

1.9. The Griffiths Mental Developmental Scale (Second level) (1970) (26) (for first level see 1.3), a comprehensive system of mental measurement for the first eight years of life.

The Griffith Scales provide a estimate of overall developmental known as the General Quotient, and scores on 6 scales of specific intellectual and behavioral functionning : locomotor, personal-social development, hearing and speech, fine-motor coordination, perceptual performances, and practical reasonning.

The Griffith is very useful because it is a very complete inventory of human abilities, from the beginning of life onwards, thus providing continuity in the assessment of children's development, which permits profile analysis.

It renders a global quotient which can be compared to more commonly used children's IQ tests.

1.10. The Mc Carthy Scales of Children's Abilities (1972) (28) is a carefully constructed individual test of human ability which provides 6 scores : verbal, perceptional-performance, quantitative, composite (general cognitive) memory, motor, individual ; it has been translated and adapted in French.

It is one of the most popular test for assessing the abilities of preschool children (Age : 2 1/2 to 8 1/2 years of age).

Proper administration of the 18 different subtests requires keen attention to detail acquired through considerable practice and careful reading of the manual, need a very good training, particulary for the motor results. To take the test lasts 60 minutes at least.

Reliability is excellent. Validity data are also encouraging : correlation with the Binet and the WPPSI are quite good (.81 and .71). Its 6 scores (general intelligence, verbal score, quantitative score, perceptual-performance score, memory score, motor score) permit pattern analysis and possible diagnostic interpretation.

1.11 The Snyders-Oomen non Verbal Intelligence Test (1958) (32) was initially constructed in Holland to assess the intelligence of deaf children, from 3 to 16 years of age, in an individual administration, orally or in pantomime, which need about 90 minutes of passation time.

It may be useful with non-deaf subjects, when verbal language is not permitted, as for the child as for the examiner (foreign language). Standardization is different in each situation.

The scale includes 8 subtests, each constituted with different items of progressive difficulty.

The complete scale is to be considered as a combinaison of 2, shorter and parallel, of 4 subtests.

P - Scale : mosaïques, pictures memory, picture arrangement, analogies
Q - Scale : geometrical drawings, Knox cubes, picture completion, seriation

Every subtest may be applied and scored separately, in mental age. The standard scores, weighted are transformable in IQ.

The neuropsycholigical examination may be performed by a clinical procedure, or a step wise approach, or by a battery approach. It tends to assess

a) the gnostic functions (how the subject recognize information as bearing special signifiance)
b) the executive functions (how the subject is able to do something with the information, by speech and praxis)

As it has been said, if the psychologist not only pays attention to the formal scores of the intellectual cognitive tests, but also to how the children act while taking the tests of intelligence, he can obtain a great deal of useful information.

High scores on certain parts of the test exclude the possibility of certain problems, making further examination unnecessary. Low scores indicate the direction which further examination needs to take. Such is the usual clinical procedure, a step by step approach, pointing out backwardness, slowness, retarded lateralization, body-schema defects, speech disorders, learning disabilities, drawing difficulties ... and mainly attention disorders.

In research, the methodology has to be identical for all the subjects, and it is useful to have at disposition specific tests which are more directly focused upon neuropsychological analysis.

1.12.An example of such test is the 8 hours (Halstead)-Reitan Neuropsychological Battery (1976) (31) for ages 5 to 14.

Numerous batteries have been offered as wide-band measures of the integrity and functionning of the brain. The Reitan Neuropsychological Test Battery remains the most researched and widely utilized measure in the United States : it consists of tests which have been shown to best discriminate between normal and patients with documented cortical damage.

The battery includes tests purposed to measure elements of memory, abstract thought, language, sensory-motor integration, and motor dexterity. This battery of sensory and motor tests, as well as tests of higher mental abilities, has been demonstrated to be sensistive to the locus and extend of impairments at the level of cerebral hemispheres. The examiner will yield measures of pure motor, sensory, language, abstract reasonning, psychomotor and perceptual abilities.

Shorter tests are :

1.13. The CPT (Continuous Performance Test) (34) for sustained attention (see Swanson, 1979).

1.14. Beery-Buktenica Developmenet Test of Visual Motor Integration (1982) 2 - 15 years - 15 minutes of passation time.
Measures the degree to wich visual perception and motor behavior are integrated in young children by having them copy 24 geometric forms with pencil and paper. The forms are arranged in order of increasing difficulty. The design are in majority straight-lines, angular configuration ; 5 designs have circular elements ; only 2 have discontineous details involving visual closure. The last 4 designs are three dimensional.

There is a proliferation of tests using geometric form reproduction. There is some communability among their purpose : to assess brain damage, visual memory, perceptual disturbances, motor maturation, ... The Beery is relatively simple compared to spatial perceptual tests using block designs and embedded figures. Interpretation about a child's perceptual abilities on the basis of this test is therefore tenuous.

This test is suitable for both group and individual administration. Correlations are higher with mental age than with chronological age.

1.15. The Bruininks-Oseretski Test of Motor Proficiency (1978) (30) is an individual wide range diagnostic and general test of motor development for ages 4 - 14 ; it assesses light arears of motor performances.

It gives 3 scores : gross motor composite ; fine motor composite ; battery composite. Individual passation - 45 minutes for administration (short form 20 minutes). No special training is required. It is a well developped, well standardized, usuable motor test.

The scales include 36 items, 6 subtests at each level which evaluate respectively : general static coordination, speed, simultaneous volontary movements, and syncinesias (involuntary movements).

Additional examination to detect impaired coordination may be usefully included and their interpretation may be based on results obtained in normal children by Touwen (1984) : balance, precision, control, lateral dominance, neurological soft signs, or true defects (16).

CONCLUSION

There is not a single test to measure intelligence from birth to adult age, but rather different tests in respect to

- ages of life
- needs to measure
- culture
- and nature of handicap

In the present case, tests such as any baby-scale, the Griffiths and the Wechsler, are bringing useful informations about mental functionning through its evolution.

Complementary tests are, of course, improving their issues.

It should not be a privilege to take time and money to use them, because they may perhaps be at the origin of new hypothesis. I am thinking to the Piaget tests, and to the Kaufman battery for instance.

2. CONSEQUENCES OF CH OF PRENATAL ONSET

Regarding intellectual outcome, at one year, no difference is observed between children with CH compared with controls. Later on, from about 2 years of age, children with CH show a slight but statistically significant downreward slight in global intelligence. But it must be observed that :

- the DQ and IQ of the CH children without additional risk factors are comparable to a control group of healthy children (5).

- whereas children with additional risk factors obtain lower values if IQ ; among them, a very few show scores strongly lower IQ (5).

Are those additional risks (such as environment, perinatal factors, illness or malformation) more frequent in a group of children with CH ? It is not impossible that the anxiety of the mother in respect to potential intellectual handicap does play a very important role.

In a group of normal 4 years old children, 51% of the variance of the IQ distribution is explained by socio-environmental variables, which are : maternal mental health, maternal anxiety, parental perspectives , mother interactive behaviors, mother's education level, occupation of head of household, minority group status, presence or absence of father in the home, family size, stressful life events (13).

Some other additional risks, linked to be intensity of the hypo-thyroïdism, are reported to be correlated with outcome : the level of T4 at the time of diagnosis, and the skeletal maturity.

Let us consider this last factor, which is the expression of CH of prenatal onset.

The IQs of children with more advanced skeletal maturity at diagnosis are distributed similar to those of controls ; and, in the other hand, a marked slight downwards is observed in children who lacked ossification centers in the knee at diagnosis.

This difference is observed in both the IQ distribution ; and in the specific measures of neurocognitive functionning, such as speech, fine and global motor coordination. This was observed at 1 year of age, and remains so at 2, 3, 4, 5, 6 and 7 years of age.

Specific learning disabilities in school are also described. Wolter, in Brussels, had pointed out all these aspects in 1979 already (18). Some researchers do not record similar facts.

According to Rovet (12) poorer perceptual abilities were evident in children from 4 to 6 years, and poorer neuromotor competence and memory were seen at age 6.

Rovet (12) considering as predictor : bone-age, treatment age, level of T4 at diagnosis and duration, in a multiple regression analysis, has shown that :

- skeletal maturity at diagnosis is a significant predictor of 3 measures of non verbal functionning (WPPSI - PIQ ; Griffiths' fine motor and Griffiths' perceptual scale - which, in fact, covers the same abilities)
- age at treatment onset was predictor of neuromotor proficiency

Globally, postnatal factors (eg duration of hypothyroïdism) appear to predict verbal abilities and memory, whereas prenatal factors (bone-age) appear to predict visuo spatial skills. Glorieux (4), in the same way, underlines the necessity to carefully watch "language development" and "reasonning", assessed by the Wechsler performance scale.

Comments and questions are :

1) Neuropsychological impairments, such as slowness, disabilities in fine motor coordination, perceptual disabilities ... influence the total score of the intelligence tests, because they interfere in many subtests. This should explain why the children with these troubles are exhibiting an IQ distribution of lower values, if they are more frequently of more severely handicapped than a normal group. Was that comparison made ?

2) Are the frequency and the nature of the neuropsychological impairments observed permanent or transiant with time ? In some children or in every child ? Intraindividual variations data would be contributive. If they are transiant, it signifies that the child has not yet achieved its true potential. If they are permanent, how can one be sure that they

are primarily caused by the direct influence of a lack of thyroïd hormone? Prospective studies of enthyroïd premature and small for gestational age infants indicate such delays in subsequent intellectual impairments(14).

Even in a non selected group of 6 years old children of normal intelligence (2) facing a neuropsychological battery of tests, only a few children manage to succeed in every test ; rare are the children who obtain a linear profile, belonging to the same quartile at every test ; more over, the number of failures increases with the number of subtests in the battery, and, of course, with decreasing IQ. Neuropsychological impairments look as if they mere related to perinatal risk factors and/or to a lack of stimulation and emotional deprivation during infancy (1).

3) Does a parallelism exist between skeletal maturity and neuropsychological disturbances ; in children in general, and in children with CH particularly ? It has been demonstrated in children with primary nocturnal enuresis. In all these children, T4 and TSH were found to be normal (8).

CONCLUSION

A few more years will certainly be necessary to bring explanation concerning the slight delay still observed today in the mental development of children with CH the most severely affected.

At the present time,it is of main interest to realize that psychological observations are corresponding to biological data.

3. POSSIBLE LONG-TERM CONSEQUENCES OF THE NEUROCOGNITIVE IMPAIRMENTS

If these disorders remain observable after 6 years of age, they may be the cause of school learning disabilities.
Prognosis, however, is complex, because no single variable is determinant of outcome, and cumulative effects from multiple risks factors increase the probability of school disabilities : socio-cultural level, environment, motivation, intelligence, are determining a quite different outcome in two children with the same neuropsychological defects. Very important also is the age at which those defects were detected and educational guidance provided (by psychomotricity, or language stimulation for example).

Studies conducted on 5 years old children to detect future school learning problems conclude that (3) (7) : children with high IQ generally succeed in school, whatever the number of failed neuropsychological tests, but are often suffering of emotional disturbances, expression of the quantity of psychic energy needed ; the IQ alone, when average, is not a predictor, and school learning problems increase in proportion of the quantity and the nature of neuropsychological impairments ; a low IQ and neuropsychological impairments are very precisely risk factors in school.

The connexion between future forms of reading disabilities and language disorders, visuo-spatial and auditory dysfunction is well known. Dyspraxias are linked with certain form of dyscalculia (some other forms are related to processes of thought maintained in Piaget's preoperative period, and to disturbances in the symbolic way of thinking).

Therefore, it would be contributive to add to the psychological examination
- some symbolic test and
- appropriate reading, spelling and arithmetic achievement tests (adapted to each country), taking into account the methods used in the country to learn to read, to spell, to calculate.

CONCLUSION

In human research, we cannot go backwards ; we can never repeat an experience exactly in the same conditions to evaluate again and again the role of one specific parameter. Now comes an ethic choice, which is of great importance on a methodological point of view.

Are we going, in the future, to provide remedial help to the children who will need it with respect to their neurocognitive impairments ?
If we answer in the affirmative, comparisons and analysis to extract the specific role of thyroïd hormone will become much more complex.
If we answer in the affirmative, it is because we really believe that training will modify the mental outcome of impaired children.

And is this true, then we have to analyse carefully the influence of this training and also the influence of psychological factors, which is one of the major factors that brings humain individual variability in statistical trends.

REFERENCES

(1) Cahn, R.
Défaut d'intégration primaire et inhibition des apprentissages instrumentaux et cognitifs.
Paris, PUF, Revue Franç. Psychanalyse
1972 ; XXXVI, 5-6 : 955-971

(2) Chiland, C.
L'enfant de 6 ans et son avenir.
Paris, PUF, 1971

(3) Devos, P.
Etude d'une méthode de dépistage en 3ème gardienne des futurs inadaptés scolaires.
In : "Problèmes de médecine préventive de l'enfant et de l'adolescent"
F. Geubelle, éd., 1975

(4) Glorieux, J., Desjardins, M., Dussault, J., Letarte, J. et Morissette, J.
Etude longitudinale du développement mental des hypothyroïdiens congénitaux - Dix années de dépistage.
Arch. Fr. Pédiatr., 1987 ; 44 : 709-14

(5) Illig, R. and Largo, R.H.
An eight year study of the mental and psychomotor development of hypothyroïd children detected by neonatal screening.
In : Therrell Ed. - Advances in neonatal screening
Excerpta Medica
International Congress Series 741
1987 ; 85-91

(6) Kaufman, A.S., Long S.W. and O'Neal, M.R.
Topical review of the WISC-R for pediatric neuro-clinicians.
J. Child Neurol., 1986 ; 1 : 89-98

(7) Klees, M.
Les troubles instrumentaux des enfants.
Editions de l'Université Libre de Bruxelles, 1983

(8) Mimouni, M., Shuper, A., Mimouni, F., Grünebaum, M. and Varsano, I.
Retarded skeletal maturation in children with primary enuresis.
Eur. J. Pediatr., 1985 ; 144 : 234-235

(9) Njiokiktjien, Ch.
Pediatric Behavioral Neurology
SUYI Publ., Amsterdam, The Netherlands, 1988

(10) Preliminary report on patients with infantile hypothyroïdism after the third grade (1987)
NE Cong. Hypothyroïdism Collaborative
Dartmouth Medical School, Dept. Mat/Child Health
In : Pediatric Research, vol. 21, April 1987

(11) Rochiccioli, P., Alexandre, F., Roge, B.
Développement neurologique des hypothyroïdiens néonatales.
Arch. Fr. Pédiatr., 1987 ; 44 : 721-4

(12) Rovet, J., Ehrlich, R. and Sorbara, D.
Intellectual outcome in children with fetal hypothyroïdism.
J. Pediatr., 1987 ; 110 : 700-4

(13) Sameroff, A.J., Seifer, R., Barocas, R., Zax, M. and Greenspan, S.
Intelligence quotient scores of 4-year-old children : Social-Environmental Risk Factors.
Pediatrics, 1987 ; 79 : 343-350

(14) Siegel, L.S., Siagal, S., Rosenbaum, P. and al.
Predictors of development in preterm and full-terms infants : a method for detecting the at risk child.
J. Pediatr. Psychol., 1982 ; 7 : 135-48

(15) Steinhausen, H.C., Gluck, M. and Wiebel, J. (1979)
Psychological development of children and adolescents with hypothyroïdism.
Act. Paedopsychiat., 1979 ; 44 : 77

(16) Touwen, B.C.L. and Pretchl, H.F.R.
The Neurological Examination of the Child with Minor Nervous Dysfunction.
London, Heineman Medical 1970
Clinics in Developmental Medecine, n° 38

(17) Touwen, B.C.L.
Neurological development in Infancy.
Clin. in Dev. Med. n° 75, SIMP, Heinemann, London and Lippincott, Philadelphia, 1984

(18) Wolter, R., Noel, P., De Cock, P., Craen, E., Ernould, C., Malvaux, P., Verstraeten, F., Simons, J., Mertens, S., Van Broeck, M. and Vanderschueren-Lodewyckx, M. (1979)
Neuropsychological study in treated thyroïd dysgenesis.
Acta Paediat. Scand. (suppl.), 1979 ; 277 : 41

2.TESTS

(20) Buros, O.K.
Tests in Print : an index to tests, tests reviews and the literature on specific tests.
New-Jersey : The Gryphon Press (10 volumes)
Dernier volume 1985
Distributed by the University of Nebraska Press

(21) Bayley Scales of Infant Development
Bayley, N.
Manual : Bayley scales of infant development
New-York : Psychological Corporation, 1969

Bayley, N.
Development of mental abilities
In : Carmichael's Manual of Child Psychology (1970)
(P.H. Mussen, Ed) John Wiley & Sons, New York, p. 1163

(22) Beery-Buktenika Development Test of Visual Motor Integration (VMI)
Beery, K.E., Buktenika, N.
Development test of visual motor integration
Chicago, Follet, 1982

(23) Echelle de Brunel-Lezine (1951, remaniée en 1965)
Brunet, O. et Lezine, I.
Developpement psychologique de la première enfance
Paris, PUF, 1951 et éditions suivantes.

(24) Denver Developmental Screening Test
Frankenburg, W.K. and Dodds, J.B.
Denver Developmental Screening Test
Denver : University of Colorado Medical Center, 1966

Bryant, G.M., Davies, K.J. and Newcombe, R.G.
The Denver Developmental Screening Tests
Develop. Med. Child Neurol., 1974, 16 : 475-484

(25) The Gesell Developmental Schedules
or Gesell Maturity Scale
or Gesell Norms of Development
Gesell, A. and Amatruda, C.S.
Developmental Diagnosis : Normal and Abnormal Child Development : Clinical Methods and Pediatric Applications
New-York, Paul B. Hoebler, 1947

(26) Griffiths Mental Developmental Scale
Griffiths, R.
The abilities of babies : a study of Mental Measurement
London : University of London Press, 1964
New-York : Mac Grow Hill

Griffiths, R.
The abilities of young children : a comprehensive system of Mental Measurement for the first eight years of life
London : Chard, Somerset : Young and Son, 1970

(27) Leiter International Performance Scale
Russel, G., Leiter, P.H.D.
General instructions for the Leiter International Performance Scale
Stoelting, Chicago, Illinois

(28) Mc Carthy Scales of children's Abilities
Mc Carthy, D.
The Mc Carthy Scales of Children's Abilities
New-York : Psychological Corporation, 1972

(29) Nouvelle Echelle Métrique d'Intelligence (NEMI)
Zazzo, R., Gilly, M. et Verba, M.
Nouvelle Echelle Métrique d'Intelligence
Paris, A. Colin, 1966, 2 tomes

(30) (Bruininks)-Oseretski Test of Motot Proficiency
Bruininks, R.H.
Bruininks-Oseretski Test of Motor Proficiency
Minnesota : American Guidance Services, 1978

(31) (Halstead) Reitan Neuropsychological Battery
Reitan, R.M.
Neurological and physiological bases of psychopathology
Annual Review of Psychology, 1976, 27 : 189-216

Reitan, R.M. and Davidson, L.A.
Clinical Neuropsychology : current status and applications
John Wiley & Sons, New-York, 1974

Hynd, G.W. and Obrzut, J.E.
Neuropsychological assessment and the school-age child
New-York, Grune and Stratton, 1981

(32) Snyders-Oomen Non Verbal Intelligence Test
Snyders, J.Th. and Snyders-Oomen, N.
Niet Verbale Intelligentie Schaal
Swets Test Services, Lisse, The Netherlands

(33) Stanford-Binet Intelligence Scale
Binet, A. et Simon, Th.
Mesure du développement de l'intelligence chez les jeunes enfants
Paris, Bourrelier, 1957

(34) Swanson, J.M., Barlow, A. and Kingsbourne, M.
Task specificity of responses to stimulant drugs in laboratory tests
Int. J. Ment. Health 1979 ; 8 : 67-82

(35) Terman-Merrill Intelligence Scale
Terman, L.M. and Merrill, M.A.
Measuring Intelligence
Boston : Houghton Mifflin, 1937

Terman, L.M. and Merrill, M.A.
Stanford-Binet Intelligence Scale
Boston : Houghton Mifflin, 1960

(36) The Wechsler Intelligence Scales
Wechsler, D.
The Wechsler Preschool and Primary Scale of Intelligence
New-York : Psychological Corp., 1967

Wechsler Adult Intelligence Scale - Revised
New-York : Psychological Corp., 1981

DISCUSSION

ROVET (Toronto) : I have two comments : 1) Most of the tests were developed in English and have been translated into a number of languages. We do not know what was lost in the translation. 2) IQ values in general populations have increased with time since the tests were developed and, consequently, the standards for a normal population used as control are higher than when the tests were developed.

MENTAL DEVELOPMENT OF PATIENTS WITH CONGENITAL HYPOTHYROIDISM DETECTED BY SCREENING: QUEBEC EXPERIENCE

Jacqueline Glorieux

Département de Pédiatrie, Université de Montréal et Centre de Recherche, Hôpital Sainte-Justine, Montréal Québec, Canada

In Quebec 100 patients born between 1976-1983 are enroled in a longitudinal study. The focus of today's presentation is the interpretation of the I.Q. scores we have collected on the first 57 patients of our cohort when they reached the age of 7 years.

DEVELOPMENTAL ASSESSMENT FROM 1 TO 7 YEARS

A) The children form an homogeneous group of francophone background, followed on a regular basis by the same team of psychologists and endocrinologists who maintained through the years a close relationship with the families. We have excluded from the study the children who were living to far away to attend the scheduled protocol, patients with any other pathology, or subjects coming from disorganised families (adolescent mothers, addicted parents, out of the norm families) with whom compliance could not realistically be expected.

B) When the children were between 1 and 5 years of age we used the Griffiths Scales of Mental Development and we formed control groups in order to assess the level of our patients because we had no precise data from our population for comparison. The results were highly satisfactory. At the age of one year we observed no differences between the control and hypothyroid groups. At the age of one and a half year, 3 and 5 years the mean Developmental Quotient from hypothyroid patients were 5, 13 and 6 points lower than matched controls. The Griffiths Global score is the total of 5 or 6 sub-scales. We found slightly larger differences in the Language scale and the 2 Reasoning scales but at the age of 5 years the difference between the mean score of controls and hypothyroid patients was within a maximum of 10 points in every scales of the test.

Even though those differences are statistically significant, they are of such a small magnitude that we may consider them as practically trivial. We thus concluded that the screening program was successful.

C) Because of the lack of predictive power of young children's mental assessment we continued the evaluation after the age of 5 years with the classic Weschsler's test of intelligence composed in 1949 and revised in 1974. This psychometric instruments is internationally known and used. In Canada, the scientific community seems satisfied with its north-american norms even though the general consensus will consider the theoretical mean of 100 to be on the conservative side. So far we have tested 57 hypothyroid patients with the W.I.S.C.R. The children at 7 years have a mean global I.Q. of 96.6 with a range between 48 and 128. The results are consistent with our previous data and we noted the good correlation between the Griffiths Developmental Scores at 5 years and the W.I.S.C.R. I.Q. at 7 years. We can say that the preventive program works.

SCHOOL REPORT APPRECIATION

Intelligence tests are far from ideal but they remain the best instrument available for revealing an individual's mental functioning. The W.I.S.C.R. I.Q. does not reflect a global summation of the brain's capabilities but it does describe learning abilities, so it predicts school achievement accurately. An analysis of the children's school reports was done 2 years ago when the children were between the ages of7 and 9 years and was a frustrating experience. Out of 45 reports only 36 were detailed enough to permit a rough classification.

We selected two fields: Arithmetic and French and classified the scores in 3 broadly defined categories (Fig. 1).

Superior: A, excellent work, or 10 points above the mean of the class.
Inferior: D, insufficient, or 10 points below the mean of the class.
Average: All the others.

We are presenting these very preliminary results as an illustration of the predictive value of I.Q. scores on school achievement: as expected the best pupils have the highest I.Q.

We intend to conduct another investigation this summer as the same children will be in grade 4 to 6 and school reports will probably be more complete.

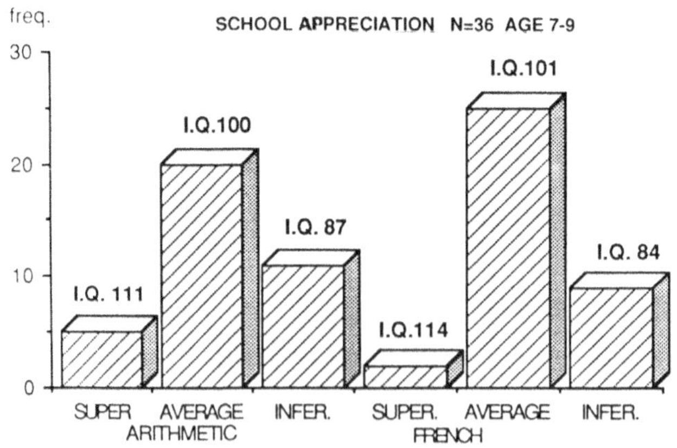

Fig. 1. School appreciation.

SOCIAL CLASS DISTRIBUTION

At the present time, social class indices, such as parents' education and occupation, predict the later intellectual development of children probably better than any other measure we may have on a young child's outcome. This holds even more if we deal with a homogeneous group of the same ethnicity, as is the case here.

The distribution of our sample shows the relatively simple educational and occupational background of our cohort (Fig. 2). Of the 50 families here represented only 2 of the fathers are professionnals and the bulk of the group belongs to the low average class calculated with the Hollingshead two factors index (1). This aspect of our research may explain to a certain degree the discrepancies with various other studies. As illustrated, we found a significant correlation between the I.Q. of our patients and the Hollingshead two factors index of their father (r = -0.42).

UNDERSTANDING THE W.I.S.C.R. GLOBAL I.Q. SCORE

A child's intellectual capacities cannot be described in a single score or even with a range of values. Wechsler groups the 12 sub-scales of his test into two sub-scores; each subtest measures some abilities that are subtest specific and some abilities that are shared with one or more other subtests: under the Verbal score we have 5 sub-scales: Information, Similarities, Vocabulary, Comprehension, they cover the knowledge of the meaning of something but also the clarity of ideation, the capacity to form a concept, to go beyond the concrete description, and Arithmetic which is something that is not related to the other scales. Under the Performance score we have 5 sub-scales: Picture completion, Picture arrangement, Block design, Object assembly and Coding. The patients reach a Verbal I.Q.: 93 sd 14 and a Performance I.Q.: 102 sd 14.

A difference of 9 points between sub-quotients is found in 25% of the normal population but these differences are equally found either in favor of Verbal I.Q. or of Performance I.Q. In our sample we found 46 patients with a better performance score and only 9 with a better

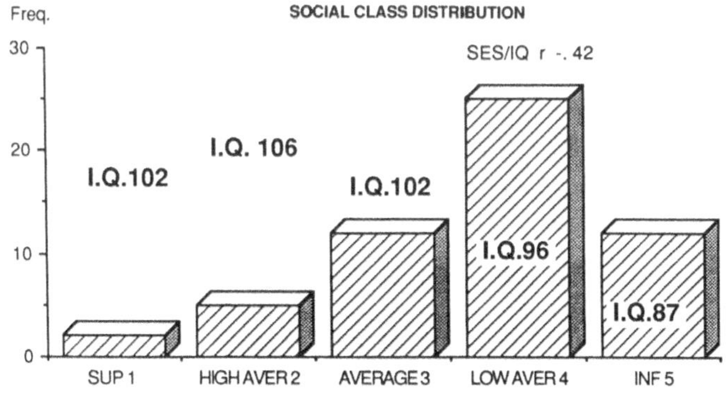

Fig. 2. Social class distribution

Verbal score. Verbal abilities are related to the familial socio-economic level. We have confirmed that point in our sample: the 9 children with a better Verbal score have also a better Hollingshead's index. On the other hand, in the 46 patients who had a lower Verbal I.Q., the mean difference between that score and their Performance score was 13. That difference is significant ($p < 0.05$). Lets us now look at the profile data of our group.

PROFILE OF SUBTESTS IN THE W.I.S.C.R. AT THE AGE OF 7

Using the presentation suggested by Kaufman (2) we grouped the sub-scales of the W.I.S.C.R. following a 3 factor analysis (Fig. 3). The first factor is verbal comprehension. The patient's mean quotient reached 95. The second factor is perceptual organisation: mean quotient reaches 106 and within both factors the sub-scores were very consistent. The third factor is difficult to describe. Kaufman

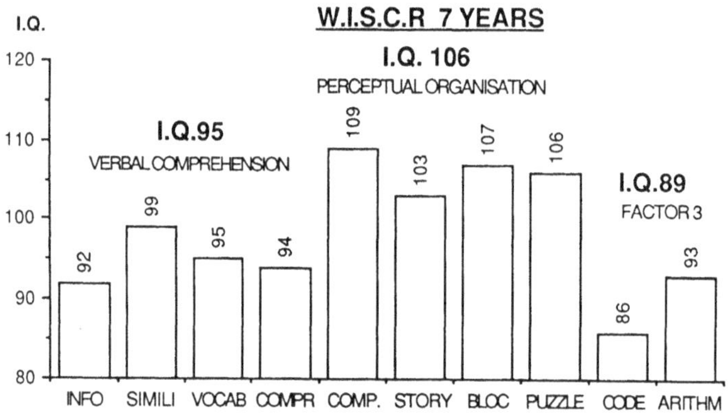

Fig. 3. Sub-scales scores.

used the term "Freedom from Distractability factor". Later on (3), he constructed a new test where he describes better the abilities required for those kinds of activities and called it: the sequential factor. Its meaning is the ability to ordinate information in a linear mode, in a succession of operation. It is an essential part of the learning process. The overall mean quotient for this factor was 89.

Low scores in these sub-scales comprising this factor are related to learning difficulties. The discrepancy of 11 points between the verbal comprehension and the perceptual organisation points in the same direction. Nevertheless, because of the adequate mean perceptual scores the majority of the children will probably manage by relying on their predominant abilities.

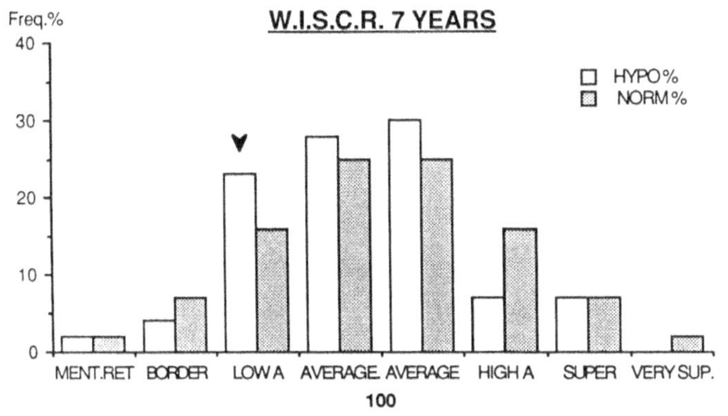

Fig. 4. Global I.Q. distribution.

DISTRIBUTION OF THE GLOBAL I.Q.

In order to add a perspective to the data, the expected normal distribution is represented in shaded tone (Fig. 4). There is a significant difference between the distribution of our sample and the normal curve.

As you can see, we may have a shortage of bright children but we have an excellent regrouping of patients in the average group, with a global I.Q. between 90 and 109, and this underlines the success of the preventive program. However we also found 13 children in the low average classification.

Because the possibility of the prevention of mental sequalae is what promoted the development of the screening procedures for a whole population of newborns, we have become interested in that low average group of children who obviously could benefit the most from early educational support. Psychological assessment at the age of 7 years can identify the patients and proper support could be provided to the patients who deserve help. But without any intellectual evaluation being built into a program can these children in need of external support be detected? We believe that it can be done and we now use two parameters that every pediatrician has available when he starts the therapy. Indeed in our sample two biological measures correlate with mental outcome: bone surfaces and T_4 levels.

In 1985 at Evian and in 1986 in Texas we demonstrated that we could discriminate between the hypothyroid patients using only the bone surface parameter: the patients with a minimal bone surface (0 or 0.05 cm^2) had a significantly lower developmental quotient in every scale at every assessment.

But the discrimination also includes a few children who were successful in the tests. If we further selected the children who had a minimal bone surface and a low T_4 level before therapy ($T_4 < 2$) the analyse becomes more accurate.

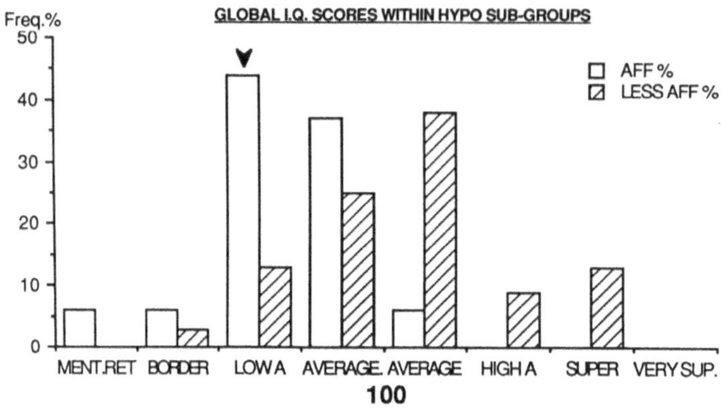

Fig. 5. I.Q. distribution for groups identified at birth.

DISTRIBUTION OF THE GLOBAL I.Q. WITHIN THE HYPOTHYROID GROUP

In 48 cases whose clinical evaluation was complete at the time of the first visit, the use of two parameters allowed us to predict that 16 children would be more affected and 32 will be less affected. In the figure 5, their results are compared and the frequency observed in each mental class is expressed as a percentage to allow for a better comparison. Out of the 16 children predicted to be more affected, we found:

- 9 children who, at the age of 7 years, had an I.Q. less than 90.
- 7 children had an average I.Q.
- none whilst could be classified in best categories.

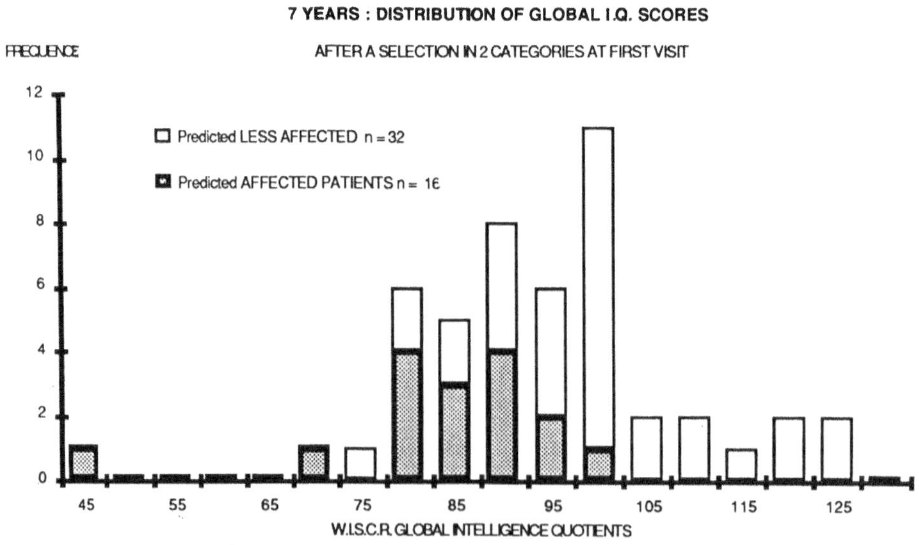

Fig. 6. Distribution of global I.Q. scores after a selection in 2 categories at the time of first visit

Their mean global I.Q. was found to be 86 with a verbal of 83 and a performance of 92. These scores are low, and differences with the norm or with the other hypothyroid patients scores are highly significant (Fig. 6) and clinically meaningful. The results of these more affected patients are below the norms, but are still higher than those reported for children who did not benefit from early therapy. The best comparison is with the group of patients described by Hulse in 1984 when he reported a mean I.Q. of only 79.5 with the same test for 99 subjects aged 6 to 16 years (4) who had delayed onset of treatment.

This categorisation has a double discriminatory effect because in the patients who had a higher T_4 level and larger bone surfaces and thus were predicted to be less affected: 27 were in the average classification and 7 belonged to the superior group, their mean I.Q. was found to be 101 with a veral I.Q. 97 and a performance I.Q. 106. The majority of the patients with higher T_4 and higher bone maturation seems free from any detectable deficit and the parents can thus expect a normal outcome, similar to their sibling's and family status expectation. Lastly, an interesting aspect of the data collected at the age of 7 years is the profile of the two subgroups on the 10 sub-scales of the W.I.S.C.R. test (Fig. 7). We observe a similarity in the and the strong points of the two groups.

Again it seems that it is the level achieved more than the pattern of abilities that differenciates the samples.

Hulse also reported that the coding sub-scale was the lowest in his sample, the highest being Block design. And he also found a difference between Verbal scores and Performance scores, in the same proportion that we did. In other words, if we consider the intellectual abilities measured with the W.I.S.C.R. test, the children born with a thyroid deficit seem to develop following a characteristic pattern that does not hamper the majority of the children but deserves a closer examination. We hope to investigate this aspect in the future.

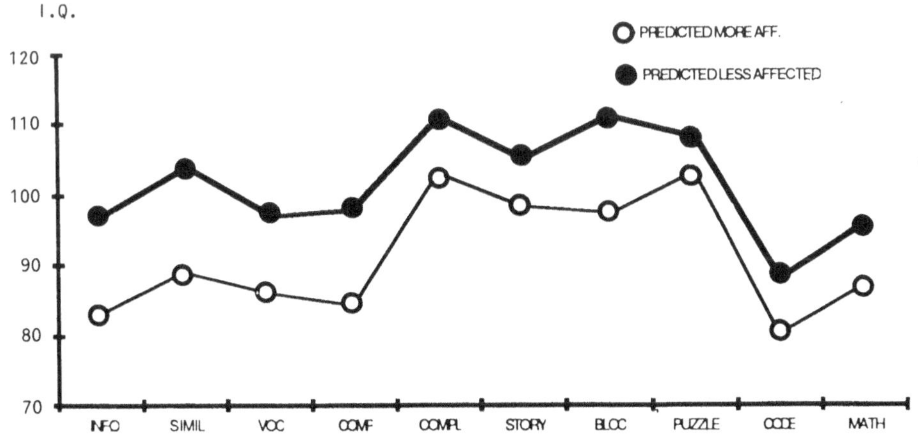

Fig. 7. W.I.S.C.R. profile age 7 for groups identified at birth

REFERENCES

1. Hollingshead AB. Two Factor Index of Social Position. Yale Station, New Haven, Connecticut, 1965.
2. Kaufman AS. Intelligent Testing with the WISC-R. John Wiley and Sons, New York, 1979.
3. Kaufman AS, Kaufman N. Kaufman Assessment Battery for Children. A..G.S. Circle Pines, Minnesota, 1983.
4. Hulse JA. Outcome for congenital hypothyroidism. Arch Dis Child 1984;59:23-30.

DISCUSSION

FUGGLE (London) : How selective was your CH patientgroup? How many children have you selected out because they belong to disorganised families ?

GLORIEUX (Montreal) : For this study, we selected families where differences between CH infants and controls, if any, should be explained only by the thyroid deficit and not by social or other factors. This is why we have not included CH infants from disorganised families, adopted children or children from immigrant families. Very few CH infants were discarded from the study because of disorganised families.

BRAVERMAN (Worcester) : Have you observed any correlation between the results of your psychointellectual evaluations and serum TSH under therapy ? Is it possible that some low IQ's are due to poor compliance ?

GLORIEUX (Montreal) : We have looked for such correlations and did not find any.

SINHA (London) : How can you apply these psychometric tests to native populations isolated in areas affected by endemic mental retardation and endemic cretinism ? How much are these tests culture free and what are the logistic problems met when psychologists, who are most usually inhabitants of large towns, use these tests in the field where most inhabitants never had any contact with urban life ?

ROVET (Toronto) : The tests that we are talking about are all culturally biased; there aren't any I know that are really applicable to various native popultations. You really have to test what is important to the culture that you are studying.

KLEIN (Hanover) : As indicated by Mrs. Klees, the LEITER test, with proper control, could probably be adapted to any population.

MORREALE DE ESCOBAR (Madrid) : The studies conducted by Prof. N. Bleichrodt from the University of Amsterdam and his colleagues in Central Java (1) and in Spain (2) were based on an adaptation of the tests to the local populations. They worked in collaboration with local psychologists to make the tests as much culture free as possible. Local personnel were then trained to use these "culture free" tests in iodine deficient and control areas.

(1) Bleichrodt N., P.J.D. Drenth and A. Querido. Effects of iodine deficiency on mental and psychomotot abilities. Am. J. Physical Anthropology 53 : 55-67, 1980.

(2) Bleichrodt N., I. Garcia, C. Rubio, G. Morreale de Escobar and F. Escobar del Rey. Developmental disorders associated with severe iodine deficiency. In : The prevention and control of Iodine Deficiency Disorders. B.S. Hetzel, J.T. Dunn, J.B. Stanbury eds. Elsevier publ., Amsterdam 1987, pp. 65-84.

NEONATAL THYROID SCREENING: NOW WE ARE NINE

New England Congenital Hypothyroidism Collaborative[1]

Department of Maternal and Child Health
Dartmouth Medical School
Hanover, N.H. 03756

Neonatal thyroid screening was begun in the hope that the resulting early treatment of infantile hypothyroidism would prevent the brain damage seen in patients treated only after a clinical diagnosis. We reported in 1979 at the Quebec meeting that the mean IQ of such children reported in the literature was under 80.[1] In addition, it was noted that Von Harnack, Zabransky, MacFaul, and (later) Wolter all reported that even those patients with normal IQs defined as above 85 or 90 were unable to progress normally in regular school classes because of specific learning disabilities, attention deficits, neurologic problems etc.[2,3,4,5,] These latter reports lacked concurrent controls and we were dubious of their conclusions. For instance, if we separate MacFaul's patients into those with IQs equal to or greater than 95 and those with IQs below 95, all but 2 of the patients with learning difficulties or neurologic problems fall in the latter group and the mean IQ of patients with learning difficulties is significantly lower than that of the remainder of the patients. Ninety-five is one standard deviation below the mean WISC-R IQ for most groups of normal children reported in recent years. Nonetheless, if these problems were real, we hoped that early treatment might also prevent them. How have our hopes been fulfilled over these years? All studies are unanimous in declaring that brain damage as measured by IQ scores has largely been prevented. There is argument only as to the degree of success obtained by different groups. Today, we would like to review briefly data on IQs in our cohort and to present new data on school progress of the patients.

Our original cohort with permanent infantile hypothyroidism, born in New England in the period of 1 January 1976 to 31 December 1980, has dwindled

[1]Current members of the Collaborative: MB Arnold, V Bapat, Y Baumgartner, A Bennett, ST Bigos, HH Bode, S Brink, J Brown, R Brown, D Carey, T Carpenter, JD Crawford, JE Crigler, M Danon, M Genel, P Gruppuso, JE Haddow, J Holmes, PR Larsen, J MacCracken, E Man, ML Mitchell, JM Orson, LA Page, S Ratzan, E Reiter, A Sadeghi-Nejad, B Senior, W Tamborlane, S Waisbren, and RZ Klein (coordinator)

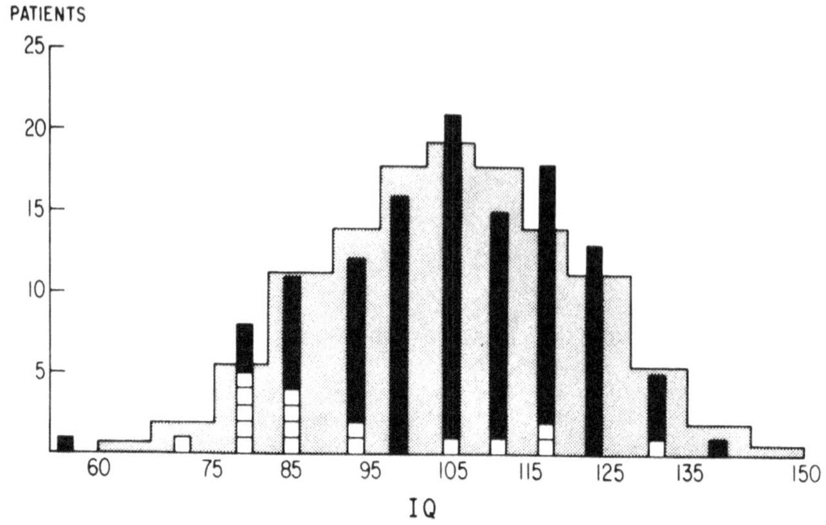

Fig. 1 Latest STANFORD-BINET Scores for 119 PATIENTS at 3-5 YEARS

from 144 patients to 93 from normal attrition and the desertion of one state health department from our collaborative. Roughly 20% of the patients had compensated thyroid disease and were at minimal risk of brain damage. Our incidence of infantile hypothyroidism is 1:4700 newborns and the incidence of those at risk for symptomatic hypothyroidism is 1:6200 which agrees with the reported incidence of clinically diagnosed hypothyrodism of 1:6000 to 7000 in Scandinavia and the Netherlands and in the elegant retrospective study of Larsson.[6,7,8,9,10] The incidence of both transient and permanent infantile hypothyroidism and compensated thyroid disease diagnosed as a result of screening in North America is lower than that in Europe presumably because of the areas of iodine deficiency extant in Europe and because of methodologic differences.

One hundred and nineteen of our cohort had one or more Stanford-Binet IQ measurements between the ages of 3 to 5 years.[6,11,12] Permission for testing was not granted for 4 children, 6 families moved from New England, and 2 others could not be located. Two children could not be tested because of severe spastic quadriplegia and two children with Down syndrome were not tested. Nine children were less than three years old when testing was discontinued. They include a third patient with cerebral palsy. The mean IQ of both the patients and controls was 105±15. Figure 1 shows the distribution of the IQ scores of the patients plotted against a normal distribution represented by the stippled background. The outlier is a girl with Williams syndrome whose score is not included in the mean calculation. If it were, the patient mean would be 104. The open squares represent patients who were indadequately treated in the first year of life as defined by repeated T4 concentrations below 8 μg/dl accompanied by TSH

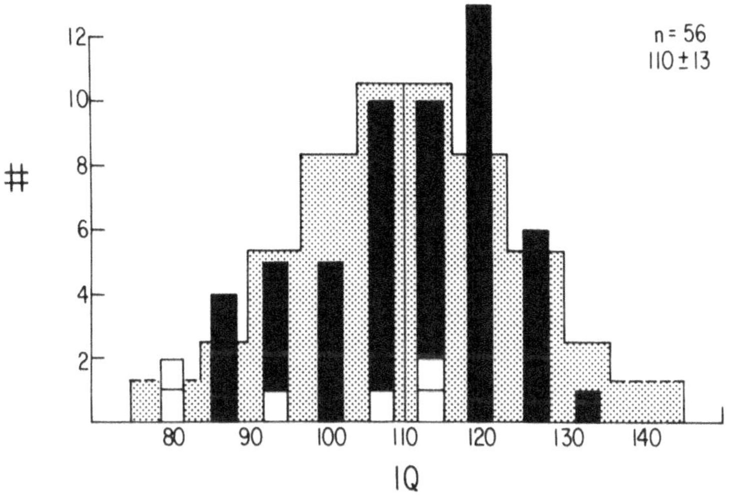

Fig. 2

concentrations above 15μU/ml after 2-4 weeks of treatment. Their mean IQ of 90 was significantly less than the group mean. Stepwise multivariate analysis demonstrated that inadequate therapy was the sole significant factor in determining IQ. Specifically, there was no significant correlation between IQ and type of hypothyroidism or other indicators of risk for brain damage such as lower T4 concentrations, retarded bone age, absence of residual functioning thyroid tissue etc. Adequacy of thyroxine replacement is essentially a function of compliance and is dependent on careful monitoring of hormonal values, meticulous parental instructions, and chance. Since patients with ectopic glands were known to have significantly higher IQs than those with atrophic glands after clinical diagnoses of infantile hypothyroidism, we felt that the lack of difference in IQ between our patient groups was most significant showing that the good results were not the result of a skewed selection of patients.[1]

At six years of age we measured WISC-R IQ scores in 56 patients.[13] This was 90% of the cohort of that age at the time our funding ceased. The results were identical to the previous results. The mean IQ for the patients was 109±13 and for the controls 110±13. Figure 2 is plotted in the same way as figure 1. At the same time we administered a battery of neuropsychologic tests. The only significant differences occurred in two tests of motor speed of small muscles, finger tapping with the non-dominant hand and the so-called "marching test" in which the children had to make a dot with a crayon in a series of preset areas. We predicted that the changes represented slower development rather than permanent deficit. This is discussed below.

Comparison of IQ and Grade Level Achievement

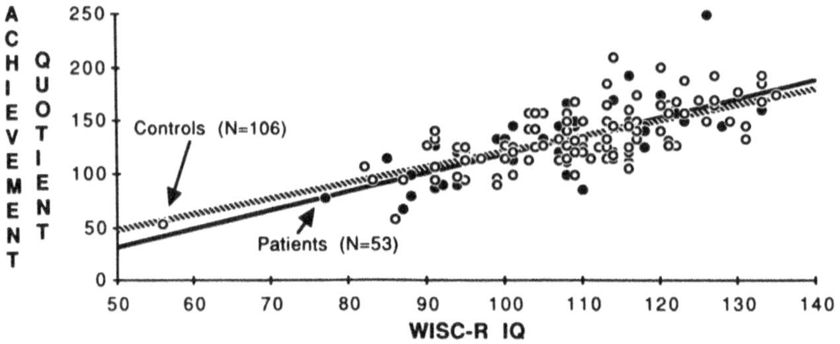

Fig. 3

We are now entering the third year of study of our patients' school progress. So far we have studied 53 patients, six of whom had finished the fourth grade and the rest the third. Their scores on a WISC-R abbrieviated to 5 items were correlated with their school achievement test scores. Twenty-four siblings and 82 classmates comprised the controls. In addition a battery of other tests looking for specific learning and neuromotor function problems was employed. The mean Hollingshead socioeconomic ranking was the same for both the patients and controls, 2.99 vs 2.84. The mean IQs were not different between patients and controls, 107±12 vs 108±16 for siblings and 110±15 for classmates. The relationships between IQs and the Peabody Individual School Achievement Tests were also the same. Figure 3 presents the regression lines for IQ and achievement scores calculated as overall grade achieved divided by grade to be entered and multiplied by 100. The lines are for 53 patients and 106 controls and are identical. The correlation coefficients are .67 and .70 respectively with r^2 values of .45 and .49. The regression lines using the individual items of the achievement tests, math, reading recognition, reading comprehension, spelling, and general information, as the dependent variables are also identical. We believe this indicates no specific learning deficits in the patients and this is verified by the results of all the tests of specific learning skills we administered which showed no difference between the control groups and the patients. These tests included the Kaufman Gestalt, Spatial Memory, and Logical Memory Tests, Boston Naming Test, Sentence Repetition Test, Children's Sentence Completion Test, Rey-Osterreith Complex Figure Test, Automatized Series Test, Beery Visual Motor Integration Test, and Achenbach Behavior Checklists.

The results of the timed motor tests revealed that the patients had caught up to the controls in speed on the marching test and finger tapping fulfilling our predictions, but they now performed more slowly on pegboard tasks. Table I shows the results for motor speed in the marching test at both 6 and 9 years of age and the pegboard test at 9 years. At no time were there any differences in accuracy in either the marching or pegboard tests.The control girls were faster at age 6 than control boys in the marching test, but the control boys caught up by age 7 and are the same at age 9. There was no difference in motor speed between male and female patients at age 6, so that only the female patients differed from their controls. At nine, both male and female patients had the same rate of speed as controls. On pegboard testing at this age, there were no differences between the sexes in rate of speed , but the differences between patients and both sibling and classmate controls were significant on ANOVA testing. The patients were not significantly slower on testing of other motor movements at 9 years.

Once again we believe the lag in small motor speed in the patients represents a delay in development of an age related skill and not a permanent deficit. One confusing factor in the interpretation of motor speed is that our patients are not as adequately treated as before. With the passage of years, both compliance and monitoring have deteriorated. We see more patients with elevated T4 concentrations and T4 values below 10 μg/dl. We even see frankly hypothyroid T4 concentrations of less than 4 μg/dl with TSHs as high as 200 μU/ml or more. Hypothyroidism at the present age of our patients should have no affect on intellectual potential but will interfere with performance on timed tests of both motor

Table I
TESTS OF MOTOR SPEED

"MARCHING"

	Patients		Controls	
At 6 Year	11 Boys	45 Girls	40 Boys	16 Girls
Dominant Hand	28±2.8*	28±1.3	28±1.3	21±0.8**
Non-Dominant	33±3.4	31±1.3	30±1.3	25±1.1**
At 9 years	10 Boys	41 Girls	20 Boys	74 Girls
Dominant Hand	18±1.6	19±1.2	18±1.3	19±1.5
Non-Dominant	22±2.7	23±1.8	21±2.7	23±1.8

PEGBOARD

At 9 Years	51 Patients	104 Controls
Dominant Hand	75±1.7***	70±1.4
Non-Dominant	85±2.5***	77±1.3

* All results are mean time in seconds±SEM
** Δ from other 3 groups, p<.01
*** Δ from controls, p<.01

Table II
SCHOOL PROGRESS THROUGH THE THIRD GRADE

	65 Patients	83 Siblings	158 Controls
NO PROBLEMS	72%	72%	73%
REPEATED A GRADE (K-3)	9%	18%	16%
SPECIAL EDUCATION	5%	5%	3%
TUTORING	14%	5%	8%

and cognitive function. Our bias is that the slow motor development is probably not related to current therapy because of the findings in our patients when younger and in patients reported by others.

Finally, we have reports of school progress through the third grade for 65 patients and 83 of their siblings, 80 of the patients' classmates and 78 siblings of these classmates. Table II gives the results. Seventy-two per cent of the patients, 72 % of their siblings , and 73 % of their classmates and the classmates' siblings went through the third grade with no difficulties. Fewer patients than controls repeated a grade but more patients needed tutoring. These differences are not significant on chi square analysis.

In summary, we think their school progress, coupled with the excellent correlation between IQ and performance on school achievement tests, is additional evidence that children with infantile hypothyroidism need not be handicapped in their progress in their real world. One caveat; these analyses of IQ, progress in school, and physical performance do not include our patients with spastic palsies. If their CNS problem were caused by their hypothyroidism, these patients would be included and the resulting results would not be as good.[14] In any event, the results we report are still somewhat better than those in some other reports. We think this is because of serendipitous, better compliance and more careful monitoring in infancy as discussed in a previous paper. [15]

Bibliography

1. Klein RZ. History of congenital hypothyroidism. In: Burrows GN and Dussault JH, eds. Neonatal Thyroid Screening. New York: Raven Press, 1980:51-59
2. Von Harnack GA, Wallis H. Zur Psychopathologie der hypothyreose in kindesalter. Monatsschr Kinderheilkd. 1960; 108:373-376

3. Zabransky S, Richter R, Hanefeld F, et al. Zur prognose der angeborenen hypothyreose. Monatsschr Kinderheilkd. 1975; 123:475-476
4. MacFaul R, Dorner S, Brett EM, et al. Neurological abnormalities in patients treated for hypothyroidism from eary life. Arch Dis Child 1978; 53: 611-619
5. Wolter R, Noel P, DeCock P, et al. Neuropsychological study in treated thyroid dysgenesis. Acta Paediatr Scand Suppl 1980; 277: 41-46
6. Klein RZ. Neonatal hypothyroidism then and now: The results of neonatal screening. Curr Probs in Pediatr 1985; 15: 1-57
7. Alm J, Larsson A, Zetterström R. Congenital hypothyroidism in Sweden. Acta Paediatr Scand 1978; 67:1-3
8. DeJonge GA. Congenital hypothyreïde in Nederland. T Kindergeneesk 1977; 45:1-5
9. Jacobsen BB, Brandt NJ. Congenital hypothyroidism in Denmark; incidence, types of thyroid disorders and age at onset of therapy in children born 1970-1975. Arch Dis Child 1981; 56:134
10. Larsson, A, Hagenfeldt L, Alm J, et al. Incidence of congenital hypothyroidism: A retrospective comparison of neonatal screening and clinical diagnosis. Pediatr Res 1984; 18:106
11. New England Congenital Hypothyroidism Collaborative. Effects of neonatal screening for hypothyroidism: Prevention of mental retardation by treatment before clinical manifestations. Lancet 1981; 2:1095-1098
12. New England Congenital Hypothyroidism Collaborative. Characteristics of infantile hypothyroidism discovered on neonatal screening. J Pediatr 1984; 104:539-544
13. New England Congenital Hypothyroidism Collaborative. Neonatal hypothyroidism screening: status of patients at 6 years of age. J Pediatr 1985; 107:915-919
14. New England Congenital Hypothyroidism Collaborative. Congenital concomitants of infantile hypothyroidism. J Pediatr 1988; 112:244-247
15. New England Congenital Hypothyroidism Collaborative. Treatment of congenital hypothyroidism. In: Delange, F. et al eds. Research in Congenital Hypothyroidism. Plenum, N.Y. 1988

DISCUSSION

DELANGE (Brussels) : School reports through the third grade indicated that the CH patients required more tutoring than the controls. What does tutoring mean in the day to day life of a child ?

KLEIN (Hanover) : Rather than holding some of the patients back for another year in the same class, they have special tutoring involving supplementary work in special classrooms of small numbers. The difference in frequency of tutoring between CH patients and controls is not significant.

FUGGLE (London) : You showed that the treatment during the first year had an effect on outcome at three years. Were there any effects on outcome at six and at nine years ?

KLEIN (Hanover) : We could not prove this because the number of patients became too small. By the time we got to 6 and 9 years, we investigated seven patients with poor compliance during the first year of life. They seem to be occupying the same area in the whole distribution, but it is not statistically significant.

DELANGE (Brussels) : Before going to the last paper of this session, we would like to ask Doctor S. Heyerdahl, child psychiatrist at the Department of Pediatrics, University of Oslo, Norway, to report on her personal experience.

HEYERDAHL (Oslo) : Our results on intellectual development of children with congenital hypothyroidism in Norway can be summarized as follows :

Our results are from a study of all known children with congenital hypothyroidism (N = 46) diagnosed during the first 3 years after screening was started in Norway.

The children were tested with Bayley Scales of Infant Development at 1 and 2 years and with Wechsler Preschool and Primary Scale of Intelligence (WPPSI) at 6 years.

The mean IQ at 6 years was lower than expected (IQ = 90) and lower than in siblings (IQ = 100). The Norwegian standardization of the test is quite recent, and if the American standardization was used the score would be about 5 points higher. The siblings were mostly older, but were tested with the same or comparable methods (WPPSI or WISC). Is this lower IQ related to hypothyroidism ?

There are some within-group differences in the population studied. Correlation between serum T4 at diagnosis and MDI (Mental Developmental Index) at 2 years is 0.32 ($r = 0.32$, $p = 0.04$); serum T4 at diagnosis and IQ at 6 years ($r = 0.29$, $p = 0.06$); age at start of treatment and PDI (Motor Developmental Index) ($r = 0.42$, $p = 0.006$). Children with athyreosis had lower IQ at 6 years (N = 13, IQ = 84) than children with ectopic or hypoplastic glands (N = 22, IQ = 100) ($p = 0.01$).

I looked at the therapeutic regimen and compared it with that of the New England group to see if this might explain the differences in the results obtained in Norway and New England. In the Norwegian study, mean age at therapy was not higher than in the New England sample, nor the dose of T4, nor the mean serum T4 during the first year of life. One problem in our study was the adequacy of therapy : we used as biochemical criteria for defining inadequate therapy a serum T4 concentration below 8 ug/dl and TSH above 15 uU/ml observed on more than one occasion.

The mean IQ at 6 years was lower in the "low treatment" group (IQ = 83, N = 13) than in the "adequate treatment" group (IQ = 93, N = 32).

In conclusion, there is a relationship between the degree of hypothyroidism and the longterm intellectual development; slightly unsufficient therapy, even if not clearly inadequate, results in a worse mental prognosis.

NEUROLOGICAL DEVELOPMENT IN CONGENITAL HYPOTHYROIDISM

Pierre Rochiccioli, Françoise Alexandre, and Bernadette Rogé

Paediatric and Medical Genetics Department, and Department of Medical Psychology, University Hospital of Rangueil, F-31054 Toulouse Cedex, France

Neonatal screening for hypothyroidism was instituted to improve the mental prognosis, which was previously unsatisfactory. Since 1975[1] early diagnosis and treatment have been possible in Europe thanks to screening. Over 10 years later, several studies[2-4] indicate that mental development is normal. However, neurological dysfunction involving the cerebellum has been reported. These abnormalities, which had been described before the institution of screening[5-8] and which constituted a supplementary handicap, seem to be present only to a minor degree in detected hypothyroid children. We therefore carried out a detailed neurological study in order to discover the importance of these neurological impairments, their correlation with other factors of gravity and the effectiveness of assistance in improving school achievement.

MATERIAL

All the hypothyroid children in our group were detected in the Midi-Pyrenees region of France during routine neonatal screening. From 1977 to 1987, 263 145 tests have been done and 69 children detected. The incidence of hypothyroidism in the region is 1/4 000. Mean age at diagnosis was 29 ± 2.4 days. The diagnosis was always confirmed by dosage of plasma thyroid hormone levels, and above all by thyroid scans, using iodine 123 in most cases and occasionally technetium 99m. The etiology was thus seen to be as follows : 39 ectopic glands, 22 cases of thyroid aplasia and 8 eutopic glands. Age at the time of neurological assessment was : 6 - 12 months, 33 children ; 12 - 36 months, 41 children ; 4 - 5 years, 37 children ; 6 - 7 years, 29 children ; 8 - 10 years, 10 children.

We also assessed a control group of normal children : 6 - 12 months, 32 children ; 12 - 36 months, 45 children ; 4 - 5 years, 38 children; 6 - 7 years, 16 children, 8 - 10 years, 15 children.

METHODS

Neurological assessment

The evaluation of Touwen[9] was used, modified in order to allow precise statistical analysis. Eight neurological apparatus were examined (9 after the age of 4).

Table 1. Neurological Profile < 3 yr (adapted from Touwen)

NAME : FIRST NAME : AGE : GROUP :

1 - Sensorimotor apparatus		**5 - Fine manipulative ability**	
Resistance to passive movement	_	Voluntary grasping	_
Range of movements	_	Quality of grasp	_
Muscle consistency	_	Releasing small objects	_
Tendon reflexes	_	Graphic ability	_
Clonus	_	Global tension	_
Total	_	Total	_
2 - Posture		**6 - Gross motor functions**	
Sitting	_	Rolling from back to stomach	_
Standing - head	_	Sitting up	_
- trunk	_	Crawling on stomach	_
- legs	_	Standing up	_
Lying down	_	Walking	_
Total	_	Total	_
3 - Balance, recovery of balance		**7 - Quality of motility**	
Pull-to-sit	_	Rapidity	_
Flat on stomach	_	Regularity	_
Sitting	_	Adequacy	_
Tests while held	_	Tremor	_
Standing	_	Clonus	_
Total	_	Total	_
4 - Coordination of the extremities		**8 - Motility of the head**	
Directed motility of hands	_	Facial expression	_
Falling forward	_	Eye movement	_
Falling backward	_	Dribbling	_
Functional tests	_	Swallowing	_
Kicking a ball	_	Language	_
Total	_	Total	_

These were sensorimotor apparatus, posture, balance, coordination of the extremities, fine manipulative ability, gross motor functions, quality of motility, head motility, and associated movements. Each subsystem was examined by 5 items scored 0, 1 or 2 according to failure, partial success or complete success. This gave a statistical result for each subsystem and also a global neurological total (table 1).

Psychomotor development

Psychomotor development was studied using Griffith's test from 0 to 6 years, giving the global development quotient (DQ) and also partial DQs : posture, coordination, language, sociability. The WISC was used after the age of 6, giving a global intelligence quotient (IQ), a verbal IQ and a performance IQ.

Study of bone maturation

The bone maturation of the hypothyroid children was studied using a personal method described elsewhere (10), giving the adjusted area of the epiphysis, the distal femoral center, the proximal tibial center, the calcaneum, the astragalus, and the cuboid bone.

Statistical analysis

Student's t test was used for comparison of the means and linear regression for the correlations.

Home stimulation programs

Specific stimulation programs were designed to be used in the home by the parents. The principles of application were explained to them and a demonstration was given. They were supplied with the necessary standardized material and with forms for noting the child's performance.

The exercises consisted of :

- manual coordination (bead-stringing, plaiting, cutting out paper, making chains with paper-clips and safety-pins, use of glue and modelling wax)
- eye-hand coordination (matching numbers and symbols)
- artwork (colouring and colouring by numbers, drawing lines, copying shapes on squared paper, joining dots, filling in drawings, following mazes).

The first stage of the program aimed at increasing the precision of the child's gestures. When this was achieved, the second stage aimed at helping the child to accomplish the tasks more quickly. Improvement was assessed at the end of the program.

Child selection

Ten children aged from 5 to 10 years were selected to follow the stimulation program. They all had specific fine manipulative disability, shown by the neurological assessment and also by Satz's test, which predicts learning difficulties due to instrumental problems, and by Stamback's test, assessing the level of manual motility development.

The child and the parents must be highly motivated. Conversely, children with important psychomotor dysfunction were directed into institutionalized rehabilitation programs.

RESULTS

Global neurological evaluation

There were no significant differences between the hypothyroid group and the control group (fig 1).

Evaluation of subsystems

There was a significant difference at 1 year for coordination, fine manipulative ability and gross motor functions, at 2 years for coordination and fine manipulative ability, at 4-6 years for fine manipulative ability and gross motor functions, at 6-8 and at 8-10 years for coordination and fine manipulative ability (fig.2).

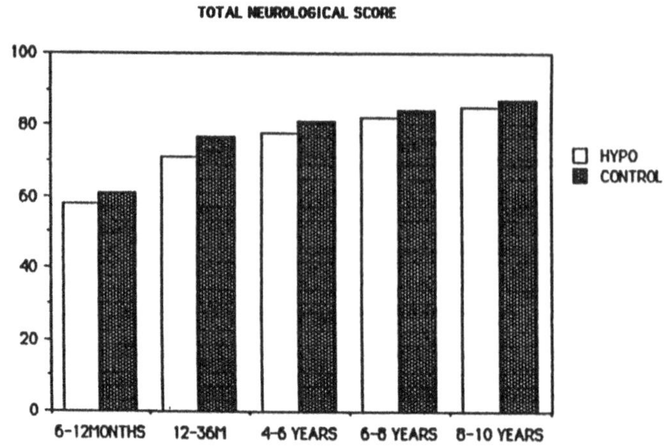

Fig. 1. Global neurological evaluation in the hypothyroid group and in the control group : no significant difference.

Follow-up

Follow-up of the evolution of 21 children over several years showed this was within normal limits for all the children except one (fig. 3).

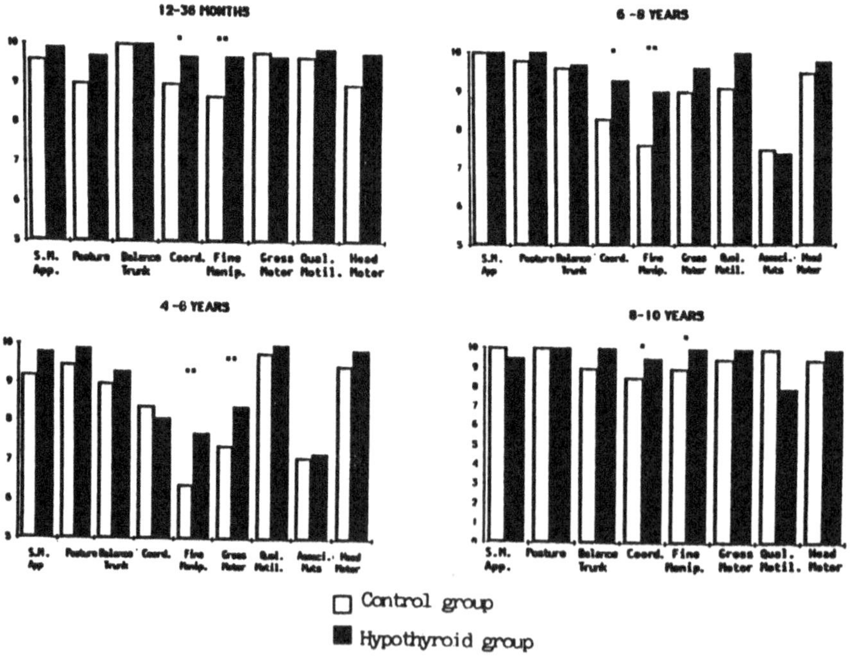

Fig. 2. Neurological profile (subsystems) at 12-36 months, at 4-6 years, at 6-8 years and at 8-10 years, in the hypothyroid and in the control group. There is a significant difference in coordination, fine manipulation and motility.

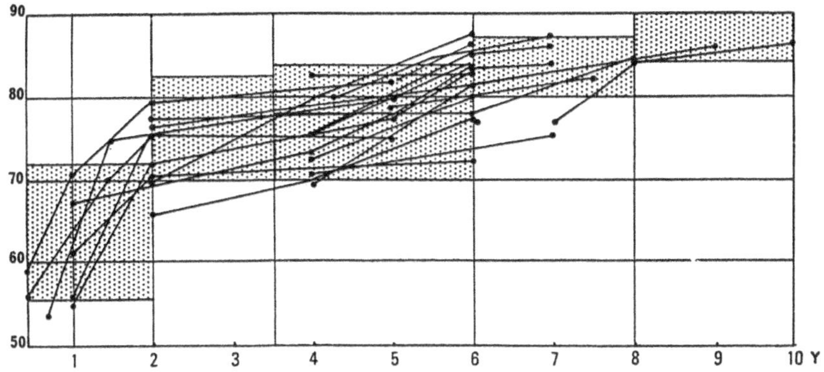

Fig. 3. Evolution of the global neurological score over several years. No significant difference with the control group.

Study of correlations

We sought various correlations: 1) with the DQ : there was a significant correlation between the global DQ and neurological score at one year and at 4 years (r = 0.73) ; 2) with the age of diagnosis : no correlation ; 3) with the socio-economic level : no correlation ; 4) etiology : we found no significant difference between children with thyroid aplasia and with ectopic glands either for the global neurological total or for the subsystems ; 5) with bone maturation : there was a significant correlation between the global neurological total at one year and the adjusted epiphyseal area of the distal femoral center at the time of diagnosis ($r = 0.30$, $p<0.01$).

DISCUSSION

The fact that neurological impairment exists in hypothyroidism has been known for some time, as the first publications date back to 1960. These described cerebellar disorders defined as "cerebellar myxoedema" by the authors. The subject was taken up again in particular in 1978[11, 12]. MacFaul[11] reported the neurological assessment of 30 subjects aged 2 to 21 years in whom hyperthyroidism was diagnosed at ages varying from 4 weeks to 9 years. He found frequent impairment : language disorders in 20% of cases, strabismus in 53%, nystagmus in 10%, minor motor dysfunction in 50%. This author analyzed the incidence of impairment according to age at diagnosis, and concluded that there was no difference in incidence in children diagnosed between the ages of 4 and 10 weeks and those diagnosed later than 10 weeks. He thus inferred that neonatal screening was of only minimal value. In the same year, Wolter[12] addressed the same problem but analyzed the results according to the pre- or postnatal onset of hypothyroidism, assessing bone age by the method of Sénécal[13]. Fifty-seven children were studied in whom hypothyroidism was diagnosed at ages ranging from 1 month to 5½ years. The abnormalities observed were hyperkinesia, impairment of spatial orientation, fine manipulative ability and speech, cerebellar ataxia, strabismus, enuresis, difficulty in learning mathematics, and dyslexia. Judging by bone age and the pre- or postnatal onset of hypothyroidism, the authors concluded that the children in whom the onset was prenatal suffered greater impairment. Moreover, if the diagnosis was made before the age of one year the abnormalities were less important and above all there was improvement as the child developed. Unlike MacFaul, Wolter[12] concluded that neonatal screening and prompt treatment reduce the neurological consequences of hypothyroidism.

Our study concerns 69 hypothyroid children detected by neonatal screening in the Midi-Pyrenees region of France. Their mean age at diagnosis was 29 days. Treatment was thus prompt and regularly followed. However, precise neurological evaluation raises many methodological problems. To avoid purely subjective appreciation, we modified Touwen's profile so that each exercise was scored. Five exercises were done for each neurological subsystem. Thus each subsystem was given a mark, as well as the neurological evaluation as a whole. Testing of a normal control group gave us a means of precise comparison. Our results showed that global neurological development was normal as compared with the control group. However, evaluation of the neurological subsystems showed significant differences between the hypothyroid children and the control group : in particular for coordination and fine manipulation at all ages, and gross motor functions at 1 year and at 4 years. However, there was no difference with the control group for all other neurological subsystems, i.e. posture, sensorimotor apparatus, quality of motility, head motility. Thus, some impairments described in the literature before screening are observed, but much less frequently and to a lesser degree.

If age at diagnosis has of course altered since the institution of screening, on the other hand the duration of hypothyroidism, its pre- or postnatal onset, has not changed. We used a personal method for calculating the epiphyseal area, which is a much more reliable criterion of the duration of hypothyroidism[10]. This is corroborated by the existence of a significant correlation between the epiphyseal area of the distal femoral center, the DQ and also the global neurological score. Moreover, analysis of the various findings according to epiphyseal area : nil, between 0 and 25 mm^2 (prenatal onset), or greater than 25 mm^2 (postnatal onset) shows a significant difference between the two groups.

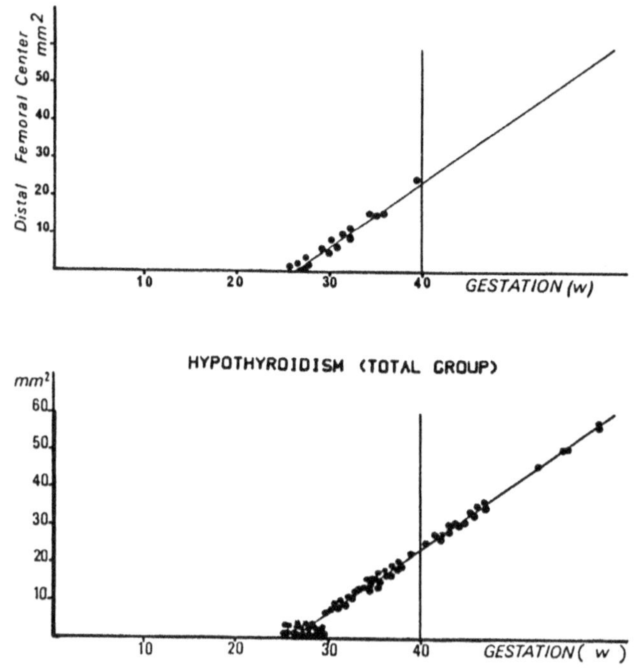

Fig. 4. Epiphyseal surface of the distal femoral center in the hypothyroid group with neurological impairment and in the total hypothyroid group. In the first group, the epiphyseal areas are of prenatal type.

These neurological impairments certainly form part of global dysfunction of the nervous system, as we found a strong correlation between the neurological total at 1 year and 4 years and the DQ at the same ages.

Analysis of the 16 cases with the highest degree of neurological dysfunction did not reveal definite etiological factors (7 cases of thyroid aplasia, 9 ectopias) nor a clear biological cause (low T4 level in 6 cases but equal to or greater than 2 μg/dl in 10 cases). However, bone maturation was retarded in all cases and hypothyroidism may be considered as being of antenatal onset, the proximal tibial center being always absent, the distal femoral center nil in 4 cases and less than 15 mm^2 in all but one of the other cases (fig. 4).

This detailed neurological examination showed that neonatal diagnosis of hypothyroidism has not resulted in the complete disappearance of the neurological dysfunction which existed before neonatal screening. However, these impairments are infrequent and seem to be restricted to disorders of coordination and fine manipulation. These impairments are more frequent where the onset of hypothyroidism was prenatal, which shows the value of study of bone maturation in this evaluation. Thus, neonatal screening considerably improves the mental and neurological prognosis. However, it does not result in the complete disappearance of certain impairments which are linked to lack of thyroid hormones during foetal life. Lastly, neurological diagnosis is important because these minor neurological abnormalities pass unnoticed and are not taken into account as necessitating psychomotor rehabilitation. They are, however, the cause of difficulties at school, related to slowness and lack of precision in gesture. Loss of interest in school work frequently follows on the original instrumental difficulty. Moreover, previous results have shown that after the age of 2 years the developmental level of the child depends more on the socio-economic environment and consequently on the quality of the stimulation received, than on the initial pathology. We therefore felt that these various elements justified the setting up of our specific home stimulation program. It has only recently been initiated and we have not yet been able to assess the results precisely. However, very rapidly, remarkable improvement in writing ability has been observed by the families, and also by teaching staff.

ACKNOWLEDGEMENTS

The authors would like to thank Ms. N. Crowte for translation of the manuscript.

REFERENCES

1. P. Rochiccioli, G. Dutau, F. Bayard, and D. Augier, Neonatal detection of hypothyroidism by radioimmunoassay of the thyroxine in the eluate of dried blood, Pediatr. Res. 9:685 (1975).
2. J. Glorieux, J.H. Dussault, and J. Letarte, Preliminary results on the mental development of hypothyroid infants detected by the Quebec Screening Program, J. Pediatr. 102:19 (1983).
3. New England Congenital Hypothyroidism Collaborative, Effects of neonatal screening for hypothyroidism. Prevention of neonatal retardation by treatment before clinical manifestations, Lancet 2:1095 (1981).
4. P. Rochiccioli, B. Rogé, and F. Alexandre, Résultats du développement psychomoteur des hypothyroïdies dépistées à la naissance, Arch. Fr. Pédiatr. 40:537 (1983).
5. R. François, M. Jeune, M. David, and A. Metton, Pronostic de l'insuffisance hypothyroïdienne congénitale, Med. Hyg. 35:3269 (1977).

6. B. Hagberg and O. Westphal, Ataxic syndrome in congenital hypothyroidism, Acta Paediatr. Scand. 59:323 (1970).
7. M. Pierson, L'avenir éloigné des enfants hypothyroïdiens, Arch. Fr. Pédiatr. 27:337 (1970).
8. J. Wiebel, Cerebellar-ataxic syndrome in children and adolescents with hypothyroidism under treatment, Acta Paediatr. Scand. 65:201 (1976).
9. B.C.L. Touwen, Examination of the child with minor neurological dysfunction, in : "Clinics in Developmental Medicine", W. Heckeman, ed., Medical Books, London, 1979.
10. P. Rochiccioli, G. Dutau and F. Despert, La surface des épiphyses du genou : index de l'ancienneté de l'hypothyroïdie néonatale, Arch. Fr.Pédiatr. 41:329 (1984).
11. R. MacFaul, S. Dorner, E.M. Brett and D.B. Grant, Neurological abnormalities in patients treated for hypothyroidism from early life. Arch. Dis. Child. 53:611 (1978).
12. R. Wolter, P. Noel and P. De Cock, Neuropsychological study in treated thyroid dysgenesis, Acta Paediatr. Scand. Suppt. 227:41 (1979).
13. J. Sénécal, M.C. Grosse and A. Vincent, Maturation osseuse du foetus et du nouveau-né, Arch. Fr. Pédiatr. 34:424 (1977).

DISCUSSION

BOYAGES (Amsterdam) : I am surprised by our observation of cerebellar deficits in CH patients because the development of cerebellum in humans is largely postnatal and because myxedematous endemic cretins, with extremely severe and untreated congenital hypothyroidism, do not have clinical or radiological evidence of cerebellar impairment.

DELANGE (Brussels) : Two questions : 1) What kind of training is required before being able to perform the neurological evaluation you described and how long does it take to make one evaluation ? 2) You have shown that there is a strong correlation between neurological score and DQ not only at one year but also at four years. At one year, such a correlation is not surprising because at that age motor performances are an important part of the evaluation of global DQ; this is not the case at four years. Why then is the correlation still so strongly positive ?

ROCHICCIOLI (Toulouse) : 1) This type of neurological evaluation requires a very long training period and experience in pediatric neurology; each examination takes about 30 minutes. 2) Most probably because the children with abnormal neurological development also have impaired intellectual development.

PART 5. ABSTRACTS OF THE POSTER SESSIONS

1.

CONGENITAL HYPOTHYROIDISM (CH) IN NORTHERN GERMANY (SCHLESWIG-HOLSTEIN, SH) : SCREENING AND FOLLOW-UP SINCE 1980

C. Brack*, P. Beyer*, G. Dinkelacker**, H.P. Müller**, P. Ball**, and R. Knuppen**

* Klinik für Pädiatrie, Arbeitsgruppe Pädiatrische Endokrinologie;

** Biochem. Endokrinologie, Zentrales Hypothyreose-Screening für das Land SH, Medizinische Universität zu Lübeck

We performe TSH dry-spot screening as "Central CH-Screening for SH" since 1980. It covers 99,3 % of all newborns in this area, the remaining are screened by more than 5 other laboratories. We follow up all cases of CH detected by our screening, either in our department or in collaboration with a local pediatrician. Since January 1980 we screened 95,657 out of 193,854 newborns in this area. There were 879 recalls for serum with a declining recall rate from 2.23 % in 1980 to 0.11 % in 1987. Therapy and clinical follow-up was begun in 65 cases, of whom 30 were later considered transient hypothyroidism, 24 of them iodine-induced. Additional 4 children had transient, 2 so far permanent hypothyrotropinemia. The remaining 30 children are still under therapy (L-T_4), therefore incidence of CH in our population is 1:3188. We saw seasonal variation in CH-incidence : 15 children were born between Jan. and April, 7 in Sept. and Oct. At about 2 years of age we reevaluate thyroid-function after disruption of T_4 substitution, including scintigraphic scan, and intellectual and psychomotor developmental testing. Start of therapy in our patients was at day 6-33 (mean 13.3). Clinical follow-up of the children depicts insufficient L-Thyroxin substitution at one or more occasions in 80 %. So far scintigraphic scan during follow-up showed 9 cases with athyreosis, 6 with ectopic thyroid tissue, and 3 with dyshormonogenesis. At age 2-4 mild deficiencies in intellectuel and psychomotor development could be demonstrated as a general trend in our patients, expressed as retardation in performance at or below the 90 % level in a standardized test[1]. Severe retardation in 6 children (28 %) as well as the milder forms was focussed on speech, perception, and less pronounced on fine motor abilities. Retardation showed no other correlation then with social status and severe additional illness.

Further studies with more detailed and frequent testing including matched pairs are necessary to elucidate prenatal neurophysiological defects in CH and the benefits of very early therapy.

---

[1] HELLBRÜGGE T. et al. Münchener Funktionelle Entwicklungsdiagnostik 2. und 3. Lebensjahr. Experimentalfassung 1977. München, Dissertations- und Forschungsdruck Frank, 1977.

2.

SCREENING FOR NEONATAL HYPOTHYROIDISM (NH) IN JUDEA AND SAMARIA

J. Sack, M. Blum, Z. Haker, T. Tulchinsky, R. Gutekunst, and A. Akstein

Pediatric Endocrinology, Institute of Endocrinology
Sheba Medical Center and Ministry of Health, Israel

The National Screening Program for NH (in combination with PKU) started in Israel in 1978. Since then, 911,000 newborns have been screened for T_4 levels, followed by TSH measurements. ALL the samples (around 95,000 per year) are processed in our laboratory. An increase of 30-50,000 samples will lead to a favorable cost benefit ratio. In Israel, 98 % of the births are in hospitals; of these 76 % are Jews, and 24 % Arabs or Druse. In Judea and Samaria there are about 32,000 births per year, however, only 63 % are in hospitals, and most mothers stay less than 24 hours in the hospital. Therefore, the screening program for PKU and NH is not feasable. To overcome this problem we decided to collect filter paper spotted blood samples in the "Mother and child health care centers" (MCH) spread all over Judea and Samaria, as early as during the first week following birth. We first undertook a study to determine the urinary iodine excretion as a measure of alimentary iodine in these areas. We found that in the Judean and Samarian cities and villages the mean excretion was between 104-155 µgI/gr. Cr., however 10 % of the children excreted less than 50 µgI/gr.Cr. As a result of this finding and in order to decrease the recall rate we began TSH screening followed by T_4 tests. Since May 1987, 10,111 newborns have been screened (age 36 hr to 3 weeks) and 5 infants were found with TSH above 100 µU/ml and T_4 less than 3.6 µg/dl. They were evenly spread geographically. Another 2 infants (aged 3 and 9 months) were also diagnosed as suffering from NH. This study suggests that in areas where a newborn screening program in the hospital is not feasable, a screening program in a MCH is possible. The incidence of NH among Arabs and Druse in Judea and Samaria is probably the same as in the rest of the country.

3.

CONGENITAL SCREENING PROGRAMME FOR HYPOTHYROIDISM IN CATALONIA

M. Albisu, N. Potau, A. Maya, M. Gussiner, E. Vicens-Calvet and the Catalan Collaborative Group*

Hospital Infantil Vall d'Hebron. Universitat Autonoma Barcelona

From January 1983 to December 1987, final diagnosis and follow-up were carried out in 47 children with positive screening test. 11 cases (23%) were normal on reassessment (transient hypothyroidism), but aetiology was known in only two cases (iodine overloading). T_4-spot values of transient hypothyroidism did not differ statistically from those of definitive hypothyroidism. TSH-spot levels differentiated both groups. Gammagraphic study with Tc99 of 36 definitive hypothyroid children showed ectopic gland in 47 %, aplastic in 36 %, and ectopic in 8 % (2 enlarged). Cervical echography was performed in 6 cases and results were concordant with those of the scan in only one case. The values of serum TSH, bone maturation (mm^2 surface of distal femoral and proximal tibial), and neonatal hypothyroidism index (NHI-Dussault) from the aplastic group, were statistically different from those of the ectopic group but serum T_4 values were not discriminatory. Treatment was started before 20 days of age in 7 children, between 20 and 40 days in 23, and after 40 days in 10. In the last two years, the onset of treatment was much earlier (28 days) than in the first years (45 days).

COMMENTS : In our area, a high incidence of transient neonatal thyroid dysfunction of known aetiology has been observed. Cervical echography in neonates is not helpful in the diagnosis of congenital hypothyroidism. There is a correlation between NHI, bone maturation, and the gammagraphic image.

---

* A. Ballabriga (Pediatr.), P. Gallofre (Scanning), M. Roig (Neurol.), J.M. Torres de Gasso (ORL), J. Tomas (Psych.), J.J. Gil (Ophtal.).

4.

NEONATAL SCREENING FOR CONGENITAL HYPOTHYROIDISM IN THE NETHERLANDS; 1986

G.J. Vaandrager, T. Vulsma, M.H. Gons and J.J.M. De Vijlder

National Institute for Preventive Health Care, Leiden, and
Department of Pediatrics, University of Amsterdam

Neonatal screening for CH in 99.5 % of all liveborn children (183.563) in the Netherlands in 1986 has been evaluated in detail per september 1987. Since 1981 a combined T_4-TSH screening is employed on the 6th-8th day of age : T_4 has been determined in all samples, TSH in the lowest 20 % of the total range. Values of $T_4 \leq -3.0$ s.d. and of TSH $\geq$ 0.50 mU/l are leading to referral to the pediatrician, values between -2.1-2.9 s.d. and 0.26-0.49 mU/L to recall. In preterm infants, since medio 1982, only TSH values as criterion for recall and referral are used. The recall percentage in 1986 was 1.54 %, referral percentage 0.41 %.

After confirmatory tests 58 children were diagnosed as permanent CH (hormone therapy necessary during at least one year) : 55 of the 58 were diagnosed as primary hypothyroidism (incidence 1 : 3300; 38 dysgenesis, 6 dyshormonogenesis, 11 subclassification not yet known) and 3 as secundary/tertiary CH (1:60.000). Besides 4 children with low T_4 and elevated TSH had no classification : all 4 with multiple anomalies; 3 of them died before the third month.

28 infants had transient CH (diagnostic test values of $T_4 \leq$ 70 nmol/L and/or TSH $\geq$ 10 mU/L); most of them were preterm infants : 5 with transient low T_4 and high TSH, 9 with transient elevated TSH and 14 with low T_4 values, TBG-deficiency excluded. 78 children were known with TBG-deficiency.

In 54 of the 58 children with permanent CH, TSH screening values were positive. The 4 exceptions were : all 3 infants with secundary/tertiary CH and 1 child with trisomia 21 and multiple congenital anomalies in which the etiology of the CH is not yet known. To our knowledge, 1 case of permanent CH was missed, because of late TSH rise : diagnosis partial organification defect.

Conclusion : As since the introduction of the nationwide screening less than 1 % of the permanent primary CH has been missed, the combined T_4-TSH screening (TSH in 20 %) will be continued. A less active policy with regard to the recall of children with low T_4 and normal TSH is under consideration.

5.

THE ETIOLOGY OF CONGENITAL HYPOTHYROIDISM (CH) IN THE NETHERLANDS

T. Vulsma, G.J. Vaandrager, M.H. Gons, and J.J.M. De Vijlder

Department of Pediatrics, University of Amsterdam; National Institute for Preventive Health Care, Leiden, The Netherlands

In the Netherlands, the national thyroid screening (measuring one week after birth T_4 and in the 20 % samples with lowest T_4-levels also TSH) was started in 1981. The children born in 1981 with abnormal screening results were evaluated 4 till 6 years later. Of the 176,311 screened children (98.7 % of all neonates) 1637 (0.93 %) had been referred to a pediatrician. Especially the cases with elevated TSH levels (> 25 mU/l in dried blood spots and/or > 10 mU/l in venous samples) were re-examined. In case of a (supposed) permanent CH of unknown cause the T_4 replacement therapy was interrupted for investigation. A permanent primary CH was found in 51 cases (incidence 1 : 3500) : 14 ageneses, 3 cryptic thyroids, 27 sublingual thyroids, 4 total organification defects, 1 Pendred syndrome, and 2 other dyshormonogenesis. Secondary/tertiary CH was found in 6 cases (1:30,000) : the 3 having an abnormal screening result were already under pediatric control since birth; in the other 3, having normal screening values, the diagnosis TSH deficiency was made after referral for growth retardation. Until now, for 1981, no false negative cases of biochemically proven primary CH have been reported.

The 89 cases (1:2000) with a transient primary CH (temporary drop of T_4 with rise of TSH) outnumber those with permanent CH. In 45 cases we found a plausible etiology retrospectively : one use of PTU in pregnancy, one congenital nephrotic syndrome and 43 in which the fetus or neonate was exposed to high doses of iodine. Severity and duration of the CH in the remaining 44 cases with unknown cause were comparable.

During re-examination we discovered 13 obvious false-positive cases, which were not recognised in the neonatal period. Those children were treated with T_4 and/or observed in the pediatric policlinic for years.

In conclusion, the overall incidence of CH in the Netherlands amounts to 1:1200 neonates. The preliminary results of the evaluation of 1982 are essentially the same. Knowledge of the precise etiology of CH has important consequences for therapy, prognosis and genetic counseling. Etiologic classification is also indispensable for epidemiologic studies and the evaluation of screening methods.

6.

POSSIBLE CAUSES OF BORDERLINE TSH : A SUMMARY OF OUR EXPERIENCE

C. Colon, J.R. Alonso-Fernandez, D.E. Castineiras, M.E. Romero, J.M. Fraga, and J. Pena

Laboratorio de Alteraciones Metabolico-Geneticas y Nutricionales. Dep. de Pediatria. Hosp. G. Galacia. Universidad de Santiago de Compostela. Xunta de Galicia, Conselleria de Sanidade e Seguridade Social. Programa Galego de Prevencion da Subnormalidade (SPAIN)

Congenital hypothyroidism began to be screened for in Galicia (N.W. Spain) in 1978. The analytical method originally used, measurement of TSH by RIA, was replaced in late 1986 by the DELFIA system.

In 1987, 22,099 samples were analysed; in 541 (2.4 %) of these cases, second samples were requested for the following reasons :

REASON FOR REQUESTING 2nd SAMPLE	No. CASES	NOT RECEIVED
TSH level higher than 20 UIU/ml serum	219 (0.99 %)	27 (0.12 %)
Improperly impregnated sample	151 (0.68 %)	29 (0.13 %)
Sample did not elute	171 (0.77 %)	38 (0.17 %)
TOTAL	541 (2.44 %)	94 (0.42 %)

Seven cases of congenital hypothyroidism and ten cases of transitory hypothyroidism were diagnosed.

In 202 of the samples prompting request of a second sample, the level of TSH (confirmed by re-analysing the same sample in duplicate) was between 20 and 50 UIU/ml, but was normal in all the 175 second samples received. Most of these samples proved to be from children born in hospitals where disinfection of the navel with iodine was recommended. With a view to investigating whether other factors might influence the observed slight, temporary rise in TSH level, we sent out home questionnaires to determine the mother's smoking habits, whether there was any family history of thyroid disorders, and whether iodine-enriched salt was used in the home.

This communication presents the results of our study and our conclusions as to the causes of the observed temporary rise in TSH level.

7.

MONOCLONAL-ANTIBODY-BASED THYROTROPIN ENZYME-IMMUNOASSAY IN DRIED BLOOD SPOTS FOR NEONATAL THYROID SCREENING

G. Ackermann[1], W. Ackermann[4], H. Willgerodt[2], H. Fiebig[3], and H.J. Thiele[5]

[1]Institute of Clinical Chemistry, [2]Paediatric Clinic (School of Medicine), [3]Laboratory of Immunology (Section of Biosciences), Karl-Marx-University, Leipiz, G.D.R.; [4]VEB Sächsisches Serumwerk, [5]Research Institute of Medical Diagnostics, Dresden, G.D.R.

We describe a two-site enzyme-immunoassay using monoclonal antibodies, suitable for the measurement of thyrotropin (TSH), from wholeblood spots on filter paper, to screen for congenital hypothyroidism. The capture antibody, immobilized on the surface of microtiter plate wells, is directed against an epitope of TSH-beta-chain, whereas the peroxidase-labeled indicator antibody recognize an alpha-subunit epitope. The sensitivity of the assay was 5 mU/L, and the intra- and interassay CVs were $\leq$ 15 % in the relevant concentration range (cutoff value : 20 mU/L). Cross reactivity with lutropin, follitropin or choriongonadotropin was negligible.

Features which make this assay advantageous to the CH-screening include simplicity of the photometric end point determination, long stability of enzyme tracer, rapid turnaround time, and absence of requirements for radioactive materials.

Besides the screening variant a highly sensitive immunoenzymometric assay using the same immunological reagents is available.

8.

COMPUTER ORGANIZATION IN THE HYPOTHYROIDISM SCREENING AT THE QUEBEC SCREENING NETWORK

J. Morissette, N. Bélanger, P. Fiset, J.H. Dussault and C. Laberge
C. Quebec Screening Network, Laval Medical Center, Ste Foy, Quebec, Canada

Since the autumn of 1987, the Quebec Screening Network uses a three level organization in its computerized system. At the first level is the reduction and accumulation of data by micro computers interfaced with laboratory instruments. An Apple II linked with an Arcus LKB (Delphia-TSH) allows the accumulation of the screening data for hypothyroidism.

A mini computer, with the UNIFY data base, does all the processing of all active screened infant records : screening formulae for the data entries, data merging between the infant records and laboratory results, on line inquiries about active infant records, sorted lists of records and finally screening activity reports. Moreover this system controls the filter paper forms in all participating nurseries.

On the last level, all infant records are transmitted to a main-frame computer, an IBM 4341, and stored on magnetic tape. Thus some statistical analysis on a tremendous amount of screening sata are readily available.

We believe that this system is essential to the quality insurance in a mass screening program.

9.

A SCREENING PROGRAM FOR THYROID FUNCTION IN PREGNANCY BY USING DRIED BLOOD SPOTS

M. Fukushi[1], O. Arai[1], Y. Mitzushima[1], N. Takasugi[1], M. Endo[2], N. Matsuura[3] and S. Fujimoto[4]

[1]Sapporo City Institute of Public Health, [2]Hakodate Central Hospital, [3]Department of Pediatrics, Hokkaido University School of Medicine and [4] Department of Obstetrics, Hokkaido University School of Medicine

It is well known that thyroid dysfunctions in pregnant women cause toxicosis of pregnancy, abortions, premature births and thyroid dysfunctions in newborns; furthermore thyroid diseases are aggravated after delivery and thyrotoxicosis and hypothyroidism are frequently observed. In order to prevent complications in fetus, newborns and mothers a screening program for thyroid dysfunction in pregnancy has been performed since June 1986 in Sapporo City. Blood samples were taken at the medical consultation of pregnants and spotted on filter paper. Analysis of the blood spots of free thyroxine (FT4) and thyrotropin (TSH) was made by radioimmunoassay. Anti-thyroglobulin antibody (ATG) and antithyroid microsomal antibody (AMC) was measured by indirect agglutination using artifical gelatin particles sensitized with thyroglobulin and thyroid microsomal antigen, respectively. By the end of December 1987, 4,806 pregnant women were screened. The positive rate of ATG and AMC were 3.17 % and 8.85 %, respectively. 69 cases (1.44 %) were suspected to have thyroid dysfunction and performed endocrinological examination in detail. As the result of the examination 17 of Graves' disease, 7 of chronic thyroiditis, 1 of hypothyroidism, 21 of transient hyperthyroxinemia at early pregnancy, 2 of other anomalies and 21 of normal cases were diagnosed. All of the pregnants with thyroid disease had no clinical signs and symptoms and one third of them need treatment. Pregnant women whose thyroid function were well controlled gave birth to normal newborns but neonatal thyrotoxicosis and transient hypothyroidism were frequently observed in infants whose mother's thyroid function were not well controlled during pregnancy.

In conclusion, early diagnosis and treatment of pregnant women with thyroid dysfunction are important and the screening program for thyroid function in early pregnancy has practical significance.

10.

MATERNAL HYPOTHYROXINEMIA IN EARLY PREGNANCY : IRREVERSIBLE BIOCHEMICAL EFFECTS IN BRAIN REGIONS OF RAT PROGENY

A.K. Sinha, A. Al-Mazidi, M.C. Ruiz de Elvira, C.G.D. Brook*, and R.P. Ekins

Department of Molecular Endocrinology, *Paediatric Endocrinology Unit, University College and Middlesex School of Medicine, Mortimer Street, London W1N 8AA

Severe hypothyroxinemia in the first trimester of pregnancy in biochemically and metabolically euthyroid subjects has been correlated with a spectrum of irreversible disorders ranging from neurological cretinism to partial paresis, auditory deficits, loss of motor coordination and cognitive dysfunction in the offspring. However, the underlying biochemical lesions in brain regions and/or cellular and subcellular compartments remain uncertain. We present data concerning the relationship between several key biochemical parameters in the CNS of the progeny and of early maternal hypothyroxinemia.

Partially thyroidectomised euthyroid rats were maintained on a small animal laboratory diet and water ad libitum at 37°C on a 12 hour light/dark cycle. The average iodine intake was 34 µg/day. The animals were mated and the progeny allowed to grow to adulthood before detailed investigation. Proteins, DNA, RNA, inorganic phosphate, B-D galactosidase, lactate dehydrogenase, aryl sulphate, succinate dehydrogenase, cholin acetyl transferase, calmodulin and calcineurine were measured in various brain regions and compared with controls. Our observations indicate a reduction in protein content in all regions by 22 % but significantly in cortex and midbrain ($p < 0.02$). DNA concentration was also reduced significantly in cortex, medulla and midbrain by 30 %, 38 % and 29 % respectively ($p < 0.05$). Inorganic phosphate concentration was diminished by 42 % and 21 % respectively in cortex and cerebellum ($p < 0.01$ and 0.05). In all brain regions there appeared to be a severe reduction in B-D galactosidase ($p < 0.05$), analogous to acid phosphatase which was also diminished by 39 % and 53 % respectively in paleocortex and midbrain ($p < 0.01$). Aryl sulphatase activity was diminished significantly in the paleocortex, medulla and midbrain ($p < 0.01$).

In summary, our results suggest that significant and irreversible biochemical and enzymic dysfunction occurs in several brain regions in the adult progeny as a consequence of early maternal hypothyroxinemia.

11.

THE VALIDITY OF 31-P MAGNETIC RESONANCE SPECTROSCOPY (31-P MRS) AS A TOOL FOR THE STUDY OF CONGENITAL HYPOTHYROIDISM IN NEONATE RATS

M. Rubinstein, O. Jeghers, R. Mathur-De Vre, Cl. Maerschalk and A.M. Ermans

Department of Radioisotopes, St. Pieter Hospital and MRS Research Group, Free University of Brussels

Neurological complications of congenital hypothyroidism (CH) have been shown to be related with a poor differentiation of neuroblasts in the very first days of life.

We applied MRS to investigate brain metabolism in the neonatal period of three groups of pups. One was born from normal mothers and two from mothers which were given PTU or iodine free diet (IfD) respectively four and six weeks prior to conception. In-vivo 31-P brain spectra were obtained from day 1 to day 45 using a 4.7 Tesla magnet and a 1 cm surface coil (Bruker MSL 200) under nitrous oxyde anaesthesia. Areas under the peaks were calculated using a Lorentzian lineshape simulation program. In normal control animals, Pcr/ATP and Pi/ATP showed a sharp increase, mostly after the second week. We observed no change in PME/ATP or PDE/ATP. Both PTU and Ifd showed identical evolution of the high energy phosphates: Pcr/ATP had a significantly lower progression than in the control group ($p < .005$), essentially after the second week, and Pi/ATP was found unchanged throughout the study.

In conclusion, our results are consistent with the known pathological data of brain dysmaturity in CH butthe versatility of 31-P MRS allows longitudinal follow-up of in-vivo parameters and can serve as a powerful tool for the understanding of cretinism.

12.

CONVENTIONAL (Tab, Mab, TBIAB) AND CYTOTOXIC (ADCC) THYROID AUTOANTIBODIES IN CONGENITAL HYPOTHYROIDISM (CH)

A. Grüters, U. Bogner, H. Schleusener and H. Helge

Depts. of Pediatrics and Internal Medicine, Free University, Berlin, FRG

Antibody-dependent cell mediated cytotoxicity (ADCC), which has been found in autoimmune thyroiditis in adults, as well as microsomal-(Mab), thyroglobulin-(Tab) and TSH-binding-inhibitory antibodies (TBIAB) were determined in 61 children with permanent CH and 46 of their mothers. Tab and Mab were measured by commercially available hemagglutination methods, TBIAB was determined using a radio-receptor ligand assay and ADCC in a 51 chromium release assay with human thyroid cells as targets. 37 of the patients with CH were newborns and 24 older patients with CH substituted with thyroxine. 12 newborns showed an increased specific lysis of 20.1 % (13.7-32) (median and range) compared to 4 % (0-11) in 50 healthy newborns. 7 of 29 of their mothers had a likewise increased lysis of 21.3 % (16.9-45), while the remaining mothers showed a lytic activity of 3.1 % (0-9.2). Also in 6 of the older patients an increased cytotoxicity of 17.9 % (12.3-22.8) was found, being 5.3 % (1.1-9) in the other children. Again 4 out of their mothers were positive for ADCC with 20.6 % (16.3-35). In 96 % results in mothers and children were concordant. An increased cytotoxicity was found in 30 % of the patients with "athyreosis", in 50 % of the patients with ectopic thyroid tissue, but only in 1 of 14 patients with a normal gland on scan. Since conventional antibodies were found only in a few patients (Mab in n = 3, Tab in n = 2 and TBIAB in n = 5), we conclude that the determination of these autoantibodies does not rule out an autoimmune pathogenesis of CH. Moreover, our results indicate, that cytotoxic antibodies could be involved in a destructive process in the fetal thyroid gland resulting in thyroid dysgenesis.

13.

AUTOIMMUNITY IN CONGENITAL HYPOTHYROIDISM DETECTED BY NEONATAL SCREENING

K.J. Hsiao, M.J. Shieh, S.H. Chiang, and F.Y. Su

Clinical Biochemistry Research Laboratory, Dept. of Medical Research; Veterans General Hospital, Taipei, Taiwan 11217, Republic of China

For understanding the relationship between autoimmunity and congenital hypothyroidism (CH), the autoimmune antibodies (Abs) in the maternal and neonatal serum from the CH cases detected by neonatal screening were studied. From 1986.1.1. to 1987.12.31, 95,798 neonates from all over Taiwan were screened for CH by TSH enzymeimmunoassay using dried blood spot samples. Among the 48 cases confirmed by serum TSH and T_4, 41 cases were scanned with ^{99m}TC (6 athyroid, 2 hypoplasia, 10 ectopic thyroid, 14 enlarged thyroid, 9 normal). Autoimmune Abs (ATA, AMiA, TBII) in maternal and/or neonatal sera from these 41 cases were studied. ATA and AMiA were determined in 36 cases and 9 of them were found to be positive at least one of the Abs (2/4 in athyroid, 0/10 in hypoplasia and ectopic thyroid, 6/14 in enlarged thyroid, 1/8 in normal). But no positive TBII result was found from 33 cases studied. (However, a positive TBII result was found in a case with transient neonatal hyperthyrotropinemia). The data showed that autoimmunity may play an important role in the CH cases detected by neonatal screening. The positives Abs cases with normal or enlarged thyroid scanning results indicate the possibility of transient CH, which should be further investigated at later age of these patients. Although the titers of the antibody were low (AMiA : 1/100, ATA : -) in the two positive athyroid cases, the pathophysiological role of the autoimmunity in agenesis of the thyroid may be very important and remains to be elucidated.

14.

INCIDENCE AND FUNCTIONAL EFFECT OF THYROID AUTOANTIBODIES IN NEWBORNS RECALLED FOR SUSPECTED CONGENITAL HYPOTHYROIDISM DURING A SCREENING PROGRAM

L. Chiovato, F.L. Giusti, F. Santini, P. Bassi, P. Vitti, C. Marcocci, M. Tonacchera, M. Ciampi, G.F. Fenzi and A. Pinchera

Cattedra di Endocrinologia, University of Pisa, Pisa, Italy

A role of maternal thyroid autoantibodies (TA) in the prediction and/or development of congenital hypothyroidism (CH) was suggested by several reports, but remains to be clarified. To this purpose, TA were studied in sera of 92 consecutive newborns recalled for examination (RN) during a screening program and in sera from 27 mothers whose children were affected by CH (MHC). Microsomal (MAb) and/or thyroglobulin antibody (TgAb) were detected by hemagglutination in 4/92 (4.3 %) sera from RN. 3/4 (75 %) RN positive for TA had CH (11 % of the whole CH group), which was due to gland agenesis (2) or thyroid hypoplasia (1). Antibody titers in 3 affected children ranged from 40^2 to 640^2 for MAb and from neg to 40^2 for TgAb. Their mothers had a previously unrecognized Hashimoto thyroiditis (HT) : 2 were euthyroid and 1 became hypothyroid in the post-partum. IgG from the latter mother produced an inhibition of TSH-induced cAMP production, I-uptake and 3H-thymidine incorporation in FRTL-5 cells, but it was ineffective in the absence of TSH. IgGs from the remaining 2 mothers with HT, from 18 MHC and from 7 mothers of normal children did not produce a stimulating or inhibiting effect in FRTL-5 cells.

In conclusion, our results indicate that : 1) 4/92 (4.3 %) newborns recalled for suspected CH were positive for TA; 2) 3 of them (75 %) had permanent CH, the incidence of TA-positive cases over the whole CH group being (11 %; 3) a biological effect of maternal IgG (i.e. inhibition of TSH action) could be demonstrated in 1/3 TA-positive cases; 4) autoantibodies are able to inhibit TSH-stimulated cAMP production; I-uptake and cell growth were not detected in mothers of CH children who were not affected by HT.

15.

MATERNAL THYROID-BLOCKING IMMUNOGLOBULINS IN CONGENITAL HYPOTHYROIDISM

R.S. Brown, P. Keating, and E. Mitchell

University of Mass. Medical School, Dept. of Pediatrics, Worcester, M.A, USA

Maternal thyroid-blocking IgGs have been causally associated with congenital hypothyroidism (CH) in some babies of mothers with 1o Myxedema (1oM). Their role in sporadic CH is less clear, however,. We prospectively evaluated 24 randomly selected CH mothers identified by the New England Regional Hypothyroidism Collaborative, two 1oM mothers whose babies had CH and 20 normal controls for the presence of IgG that blocked ^{125}I-TSH binding to thyroid membranes (TBII), TSH-induced growth (TGI-block), and TSH-induced thyroid stimulation (TSI-block). TBII were measured by a modification of the Smith & Hall method; thyroid growth was assessed by ^{3}H-thymidine uptake, and stimulation by ^{125}I-uptake by a cloned TSH-dependent rat thyroid (FRTL5) cell line, sensitivity being less than 20, 2.5 and 1 µU bTSH, respectively. Only one randomly selected CH mother and 2/2 1oM mothers had potent TBII, TGI-block, and TSI-bloc compared with the remaining 23 CH mothers whose results did not differ from the normal controls. Positive IgG were studied in further detail. Blocking activity was absorbed by and could be eluted from staphylococcal protein A. It was not an antiTSH idiotype as preincubation of the cells with IgG and washing prior to addition of TSH, versus preincubating the IgG with TSH prior to the addition of the cells gave no difference in the inhibition of thyroid growth observed. Inhibition of TSH-induced growth in each case was verified by evaluating the number of FRTL5 celles after 5 days' incubation with TSH in the presence of bloccking IgG (2.0×10^{5}) versus normal IgG (3.3×10^{6}). Half-maximal inhibitory concentration in the 2 assays (TBII, TGI-block) performed to date were 7.5 and 2.5 µg IgG, respectively. Finally, microsomal antibodies were positive in 2/18 normals, 4/25 CH mothers but not in the CH mother with potent blocking activity up to a dilution of 1 in 1.9×10^{8}.

CONCLUSIONS : 1) Metrnal thyroid-blocking IgG do not play a role in most cases of sporadic CH; 2) In view of the reasonably correlation between the 2 activities studied to date, TBII, TGI-block (and probably TSI-block) may be due to the same TSH receptor antibody population; 3) Determination of TBII but not microsomal antibody is a sensitive screening test for the presence of thyroid-blocking Abs.

16.

THYROID ANTIBODIES IN CONGENITAL HYPOTHYROIDISM

A. Ilicki[1], A. Larsson[1,3], and F.A. Karlsson[2]

From the Departments of Pediatrics[1] and Internal Medicine[2], Uppsala University, Akademiska sjukhuset, Uppsala, and [3]The PKU Section, National Bacteriological Laboratory, Stockholm, Sweden

The etiology in congenital hypothyroidism (CH) was investigated by analysing sera from CH patients, detected by neonatal thyrotropin (TSH) screening, for the presence of antibodies against thyroid peroxidase (TPO-Ab), thyroglobulin (Tg-Ab) and TSH receptor (TSH-rec-Ab). TPO-Ab and Tg-Ab were determined in 144 CH patients using ELISA assays (Pharmacia, Uppsala, Sweden) and TSH-rec-Ab in 118 CH patients by a radioreceptor assay (TRAK, Henning, Berlin, West Germany). All available sera in the National Swedish screening programme since the start in 1980 were utilized. Serum was collected two weeks post partum. Positive titres of TPO-Ab were found in 6 patients (4.2 %), whereas Tg-Ab were not found in any patient. TSH-rec-Ab were found in one of these patients and confirmed by elevated titres in the mother. The antibodies blocked TSH stimulated cAMP production in thyroid membranes. This patient had transient CH and the therapy could be stopped. In another serum an exceptionally high precipitation of the TSH tracer in the TSH-rec-Ab assay was found. This was due to the presence of antibodies against TSH.

It is concluded that the vast majority of CH in neonates is not associated with the presence of thyroid specific antibodies. The pathogenetic significance of TPO-Ab in certain CH patients ramains to be determined. Analysis of TSH-rec-Ab may detect a transient form of CH in some patients and provide a basis for stopping the treatment already a few months after birth. The same assay may also reveal the cause of false negative screening results due to antibodies against TSH.

17.

FAMILIAL THYROID ECTOPY

H. Capiau (1), L. Roelen (2) and R. Clara (1)

(1) Department of Pediatric Endocrinology. (2) Department of Genetics, Algemeen Kinderziekenhuis Antwerpen, Antwerp Belgium

We report two brothers with ectopic thyroid gland. One child, 6 weeks of age, shows the stigmata of hypothyroidism. This is confirmed by laboratory data. Scintigraphy reveals a sublingual thyroid gland. On screening the other members of the family, we find an elevated TSH with normal FT3 and FT4 in a clinical normal 13 month aged brother. Again a sublingual thyroid gland is shown on scintigraphy. Ectopic thyroid gland usually occurs sporadically and in most of the familial cases multifactorial heredity is suggested. However, we believe that ectopic thyroid gland in our family has to be considered as an autosomal recessive trait. A review of literature findings supports our consideration.

18.

CONGENITAL HYPOTHYROIDISM : INCREASED INCIDENCE IN ASIAN FAMILIES

G.M. Addison, M. Rosenthal and D.A. Price
Royal Manchester Children's Hospital, U.K.

100 cases of congenital hypothyroidism were detected in the N.W.R.H.A. in a screening population of 289,697 infants of which 18,366 (6.3 %) had mothers of Asian origin (India, Pakistan & Bangladesh). 77 infants were white caucasian, 20 Asian, 1 Arab, 1 Jewish, 1 Chinese. None were Negro. The incidence of congenital hypothyroidism in this U.K. region is :

	MEAN	RANGE	F:M:
Total	1:2897	1:2399 - 1:3621	2.7:1
Asian	1:918	1:612 - 1:1670	5.7:1
Non Asian	1:3391	1:2740 - 1:4396	2.3:1

Asian v Non Asian $p < 0.001$ (chi squared)
Total v mean European $p < 0.01$ (chi squared)
Sex differences. Asian v Non Asian. Not significant at 5 % level.

13 Asian couples were consanguinous. These patients were 4.3 times more likely to have anatomically normal or enlarged, normally positioned thyroid glands on scan, but this accounted for only 5 out of 13 infants. The reasons for the increased incidence of congenital hypothyroidism is unclear but is only in part related to consanguinity and inherited metabolic defects. Diet or other specific environmental causes can not be ruled out.

19.

NEONATAL THYROID SCREENING AS AN INDEX OF ENVIRONMENTAL IODINE DEFICIENCY

M. Carta Sorcini°, S. Carta°, A. Diodato°, C. Fazzini°, M.E. Grandolfo°, L. Donati°°, C. Romagnoli+, R. Luciano+ P. Bourdoux°°° and F. Delange°°°

°Istituto Superiore di Sanità, Roma; °°Civil Hospital, Urbino; +Catholic University, Rome; °°° Université Libre de Bruxelles

During a multicentric pilot screening program for congenital hypothyroidism a comparison between the results obtained from Urbino, an Italian area characterized by low iodine supply and endemic goiter, and Rome, a non endemic area, has been made. The evaluation of neonatal ioduria in the two areas showed significantly ($p < 0.05$) lower iodine urinary excretion levels in Urbino than in Rome. A shift of TSH at screening toward significantly ($p < 0.01$) higher values as well as a higher percentage of recall in Urbino area than in Rome was observed. This finding that well correlates with a low environmental iodine supply stresses the importance of screening for congenital hypothyroidism as a suitable index of the presence and action of goitrogenic factors in the environment.

This concept was established in areas of extremely severe or severe endemic goiter. The present paper provides evidence that the concept is also true in moderate endemic goiter areas.

20.

EPIDEMIOLOGIC STUDY OF TRANSIENT HYPERTHYROTROPINEMIA IN GREECE

Ch. Mengreli and S. Pantelakis
Institue of Child Health, Athens, Greece

Systematic screening for congenital hypothyroidism (CH) in Greek newborns was started in 1980. As the screening method TSH measurement in dried blood spots collected between the 3rd and 7th day of life is used. The cut-off point in our method is 30 mU/l of TSH in whole blood. By the end of 1987 a total of 865,425 newborns have been examined and in 263 of them the diagnosis of CH was confirmed (incidence 1:3290). The overall recall rate in our screening programme is 0.25 %, which lays in the higher range of the recall rate in Europe according to different reports. Transient hyperthyrotropinemia (TH) and transient hypothyroidism expressed by higher recall rates have recently been reported as a possible index of iodine deficiency. Analysis of our recall results for a three years period (336,610 newborns) showed a wide variation in the 52 geographic regions of the country, ranging from 0.04 % in Athens area to 1.7 % and 2.4 % in some regions of the north and central part of Greece. The finding that the region of Athens, in which one third of the Greek population leaves and consumes iodinated salt, has a very low recall rate, in contrast to other parts of the country, where in a great percentage non-iodinated salt locally produced is consumed, supports the hypothesis that TSH may indicate iodopenia. Iodine measurements in mothers milk and newborns urine during the first days of life could confirm this hypothesis.

In conclusion, we believe that national screening programmes for CH, in addition to the detection of CH, may reveal regions of iodopenia. Each state can use such maps of iodopenic regions in order to take active measures of intervention (i.e. distribution of iodinated salt) and eliminate iodopenia, which can lead to brain impairment in the newborn or predispose in thyroid cancer in case of nuclear accident.

21.

JUVENILE HYPOTHYROIDISM : NUTRITIONAL MARKER OF GOITROGENICITY

B. Swennen, J. Vanderpas, P. Van Den Abeele, J. Gillies and C. Thilly

Brussels University, Belgium

In a severe goitrous endemia in Zaire (Z), high prevalence of cretinism and thiocyanate overload (a potent goitrogenic factor) acting after weaning are described. In another endemia in Malawi (M), we recently observed that few cretins are present and that undetoxified cassava, from which thiocyanate is produced, is not consumed. In both endemia (Z versus M) we compared the thyroid function of children in different age groups. We concluded that 1) in Zaire, there is a late juvenile degradation of the thyroid function, starting after one year of age which is not observed in Malawi. 2) This juvenile rather than perinatal hypothyroidism is probably responsible of the higher frequency and of the stunded growth of cretins and cretinoïds in Zaire compared to their low frequencies and the lack of dwarfism in Malawi. 3) The range of abnormalities observed in Zaire is not only linked to the iodine deficiency but also to the thiocyanate overload.

AGE (YEARS)	<1		1 to 4		4 to 7		7 to 15	
	Z	M	Z	M	Z	M	Z	M
SERUM T4 (μg/dl)	9.6	8.6	5.8	6.1	4.8	5.8	4.1	6.3
SERUM FT4 (ng/dl)	0.81	1.00	0.64	0.74	0.60	0.73	0.48	0.80
SERUM TSH (μU/ml)	10.4	3.25	10.1	3.44	24.3	3.01	24.1	4.49

22.

TRANSIENT CONGENITAL HYPOTHYROIDISM WITH AMIODARONE ADMINISTRATION

J. De Schepper, D. De Wolf, and L. Smits

Department of Pediatrics, Academisch Ziekenhuis V.U.B., Brussels, Belgium

Amiodarone, a drug used in arrythmias, contains 75 mg of iodine/200mg active substance and is a frequent cause of thyroid dysfunction in adults. No adverse effects on fetal thyroid function have been reported with maternal ingestion of Amiodarone during pregnancy.

200 mg Amiodarone daily from the 13th week of gestation until delivery was taken by a 29 year-old-woman. Bimonthly thyroid function tests during pregnancy were normal. She delivered at 38 weeks a 2,450 kg weighing boy with a goiter and clinical signs of hypothyroidism. Cord blood TSH was > 100 mU/l and T4 was 35,9 µg/l. Iodine content of the urine was 144 µg/dl. Levels of Amiodarone were still detectable in the infant's plasma after one month. The goiter disappeared after 3 months of substitution therapy with levothyroxine, which could be stopped definitively at 20 months. Psychomotor development showed a 4 months retardation at that time. Follow-up thyroid function tests (including TRH-test) remained normal.

This observation calls for cautious use of Amiodarone during pregnancy and for early assessment of neonatal thyroid function to avoid the risk of neurologic impairment.

23.

IODINE OVERLOAD AND TRANSIENT HYPOTHYROIDISM DURING NEONATAL SURGERY

J.P. Chanoine, M. Steppe, P. Bergmann, D. Biarent, M.H. Delaet, J.M. Bouton, and F. Delange

Departments of Pediatrics, Radioisotopes and Pediatric Surgery, Free University of Brussels, Belgium

The use of iodianted contrast media in radiology and iodinated skin disinfectants in surgery result in iodine overload. Patients admitted in intensive care units (ICU) are routinely submitted to both procedures. The thyroid gland of infants is highly sensitive to the antithyroid effects of iodide excess. Therefore, we determined whether iodine overload could be a significant cause of hypothyroidism in severely ill young infants admitted in a surgical ICU. We investigated 7 newborns (gestational age : mean $\pm$ SD : 38.3 $\pm$ 1.8 weeks) aged 1 to 9 days (gr I) and 7 infants aged 1 to 5.5 months (gr II) admitted to the ICU for cardiac or abdominal surgery and submitted to iodine overload following IV injection of iohexol (Omnipaque R) for cardiac catheterization and/or application of povidone iodine (Isobetadine ^{R}) for skin disinfection. We measured serum TSH (µU/l, mean $\pm$ SD), T_4, FT_4 and urinary iodine before and during 10 days following iodine contamination.

Results : (TSH values compared by student's test for paired values; $p < 0.01$: ** as compared to basal TSH)

	Basal TSH	Maximum TSH	Day (mean + SD)
Gr I	3.4 $\pm$ 5.7	29.8 $\pm$ 7.6**	6.4 $\pm$ 1.4
Gr II	1.0 $\pm$ 0.8	3.4 $\pm$ 1.0**	4.8 $\pm$ 1.4

T_4 and FT_4 presented a transitory decrease as compared to basal values both in gr I and gr II (Maximum decrease : 50 and 63 % respectively, on day 2.5 $\pm$ 1 after iodine overload). Mean urinary iodine was 5,827 $\pm$ 3,715 (SD) µg/dl after contamination. It returned to normal values (< 15 µg/dl) within 5.2 $\pm$ 1.1 days.

Conclusion : Iodine overload in severely ill newborns results in a Wolff-Chaikoff effect with a transient and sometimes severe primary hypothyroidism, as evidenced by the decrease in T_4 and FT_4 values followed by a 9 times increase in serum TSH. In infants older than one month, serum TSH also increased but remained within the euthyroid range. Consequently, the use of iodinated agents in critically ill newborns, if necessary, should require close monitoring of the thyroid function.

24.

SLEEP DISTURBANCES IN CHILDREN WITH CONGENITAL AND ACQUIRED HYPOTHYROIDISM

M. Hayashi, M. Shimohira, S. Saisho, H. Suzuki, K. Shimozawa and Y. Iwakawa

Department of Pediatrics, Tokyo Medical and Dental University, School of Medicine, Tokyo, Japan

All night polysomnography was examined in 8 patients (2 males, 6 females). Five of them were congenital hypothyroidism, two were patients of inappropriate TSH secretion (ITSH) and one was chronic thyroiditis.

In four cases of congenital hypothyroidism detected by neonatal screening, the frequency of body movements (BMs) during sleep was disturbed and abnormal distributions of BMs against each sleep stage were also observed. Those changes did not normalize even after long-term L-thyroxine replacement therapy. In addition, two cases also showed a decrease of the sleep stage of rapid eye movements (SREM) at neonates. In a girl of congenital hypothyroidism which was found at the age of four because of her dwarfism, rapid eye movements (REMs) during SREM were hardly noticed, while she exhibited only a mild decrease of BMs. On the other hand, three children of ITSH and chronic thyroiditis showed no significant changes in BMs and REMs.

Both BMs and REMs are considered to be closely related to monoaminergic neuron systems in the brainstem. And it is well known in animals that congenital hypothyroidism can impair brain catecholaminergic and serotonergic systems. The present investigation suggests that hypothyroidism might also insult those systems during the perinatal period in humans and L-thyroxine replacement therapy could not improve those brainstem disturbances although thyroid functions were normalized. It is important to evaluate the sleep development of children with congenital hypothyroidism and define the critical period of brain involvements.

25.

DETECTION OF HYPOTHYROIDISM IN DOWN SYNDROME CHILDREN IN THE ABSENCE OF ELEVATED THYROID ANTIBODY LEVELS

R. Bellisario, L.J. Freligh, A.M. Wiley, M.L. Cowger and M.R. Pandian

NYS Department of Health, Albany Medical College, Albany, NY and Nichols Institute, Los Angeles, CA

Adolescents and adults with Down syndrome have an increased prevalence of thyroid dysfunction, ranging from 13-54 % in different studies. The thyroid dysfunction has been reported to be positively correlated with elevated thyroid antibody levels. A longitudinal study was carried out to determine whether the incidence and onset of hypothyroidism coincided with the elevation of thyroid antibodies in Down syndrome children. From a group (N = 191) of Down syndrome newborns screened in the New York State Newborn Screening Program, 12 (6.4 %) were recalled for follow-up testing because of abnormal thyroid function test results compared with a 1 % recall rate for all newborns. Two of the infants recalled were subsequently diagnosed with primary hypothyroidism, suggesting an incidence of hypothyroidism of about 1:100 (2/191) in Down syndrome children in agreement with a previous report (Fort et al. (1984) J. Pediatr. 104 : 545-8). Diagnoses of Down syndrome and hypothyroidism were obtained from NYS cytogenetic laboratories and hypothyroidism treatment centers, respectively. Measurement of antimicrosomal and antithyroglobulin levels of 46/191 follow up serum specimens taken from this group up to 3 years of age showed that 9 % (4/46) exhibited antimicrosomal titers of 1:100 and none of the antithyroglobulin levels were elevated. None of the 11 Down syndrome children tested axhibited elevated levels of TSH Binding Inhibitory Immunoglobulin (TBII) levels.

Measurement of antimicrosomal and antithyroglobulin antibody levels of another group of Down syndrome children, clinically diagnosed with hypothyroidism by 3 years of age, showed that 0/11 exhibited elevated antimicrosomal or antithyroglobulin antibody titers. None of the 3 hypothyroid Down syndrome children tested exhibited elevated TBII levels. It appears that in the first 3 years of life about 1 % of Down syndrome children are hypothyroid without significant titers of elevated thyroid antibodies.

26.

NEONATAL HYPOTHYROIDISM : COMPARISON OF RADIOISOTOPE AND ULTRASOUND IMAGING IN 37 CASES

R. de Bruyn, NG, and D.B. Grant

The Hospital for Sick Children, Great Ormond Street
London WC1

Thirty seven neonates identified by the North East and North West Thames Regional hypothyroid screening programme between January 1985 and June 1987 were investigated with radioisotope and ultrasound scans of the thyroid before treatment with l-thyroxine was commenced. Compared with the radioisotope scan, ultrasound identified normally sited thyroid tissue in 7 out of 9 patients, and usually indicated when no normally sited thyroid tissue was present (25 out of 28 cases). However, ultrasound did not reliably detect ectopic sublingual thyroid tissue (5 out of 16 cases). In spite of this shortcoming, ultrasound examination could play a larger role in the investigation of congenital hypothyroidism, particularly in identifying patients with defects of thyroid hormone synthesis which carry a high risk of recurrence.

27.

TREATMENT AND FOLLOW-UP IN 25 NEONATAL HYPOTHYROID (NH) INFANTS DETECTED BY THE ISRAELI NATIONAL SCREENING PROGRAM

D. Eldar, O. Amado, and J. Sack

Pediatric Endocrinology, Institute of Endocrinology
Sheba Medical Center, Israel

The National Screening Program for NH started in Israel in May 1978. The incidence of NH infants detected is 1:3200. Currently the required time of final diagnosis and begining of treatment is 23 ± 2 days (mean ± SEM) (however, at the start of the screening program the required time was longer). This study was undertaken in order to investigate the relatively large number of NH infants with prolonged hyperthyrotropinemia. In 25 NH infants identified by the screening, treated and followed by us, thyroid function was assessed each month. The infants were started on T_4 15 μg/kgBW/day. This dose was reduced to about 10 μg/kgBW/day one month later. The infants were divided into two groups : those who suppressed their serum TSH values to normal within 3 months of treatment (5 ± 1 weeks) and those where suppression was delayed (15 ± 3 weeks, mean ± SEM). The initial filter paper T_4 was higher in the first group (2.8 ± 0.5 vs. 2.1 ± 0.3) however, not significantly. There were no differences between the two groups in the etiologies of NH nor were there differences in bone age or growth and development. During the 24 months follow-up the T_4 treatment dose was the same in both groups : 5.7 ± 0.4 vs. 5.5 ± 0.3 at the age of 6 months, 4.5 ± 0.5 vs. 4.9 ± 0.2 at 12 months and 5.1 ± 0.3 at 24 months. Despite equal mean serum T_4 (range 11.8-12.6) and T_3 (range 146-207) concentrations during this period, the mean serum TSH concentrations remained significantly higher at 3, 12 and 18 months in the group where delayed suppression of TSH was observed. This suggests, that the thyrotroph in these children is more "resistant" to serum T_4 and T_3 concentrations. A delay in onset of treatment in the "resistant" group (5.6 ± 0.7 vs. 3.8 ± 0.6 weeks) was observed. This postnatal delay in the onset of T_4 treatment of NH infants might explain the relative large number of NH infants with "resistance" to T_4 treatment for a prolonged time.

28.

ABOUT THE FORM AND REPLACEMENT DOSE OF THYROID HORMONES IN THE TREATMENT OF CONGENITAL HYPOTHYROIDISM (CH)

F. Péter, L. Blatniczky, and H. Breyer
Buda Children's Hospital and Polyclinic, Budapest, Hungary

Nearly generally accepted that replacement therapy of CH should be done with thyroxine (T_4) not triiodothyronine (T_3). The recommended dose of L-T4 at the start of therapy is 25-50 µg/die for a term infant. On the other hand the aim of the newborn screening programs for CH is to avoid the prolonged hypothyroid state during the first weeks of life.

Infants born in half Hungary undergo filter paper TSH testing in our laboratory. The incidence of hypothyroidism detected was 50/250.000. The median age at the onset of therapy was before the end of the 3rd week of life. The majority of the infants was treated only with L-T_4 but sometimes we started with thyreoidea sicca (Thyranon, Organon) and changed over to L-T_4 (L-Thyroxine, Henning; Eltroxin, Glaxo) later.

The TSH level decreased from > 80 mIU/L to 2.2 $\pm$ 1.2 mIU/L (T_3 to 2.7 $\pm$ 1.8 ng/ml; T_4 to 11.6 $\pm$ 3.8 µg/dl) up to the first control in the cases (10) treated with Thyranon (25 mg/die) but the decrease of TSH was moderate (to 36.8 $\pm$ 12.1 mIU/L) of the L-T_4 treated infants. Furthermore the replacement dose of L-T_4 was 50 µg/die in the cases treated previously with 25 mg/die Thyranon - titrated until TSH levels became normal (ratio 2:1). A few years ago as we changed the replacement therapy of the older children with CH from Thyranon over to L-T_4 the ratio was found to be cca 1:1.

In conclusion : these results confirm the suggestion that T_3 may play a more important part than T_4 in regulating the serum TSH concentration and the ratio of conversion from thyreoidea sicca to L-T4 may change during the life.

29.

THYROID FUNCTION AFTER WITHDRAWAL OF TREATMENT IN CHILDREN WITH ECTOPIC THYROID TISSUE

D.B. Grant, W.K.Ng, S.P. Leung, and J.A. Hulse

Hospital for Sick Children, Great Ormand Street, London WC1, U.K.

Thyroid function was studied after temporary withdrawal of treatment in 30 children with sub-lingual thyroid tissue demonstrated by technetium scans at the time of diagnosis. In 27 children treatment was stopped for one week after three week's therapy with T3 (20 μg/day) after the age of one year. In 11 children with plasma thyroxine (T4) levels over 65 nmol/l at the time of diagnosis, treatment was stopped for one month between the ages of 5 and 7 years.

In all cases plasma TSH rose above 15 mU/l on stopping treatment. Nine of the 27 children tested between the ages of one and two years achieved normal plasma T4 levels over 65 nmol/l. Five of the 11 children tested between the ages of 5 and 7 years had plasma T4 levels above 65 nmol/l 4 weeks after stopping treatment.

These results indicate that useful thyroid function is preserved for considerable periods in some children with ectopic thyroid tissue. Although there was an overall correlation between plasma T4 level off treatment and plasma T4 at the time of diagnosis, the T4 at diagnosis did not reliably predict those patients with ectopic thyroid tissue who would continue to have useful thyroid function.

30.

PREDICTING DEVELOPMENTAL OUTCOME IN CHILDREN WITH HYPOTHYROIDISM IDENTIFIED ON NEWBORN SCREENING

J. Rovet, R. Ehrlich, and D. Sorbara

Depts. Psychology & Pediatrics, Hospital for Sick Children
Toronto, Canada M5G 1X8

To test whether subtle neurocognitive deficits persist in hypothyroid children despite screening and to identify the children at greatest risk, we have been following since 1981 190 cases detected on screening programs in Ontario. Our sample representing about 95 % of all cases being treated at Toronto's Hospital for Sick Children, includes 31 with athyrosis, 29 with goitre or enzyme defects and 49 with ectopic glands. Children are assessed annually until age 9 with measures covering a wide range of psychological functions. Results indicate IQs within the normal range but significantly lower PIQ scores than siblings when tested at the same age ($N = 74$ pairs, $p < .03$), perceptual, neuromotor and memory deficits, particularly after age 5 and among children with athyrosis and/or delayed skeletal maturity at diagnosis. A series of multiple regression analyses predicting subsequent specific abilities from factors associated with early disease and treatment have shown that those factors reflecting hypothyroidism in utero and severity at time of diagnosis were significant predictors of later perceptual and motor skills while factors reflecting treatment and duration of hypothyroidism predicted language abilities. Preliminary findings on school age children indicate that a significant number are experiencing mild to moderate learning and attentional problems especially in the arithmetic area.

31

RESULTS OF FOLLOW-UP AT 5 YEARS IN A GROUP OF HYPOTHYROID AUSTRALIAN CHILDREN DETECTED BY NEWBORN SCREENING

A. Rickards, J. Coakley, I. Francis, S. Armstrong, H. Medson, and J. Connelly

Departments of Child and Family Psychiatry and of Clinical Biochemistry, Royal Children's Hospital, Melbourne, and State Screening Laboratory, Mont Park

All cases of congenital hypothyroidism (CH) identified by newborn screening and referred to the Royal Children's Hospital, Melbourne are being reviewed prospectively using a set protocol. Psychological assessments are performed by 1 psychologist (AR) using the Bayley procedure at 6 months, 1 and 2 years, Stanford Binet I.Q. at 3-5 years, WPPSI at 5 years and WISC-R at 8 years, with social background and temperament also recorded. A control group of children of comparable ethnic and social background selected by a systematic random method have also been assessed at 2 and 5 years with a further review planned for 8 years. For statistical comparison the non-parametric Mann-Whitney U test has been used.

Results of comparison at 2 years were reported to the 6th International Screening Symposium in Austin Texas in November 1986. These showed that while most CH children were functioning within the normal range their MDI's and PDI's were statistically lower than the control group.

Review of the control children at 5 years has recently been completed and their WPPSI results, compared with the 5 year old CH children, most of whom had been assessed at 2 years, are as follows :

	VERBAL			PERFORMANCE		
	No.	$\bar{x}$	SD	No.	$\bar{x}$	SD
Hypothyroid	39	96.13	13.12	40	107.30	15.44
Control	44	111.25	15.63	45	113.44	15.22
		$p < .0001$			$p < .05$	

Results therefore, show a statistical delay in both parameters but greater in verbal skills. Studies are in progress to search for deficiencies in specific sub tests and to identify risk factors, both thyroidal and extra thyroidal, influencing outcome. Results available in time for the workshop will be reported.

32.

CONGENITAL HYPOTHYROIDISM : PSYCHOLOGICAL OUTCOME AT THREE AND FIVE YEARS

P. Fuggle, G. Murphy, G. Thorley, and D.B. Grant
Institute of Child Health, London

Psychological assessments (McCarthy Scales) were carried out at three years (Murphy 1986) and at five years (Wechsler Scales; WPPSI) on children with congenital hypothyroidism (CH) born in the North Thames Health Regions and started on l-thyroxine therapy at a mean age of 27 days. Children matched for age, sex, social class, language background, and ethnic origin were tested as controls.

At three years 58 CH children and 39 controls were assessed. No significant differences between matched patients and control children was found (Patients : GCI 103. SD 13.7; Controls : 105 SD 11.3) but significantly lower mean test scores were obtained in CH children with initial bone ages below 30 weeks or initial plasma T4 levels below 20 nmol/l.

At five years 59 CH children and 52 matched controls were assessed. Mean scores for matched patients and controls were not significantly different (Patients : IQ 109. SD 17.0; Controls : 111 SD 15.5) but again significantly poorer scores were obtained in the CH children with initial plasma T4 levels below 20 nmol/l compared with CH children with T4 levels above 60 nmol/l (mean difference 19 IQ pts. p 0.01). There was also a significant correlation between initial T4 level and performance IQ ($r = 0.36$; p 0.006) but not with verbal IQ.

At both three and five years there was no significant correlation between age at start of treatment and psychological outcome. The data suggest that, although CH children as a group fall well within the normal range of cognitive ability, initial severity of hypothyroidism continues to be associated with poorer outcome.

33.

MENTAL DEVELOPMENT IN 75 CHILDREN WITH CONGENITAL HYPOTHYROIDISM DETECTED BY NEONATAL SCREENING

L. Moschini[+], M. Carta Sorcini°, P. Costa[+], I. Antonozzi[++], A. Paolella[-], G. Porro[-], and S. Carta°

[+]3Catt. Puericultura; [++]Dip. Med. Sper., [-]Ist. Neuropsc. Inf. Univ. "La Sapienza" Roma; °Istituto Sup. Sanità, Roma

Seventy-five infants (48 girls and 27 boys) with primary CH detected by neonatal screening were included in our study. Screening program was based in T4 and/or TSH determination on filter paper spot at 3-5 days of life. At time of diagnosis, measurements of serum TSH, T4, FT4, T3 and determination of bone age were performed. Technetium 99mTc scan on 58 subjects revealed athyreosis in 25 patients, ectopic glands in 28 and a gland in normal position in the remaining 5. The mean age at onset of therapy with L-thyroxine was 30 ± 7 days in 70 and 60 ± 6 in 5 infants. Evaluation of mental development was assessed using Brunet-Lezine's method from 6 to 24 months, Stanford-Binet at 3-5 years and WISC at 6 years of age. At the time of the screening mean T4 concentration was significantly ($p < 0.001$) lower, 1.11 ± 1.36 µg/dl in agenesia than in ectopic glands, 4.37 ± 3.22 µg/dl. Mean serum levels of both T4 and TSH were significantly different (respectively) $p < 0.001$ and $p < 0.01$) in agenic with respect to ectopic thyroids at the time of final diagnosis. Assessment of mental development showed mean global score 88 ± 11 at age 6 months, 88 ± 12 at 1 year, 90 ± 14 at 3, 97 ± 10 at 4, 97 ± 8 at 5 and 100 ± 15 at 7 years. There were no differences among the patients when they were classified according to etiology of the disease (athyroidism, ectopic thyroid). In agreement with previous reports, the mental development of our CH children is on average within the normal range. However in a few cases it could have been influenced by many factors of pre- and perinatal risks.

34.

FACTORS RELATED TO THE INTELLECTUAL DEVELOPMENT OF CHILDREN TREATED FOR CONGENITAL HYPOTHYROIDISM

J.E. Toublanc, S. Rives, and J.C. Job
Hôpital Saint Vincent-de-Paul, Paris, France

Follow-up was obtained for at least 4 years in 52 children (40 F, 12 M) who had congenital hypothyroidism, detected by TSH screening at 5 days and persisting at the 3th-5th postnatal week. 8 had athyreosis, 33 an ectopic thyroid gland and 11 an anatomically normal thyroid. Treatment began at day 29 $\pm$ SD 12 with l-T4 solution, 5-7 µg/kg/day and was followed in order to obtain immediately T4 and FT4 blood levels in the high normal range, with normal TSH at least from age 6 mth. Mental development was evaluated at age 0.5, 1 and 2 years with the Brunet-Lézine scale (BL), and at 4 years with the Termann-Merrill test (TM). 20 of them were re-evaluated at age 7 years with the WISC R test.

The results of these tests were all in the normal range, with individual differences, the range being 85 to 145 with TM at 4 yrs and 95 to 133 with WISC R at 7 yrs, not different from those obtained in age-matched controls and in the normal siblings of the patients. The WISC R test at 7 yr did not show difference between verbal and performance levels. Its results were correlated with those of the TM test performed at age 4 yr in the same 20 patients : $r = 0.75$, $p < .01$. Scholarship at 7 yr was normal for age in these 20 hypothyroid children.

Among the pre-treatment data, only the presence of the two epiphyses of the knee had an influence on the outcome ($p < .025$). Neither clinical signs, pre-treatment T4 and TSH, or the time of starting the treatment within the 2 first months of life, correlated with the results of the tests. Three main factors had a significant influence on the mental outcome. 1) The socio-economic level of the family, evaluated according to the score of Graffar : $r = - 0.50$ and $p < .005$ at 4 years, $r = - 0.55$ and $p < .025$ at 7 years; 2) The quality of hormonal control during the first year of treatment ($p < .025$) and the following years ($p < .05$). 3) The existence of psychological disturbances during infancy and early childhood, which were predictive of poor outcome : at 4 years, mild problems $p < .005$, and serious problems $p < .0005$, at 7 years $p < .001$ with $r = - 0.76$.

These data show that medical care and familial support are more important for a good mental development of the congenitally hypothyroid children than the cause of the pre-treatment data.

35.

FOLLOW-UP OF CONGENITAL HYPOTHYROIDISM (CH) DETECTED BY NEONATAL SCREENING IN CATALONIA

L. Ibanez, M. Albisu, N. Potau, E. Vicens-Calvet and the Catalan Collaborative Group*

Hospital Infantil Vall d'Hebron. Univeristat Autonoma Barcelona

The follow-up of 36 confirmed cases of CH, diagnosed by neonatal screening between January 1983 and December 1987 is presented. Clinical and biological characteristics of these patients have been described elsewhere. The clinical follow-up consists of : a) periodic endocrinological, clinical and anlytical control, b) ophthalmological examination : ocular motility, fundus oculi and ocular refraction (6-12 mo.); c) neurological study : Denver Developmental Screening test (DDST)(6-12mo.); d) hearing tests : brainstem (BAEP) and cortical (CAEP) auditory evoked potentials (from 6mo), Warblet-Rudmose test (6-24mo.), Suzuky test (24mo-5yrs), and tympanometry when considered necessary; e) psychological study : Gesell test (18-36mo), and/or Terman-Merrill test (> 3 yrs).

Ophthalmological examination (n = 36), revealed strabismus in 6 patients (17 %), whom 2 had associated hypermetropia. Another patient (3 %), had megalocornea, and yet another (3 %), had anisocoria. Neurological study (n = 36), revealed no evaluable anomalies in any of the cases. Hearing tests (n = 28), detected sensorineural hearing loss to some degree in 7 cases (25 %), and mixed loss in 2 cases (7 %). On first examination, in 8 cases (28 %), an integration deficit at cortical level (CAEP) was observed, which improved with age on subsequent controls. Psychological examination (n = 14), disclosed an overall IQ < 90 in 3 patients (21 %), with psychomotor and language area involvement. In a further 4 patients (28 %) with overall IQ's in the normal range, a relative delay in language with respect to other areas examined was observed. The incidence of anomalies in the different areas studied was somewhat higher in patients with thyroid agenesis (42 %), than in those with thyroid ectopia (36 %), although short follow-up in some cases does not allow for definite conclusions to be drawn at present.

---

* A. Ballabriga (Pediat.), P. Gallofre (Scanning), M. Roig (Neurol.), J.M. Torres de Gasso (ORL), J. Tomàs (Psych.), J.J. Gil (Ophtal.), A. Maya (Screen.).

36.

LONG TERM NEUROLOGICAL PROGNOSIS IN EARLY TREATED CONGENITAL HYPOTHYROID CHILDREN

G.F. Rondanini, A. Bollati, V. Della Porta, R. Cerabolini
C. Lenti, L. Rovey, A. Manzoni, and G. Chiumello

Scientific Institute H. San Rafaele, Dept. of Pediatrics
Endocrine Unit, University of Milan, Italy

A prospective study of the mental development of hypothyroid infants detected by neonatal screening began 8 years ago. Thirty infants were followed (group A); the mean age at initiation of l-T4 therapy was 36 days. Brunet-Lezine and Stanford-Binet IQs were assessed. Neurological examination included Prechtl & Saint Anne Dargassies tests (in newborns), integrated by Vojta tests (in the first year), and Touwen & Prechtl test and Oseretsky test in the following years. Results were compared to those obtained with the same tests in a group of hypothyroid children diagnosed at a mean age of 170 days, before the beginning of the screening program (group B).

The IQ is within the normal range in group A; at six months of age it is in the lower normal range, but it shows a significant recover before the first year. In group B the IQ is well below the normal range, even if at 5 years it shows a significant increase (but not a normalization), demonstrating the usefulness of a follow-up program. A neurological delay is present in 55 % of group A infants and in 73 % of group B infants in the first year of life. During the following years we observe a progressive and complete neurological normalization in group A, while in 60 % of group B children a neurological impairment is still present at 5 years.

The results indicate that early treatment of congenital hypothyroidism and appropriate follow-up programs permit hypothyroid children to achieve a normal neuro/psychological development.

37.

NEUROPSYCHOLOGICAL DEVELOPMENT OF CONGENITAL HYPOTHYROID CHILDREN

S. Bargagna, L. Cittadoni, G. Falciglia, G. Ferretti, M. Marcheschi, *L. Chiovato, *F.L. Giusti, *G.F. Fenzi and *A. Pinchera

Istituto di Neuropsichiatria Infantile, and *Cattedra di Endocrinologia, University of Pisa, Pisa, Italy

In this study we evaluated the neuropschological development of a group of congenital hypothyroid children diagnosed by neonatal screening and treated early after birth. 17 cases were followed-up. Examination was performed at 3, 6, 9, 12, 18, 24, 30, 36 months and then once a year using the following tests : from 0 to 2 yr : neurological examination (Touwen), ordinary scales of Uzgiris-Hunt, Brunet-Lezine test, direct observation of mother-child interaction according to Kreisler; from 2 to 6 yr : standardized neurological examination, Termann-Merrill scales, projective tests, psychological interview with child and family; at any age : evaluation of psychological factors (home inventory). Postural attainements had a normal development, but were slightly slower than in normal children (mean walking age = 13.8 months). Neurological examination did not show specific deficits in 14 children, mild hypotonia being found in one and tremor of distal ends in 2 cases. The development quotient was : 100 ± 7.5 (mean ± SD) at age 12-month (n = 17) and 102 ± 5.3 at age 18-month (n = 13). The IQ was 99 ± 13.7 at age 36-month (n = 6). Cognitive strategies were normal in relation to age. A generalized delay of language development was found in 3 cases. The behaviour was characterized by irritability and opposition at age 2-yr, while inhibition and lack of self confidence prevailed in older children. Relational attitudes were normal but parents-child intercation was disturbed by hyperstimulation and hyperprotection due to anxiety derived from diagnosis.

In conclusion, our preliminary results indicate that the neuropsychological development of the 17 children was generally within normal limits although mild and possibly transient neurological and psychological defects were found.

38.

EARLY TEMPERAMENTAL CHARACTERISTICS OF NEONATALLY IDENTIFIED HYPOTHYROID CHILDREN

D. Sorbara, J. Rovet, and R. Ehrlich

Departments of Psychology and Pediatrics, Hospital for Sick Children, Toronto, Canada M5G 1X8

Atypical temperamental characteristics have been noted in preschoolers and school-age children with congenital hypothyroidism (CH) identified by newborn screening in early infancy. We now report on the temperaments of 50 six month old CH infants detected via screening in Ontario. Temperament was evaluated by parental report using Carey's Infant Temperament Questionnaire. Results indicate : (i) a greater than usual incidence of difficult infants; (ii) a profile pattern involving a lower response threshold and more intense reactions to sensory stimulation; (iii) higher T3 and T4 levels for more difficult than easier children; these were evident for T3 (nmol/l) at 3 mos (2.93 vs 2.45, $p < .03$) and for T4 (nmol/l) at 1 (14.4 vs 9.7, $p < .01$), 2 (16.6 vs 11.6, $p < .005$), 3 months (15.0 vs 11.3, $p < .01$) and 4 months of age (196.7 vs 162.4, $p < .10$), but not at diagnosis (55.0 vs 71.7); (iv) minimal stability with subsequent temperament ratings (to age 3); (v) no correlation between either hormone levels in infancy or 6 month temperament ratings and subsequent intellectual outcome. We conclude that while higher levels of T4 in infancy may increase temperamental difficulty, this does not have long-lasting effects.

PART 6. SUMMARY AND RECOMMENDATIONS

SUMMARY AND RECOMMENDATIONS

I. SUMMARY

PHYSIOPATHOLOGY SECTION

Dr. Braverman and Dr. Emerson reviewed with us placental function as it relates to the thyroid gland reminding us that the placenta is permeable to iodide, TRH and antithyroid drugs. The human placenta transports immunoglobulins to the fetus during the third trimester including antithyroid antibodies. Placental outer ring iodothyronine deiodinase converts T4 or rT3 and T3 to T2 and T1. The placental enzyme is inhibited by iopanoic acid but there is no effect of hypothyroidism or fasting.

Dr. Morreale and Dr. Obregon reviewed the now clear evidence for placental transport of thyroid hormone, showing significant levels of T4 in embryonic and trophoblastic tissue in the rat. These levels can be reduced by maternal thyroidectomy and by a low iodine (maternal) diet. They described the presence of nuclear T3 receptors in fetal rat tissues including brain and the presence of outer ring iodothyronine deiodinase in the fetal brain tissue. Thus the fetal brain can convert T4 to T3 and this activity is increased by hypothyroidism. T3 is relatively concentrated in fetal brain with 60-80 % saturation of brain T3 receptors, by T3 levels approximating 1 ng per gram tissue. Liver T3, by contrast, parallels the lower serum T3 values. Brown adipose tissue also can convert T4 to T3 and has the highest prevailing tissue T3 concentration (about 2 ng/g tissue). Brown adipose tissue thermogenin is thyroid hormone responsive by 20-22 days and increases markedly in the newborn, supporting an active role for brown adipose tissue in neonatal thermogenesis.

Dr. Ekins pointed out that T4 dissociates from plasma binding proteins at a finite rate, and this rate constant for a given protein determines the diffusion rate of T4 to tissues. Thus, binding proteins in mammals and TBG in humans assists T4 transfer to tissues. The increase in TBG in human pregnancy, in his view, supports rather than inhibits maternal to fetal T4 transfer.

Dr. Fisher reviewed thyroid system maturation in rats, sheep and man focusing on the ontogenesis of fetal T3 nuclear receptors and various postreceptor effects. Thyroid hormones appear during the second and third trimesters of system development and thyroid actions generally appear during the third trimester or later. There are relatively early effects on brain, bone and skin and later effects on liver. There also are early effects of T4 on brain outer ring monodeiodionase and extrahypothalamic TRH concentrations. These systems seem oriented to the maintenance of brain thyroid hormone T3 levels in the event of T4 deficiency.

Dr. Dussault reminded us of the complexity of the brain from the early induction of the neural plate through progressively regionalized and localized neural proliferation, migration and connectivity. He reviewed the wide distribution of brain T3 receptors and reminded us of the host of brain biochemical events affected by thyroid hormones, including the critical actions on cell migration and connectivity.

In summary, these physiopathology data support the concept of some significant maternal-fetal thyroxine transport and sequestration in fetal tissues. Fetal brain T3 receptors and T4 to T3 conversion occur relatively early in thyroid system ontogenesis. The timing of the thyroid effect on brain development varies with species. Brain development in fetal sheep is thyroid hormone dependent at least by 80 to 90 days of the 150 day gestation period and the precise timing of the thyroid hormones effects on brain maturation in rats and humans is not clear. Further information is essential to clarify the significance of thyroid hormones on fetal brain maturation in these species.

AUTOIMMUNITY SECTION

Dr. Matsuura reviewed the Hokkaido infants with congenital hypothyroidism (CH). Ninety-five infants with permanent CH were identified from 556,000 infants screened, and an additional 22 had transient disease. In 18 mother-infant pairs from the transient group 4 were positive for TSH binding inhibiting immunoglobulin (TBII) and thyroid blocking antibody (TBA); 2 of these had low IQ values.

Dr. Dussault reported 5 infants with permanent CH and antithyroid antibodies from 85 mothers screened. These included 2 families with 2 involved infants. One of these included 2 infants of 3 with permanent CH and another with 2 infants of 3 with transient CH.

Dr. Pinchera reported that 10 % of Italian infants with CH have positive antithyroglobulin antibody (ATA and/or antimicrosomal antibody (AMA) titers. From this group he identified 3 mother-infant pairs with high titers of TBA and CH.

These patient data suggest that some 5-10 % of CH infants have positive AMA and/or ATA representing passive transfer of thyroid autoantibodies. Some 2-3 % have transient or permanent CH with TBA measured using the FRTL-5 cell line.

Dr. Drexhage reviewed data derived using a cytochemical bioassay for thyroid growth blocking antibody (TGBA). This was present in some 50 % of CH infants and mothers. The colloid antibody of Balfour (CA2) is present in 36 % of mothers and 9 % of infants. Suprisingly, these titers persisted for many years.

Dr. Boyages reviewed information from the Chinese Quinghai endemic. This included both neurological and myxedematous cretins with an average age of 25 years. The neurological abnormalities and mental retardation were similar in the two groups. Suprisingly, again, these cretins had thyroid autoantibodies. TBII levels correlated with TSH results and 7/8 myxedematous cretins had significant TGBA titers; 40 % of the cretins had significant CA2 titers.

These autoimmunity data and the ensuing discussion suggest that there is no correlation of TGBA and thyroid blocking antibody (TBA) measured using the FRTL-5 cell line. The TGBA probably represents antibody to another (non TSH) receptor involved in thyroid growth. There was reasonable concensus that 5-10 % of CH infants have ATA and/or AMA; TBA or TBII are seen in 1-3 %. In addition wome 40-50 % manifest significant TGBA (cytochemical bioassay) and CA2 antibody. The former antibodies are not involved in the pathogenensis of CH in most infants with sporadic disease. The significance of TGBA and CA2 in pathogenesis of CH remains unclear.

GENETICS ENVIRONMENT SECTION

Dr. Lazarus reviewed an English series of CH infants and questionnaire data from 1538 CH from several countries for the presence of congenital anomalies (CA). CA varied from zero to more than 20 % with an average of 8 %. 30 % of the anomalies were cardiac and 12 % had chromosomal abnormalities. The overall 8 % prevalence probably is greater that the expected 2-4 % in a normal population. However, 2 countries accounted for most of the CA infants. Clearly more data are necessary to clarify this issue.

Dr. Refetoff reviewed thyroid resistance syndromes noting that there are now 142 reported cases, 139 of which were familial. He described an approach to differential diagnosis of hyperthyroxinemia and the need to confirm the diagnosis of peripheral thyroid hormone resistance by thyroid loading tests. He described a 9 day thyroid loading test with 5, 10 and 20 ng T3/kg daily (for 3 days each) with before and after measurements of body weight, sleeping pulse, BMR, serum steroid hormone binding globulin (SHBG), ferritin, and cholesterol, and urine magnesium, carnitine and urea nitrogen all of which are responsive in normals and unresponsive or relatively so in patients with peripheral resistance. The need now is to simplify the test procedure.

Dr. Delange reviewed the problem of endemic cretinism and the ICCIDD program now extending worldwide. He pointed out the significant decrease in mean IQ in the noncretin cohorts in Ecuador and India. He also emphasized the several areas of Europe where iodine excretion still is less than 100/µg/day and iodine content of breast milk is low. There is a threshold of iodine intake (50 µg/day) and urine iodine (4.5 µg/dl) below which cord TSH levels are elevated. Iodine deficient infants also are susceptible to Wolff Chaikoff blockade with iodine exposure.

DIAGNOSIS SECTION

Dr. Ermans reminded us of the utility of thyroid scanning in differential diagnosis of neonatal CH : 125I is preferred and 99m Technetium a satisfactory substitute. There is a low and insignificant radiation risk with good technique an equipment.

Dr. Hassan and Dr. Spehl reviewed their experience with echography concluding that current equipment and methodology are inadequate to reliably detect ectopic thyroid tissue. There is little clinical utility for ultrasound in the differential diagnosis of CH at present. However, ultrasound always detects thyroid tissue located in normal position and, consequently, will contribute to more selectively decide when to perform thyroid scanning.

Dr. Czernichow reviewed recent experience with serum thyroglobulin (Tg) measurements in differential diagnosis of CH, relating results to scan data. He reported both false positive results in athyreotic infants, and false negative results in infants with ectopic thyroid glands concluding that a Tg level above 10 ng/ml indicates the presence of a thyroid gland. A value less than 10 ng/ml is not, however, diagnostic of thyroid agenesis. He also reported detecting Tg in children with CH off treatment after 7 years indicating prolonged survival of the gland remnants.

Dr. Delange reported a collaborative study from 6 European centers by members of the ETA Screening Committee. There were 181 infants scanned before treatment in whom T4, Tg and bone age data were related to the scan results. T4 values ranged from 0-4 µg/dl and Tg values from 0-30 ng/ml in athyroid infants. Bone age was correlated with initial T4 values. Delayed bone age maturation was observed in infants with T4 values less than 6 µg/dl, but the retardation was variable and poorly correlated with T4 below that level.

He concluded that there was a correlation between increased clinical or X-ray abnormalities and thyroid agenesis by scan. The laboratory results (T4 and Tg) in athyroid infants are most compatible with some placental T4 transfer, but some small amount of undetected tissue could not be excluded.

THERAPY SESSION

Dr. Mitchell summarized treatment results from the New England newborn thyroid screening program. He characterized 75 % of infants with adequate and 25 % with inadequate treatment in retrospect. After 8 weeks of treatment the adequate group had a mean serum T4 of 13.1 and a TSH less than 20. The mean T4 treatment dose was 7.5 µg/kg/day. The inadequate group had a mean T4 of 9.7 µg/dl and TSH over 20. The mean treatment dose was 5 µg/kg/day. Ten percent of the infants has a TSH over 20 after 1 year. He proposed more frequent measurement of T4 and TSH in treated infants and the need to suppress serum TSH levels to less than 20 µU/ml.

Dr. Van Vliet reviewed treatment data from Brussels where the T4 dose ranged from 25 to 50 ug/day and follow up attempted to maintain a free T4 level of 2 ng/dl. Mean age of initiation of treatment was 14 days and mean bone age 38 weeks. At one year the head circumference was about 1 SD increased and the TSH above normal in all infants. Developmental scores versus age showed lower slopes for CH infants (versus controls) for motor, social, adaptive, and language skills. Bone age remained relatively retarded in the delayed group for 3 years. He concluded that the data were compatible either with a prolonged or irreversible effect of prenatal hypothyroidism or inadequate treatment.

Dr. Dhondt reported results in 94 CH infants from France. Sixty-five percent had ectopic glands and treatment was begun at an average 25 days. He observed a correlation between low IQ values and low initial serum T4 or bone surface area at 1 and 2 years as well as a correlation between low bone surface and low T4 at birth. Infants with ectopic glands had bone surface scores higher than those with agenesis. Also there was a tendency to higher IQ values in infants with higher initial bone surface scores. These results during treatment, including low T4 values during treatment in some infants, elevated TSH values in many infants, increased head circumference, and reduced bone age at one year support the view that a significant proportion of treatment CH infants are not treated adequately. Thus inadequate treatment might be related to the low IQ values observed in treated CH children in some programs.

PSYCHOINTELLECTUAL ASSESSMENT SECTION

Dr. Klees reviewed the testing modalities available, their application and their significance in the evaluation of psychointellectual development of children.

Dr. Glorieux reviewed the Quebec data describing IQ test results in treated CH children relative to controls at 1, 3, 5, 7 and 9 years. The global IQ scores of CH children were low at 1, 3 and 5 years. The 7 year mean WISC result was 96 (range 48-128). Preliminary data on School Appreciation Evaluation confirmed the WISC predictability. Children with higher WISC scores do better in school. Results were lower in lower social class children with CH as expected. There were relatively low WISC verbal and IQ subset scores in the CH group. A separate group of children with low T4 and low bone surface scores at birth had a significantly lower mean IQ values. The profile of these children resembled that of the less affected infants but the IQ subset values were 10-15 points lower.

Dr. Klein presented the New England collaborative data describing IQ results in 90 infants followed through 7 years. The mean IQ was 105. There was a good correlation between IQ and grade level achievement. A subset of children with inadequate early treatment had a mean IQ of 90 (15 points low).

Dr. Heyerdahl reported the Norway experience with 46 children in Norway. The mean IQ values of CH and control children were similar. However, at 6 years the mean CH value was low in CH children (90 versus 100 for sibling controls). There were correlations of low IQ with low T4 at birth and delayed age of treatment. IQ values were lower in athyroid infants and in a low treatment group of infants (TSH $>$ 15 and T4 $<$ 8 at some time during first year).

Dr. Rochiccioli described neurological abnormalities in 69 French children from the Pyrenees area followed over 10 years. There was a normal global neurological score in these CH children relative to controls but subset scores involving coordination and fine manipulative skills were significantly retarded. These retardations correlated with IQ, with T4 at diagnosis, and with bone maturation at birth. There was no correlation with age at diagnosis, adequacy of treatment or athyreotic versus ectopic scan results.

These psychointellectual data show low IQ results (by 5 to 15 points) in subsets of CH infants presenting more severe hypothyroidism at birth, delayed bone maturation at birth, and/or inadequate treatment.

II. RECOMMENDATIONS

1. Over the longer term, further work to delineate the role of maternal T4 in fetal brain development and the timing of thyroid hormone actions on various aspects of brain maturation is desirable. Such information will enable more appropriate decisions regarding intrauterine thyroid

hormone treatment in selected instances and more knowledgeable therapy strategies for CH infants.

2. For the near term, more attention to early treatment of CH infants is desirable. More vigorous treatment and follow up during the early weeks and months of life may reverse any untoward perinatal effects of thyroid hormone deficiency on brain development.

3. Prompt raising of the serum corrected T4 (Free T4 index) or free T4 concentration to the upper normal range and suppression of serum TSH to 20 uU/ml are recommended. Careful follow-up to guarantee maintenance of optimal serum T4 and avoidance of overtreatment is advised. Regular assessments of growth rate, bone age advancement and developmental parameters are necessary.

4. Thyroid scanning with 123I or 99m Technetium is the most useful differential diagnostic procedure in the neonatal period. Serum thyroglobulin measurements may be helpful. Ultrasound is not presently contributive to the differential diagnosis of CH. Measurements of maternal and newborn TSH receptor-blocking antibody is useful to identify antithyroid antibody-induced transient hypothyroidism in infants born of mothers with autoimmune thyroid disease.

5. Continued careful assessment of psychointellectual development and neurological function in children with CH is important to define functional deficits. Continued efforts to correlate these results with neonatal data and treatment parameters, will better characterize the early risk factors and, thereby, improve treatment strategies.

SPEAKERS

BRAVERMAN, L.

University of Massachusetts School of Medicine
Worcester, MA, USA

CZERNICHOW, P.

Hôpital des Enfants Malades
Paris, France

DELANGE, F.

Depts. of Pediatrics and Radioisotopes, Univ. Hosp. Saint-Pierre
Free University of Brussels, Brussels, Belgium

DREXHAGE, H.A.

Laboratory for Clinical Immunology, Dept. of Pathology
Free University Hospital, Amsterdam, The Netherlands

DUSSAULT, J.H.

Centre de Recherche du Réseau de Médecine Génétique du Québec
CHUL, Ste Foy, Québec, Canada

EKINS, R.

Dept. of Molecular Endocrinology, University College and
Middlesex School of Medicine, University of London,
London, United Kingdom

EMERSON, C.

University of Massachusetts School of Medicine
Worcester, MA, USA

ERMANS, A.M.

Dept. of Radioisotopes, Univ. Hosp. Saint-Pierre,
Free University of Brussels, Brussels, Belgium

FARRIAUX, J.P.

Service de Génétique et Maladies Héréditaires du Métabolisme
Centre Hospitalier Universitaire, Hôpital C. Huriez
Lille, France

FISHER, D.A.

Dept. of Pediatrics, UCLA School of Medicine, Harbor-UCLA Medical
Center, Torrance, Ca, USA

GLORIEUX, J.

Dépt. de Pédiatrie, Université de Montréal, Centre de Recherche,
Hôpital Saint-Justine, Montréal, Québec, Canada

HASSAN, M.

Dépt. de Radiologie, Hôpital R. Debré
Paris, France

IRIE, M.

First Department of Medicine, Toho University
Tokyo, Japan

KLEES, M.

Dépt. de Pédiatrie, Hôpital Universitaire des Enfants Reine Fabiola, Free University of Brussels,
Brussels, Belgium

KLEIN, R.

Dept. of Maternal and Child Health
Dartmouth Medical School, Hanover, NH, USA

LAZARUS, J.

Dept. of Medicine, University of Wales College of Medicine,
Cardiff, Wales, United Kingdom

MATSUURA, N.

Dept. of Pediatrics, Hokkaido University School of Medicine,
Sapporo, Japan

MITCHELL, M.

New England Regional Screening Program, State Laboratory Institute, Jamaica Plain, MA, USA

MORREALE DE ESCOBAR, G.

Unidad de Endocrinologia Experimental, Instituto de Investigaciones Biomedicas, Faculty of Medicine,
Madrid, Spain

OBREGON, M.J.

Unidad de Endocrinologia Experimental, Instituto de Investigaciones Biomedicas, Faculty of Medicine,
Madrid, Spain

PINCHERA, A.

Cattedra di Endocrinologia, University of Pisa,
Tirrenia, Pisa, Italy

REFETOFF, S.

Depts. of Medicine and Pediatrics, The University of Chicago,
Chicago, Ill., USA

ROCHICCIOLI, P.

Dept. of Pediatrics and Medical Genetics,
University Hospital of Rangueil, Toulouse, France

SPEHL, M.

Pediatric Radiology, Univ. Hosp. Saint-Pierre,
Free University of Brussels, Brussels, Belgium

VAN VLIET, G.

Dépt. de Pédiatrie, Hôpital Universitaire des Enfants Reine Fabiola, Free University of Brussels, Brussels, Belgium

DISCUSSANTS

ADDISON, G.M.

Royal Manchester Children's Hospital
Pendleburry, England

BOYAGES, S.

Dept. of Immunopathology,
Free University of Amsterdam, The Netherlands

BROWN, R.S.

Dept. of Pediatrics, University of Massachusetts Medical Center,
Worcester, MA, USA

CHANOINE, J.P.

Dept. of Radioisotopes,
Free University of Brussels, Brussels, Belgium

CHEN, J.S.

Dept. of Clinical Pathology, University of Taiwan
Taipei, Republic of China

CHIOVATO, L.

Dept. of Endocrinology, University of Pisa,
Tirrenia (Pisa), Italy

DE NAYER, Ph.

Dept. of Hormones Metabolism, International Institute of
Cellular Molecular Pathology,
University of Louvain, Brussels, Belgium

D'HONDT, J.L.

Dept. of Genetics and inborn errors of Metabolism,
University of Lille, Lille, France

FUGGLE, P.

Institute of Child Health
London, England

GIOVANNELLI, G.

Dept. of Pediatrics
University of Parma, Parma, Italy

GLINOER, D.

Depts. of Medicine and Radioisotopes,
Free University of Brussels, Brussels, Belgium

HEYERDAHL, S.

Dept. of Pediatrics,
University of Oslo, Oslo, Norway

ORGIAZZI, J.

INSERM U 197
Lyon, France

PANTELAKIS, S.

Institute of Child Health, Aghia Sophia Childrens Hospital
Athens, Greece

RONDANINI, G.F.

Dept. of Pediatrics, Ist. Scient. H. San Raffaele,
Milan, Italy

ROVET, J.

Hospital for Sick Children
Toronto, Canada

SACK, J.

Chaim Sheba Medical Center
Tel Hashomer, Israel

SINHA, A.

Dept. of Molecular Endocrinology,
Univ. College and Middlesex School of Medicine,
London, United Kingdom

VULSMA, Th.

Dept. of Pediatrics
Free University of Amsterdam, Amsterdam, The Netherlands

WILLGERODT, H.

Dept. of Pediatrics, Karl-Marx University
Leipzig, German Democratic Republic

AUTHOR INDEX

SUBJECT INDEX

GPSR Compliance

The European Union's (EU) General Product Safety Regulation (GPSR) is a set of rules that requires consumer products to be safe and our obligations to ensure this.

If you have any concerns about our products, you can contact us on ProductSafety@springernature.com

In case Publisher is established outside the EU, the EU authorized representative is:

Springer Nature Customer Service Center GmbH
Europaplatz 3
69115 Heidelberg, Germany

Batch number: 10838292

Printed by Printforce, the Netherlands